VOLUME ONE HUNDRED AND NINETY ONE

Progress in MOLECULAR BIOLOGY AND TRANSLATIONAL SCIENCE

Human Microbiome in Health and Disease - Part A

VOLUME ONE HUNDRED AND NINETY ONE

Progress in Molecular Biology and Translational Science

Human Microbiome in Health and Disease - Part A

Edited by

BHABATOSH DAS

Molecular Genetics Laboratory, Infection and Immunology Division, Translational Health Science and Technology Institute, Faridabad, Haryana, India

VIJAI SINGH

Department of Biosciences, School of Science, Indrashil University, Rajpur, Mehsana, India

ELSEVIER

ACADEMIC PRESS

An imprint of Elsevier

Academic Press is an imprint of Elsevier
50 Hampshire Street, 5th Floor, Cambridge, MA 02139, United States
525 B Street, Suite 1650, San Diego, CA 92101, United States
The Boulevard, Langford Lane, Kidlington, Oxford OX5 1GB, United Kingdom
125 London Wall, London EC2Y 5AS, United Kingdom

First edition 2022

ISBN: 978-0-323-99786-7
ISSN: 1877-1173

For information on all Academic Press publications
visit our website at https://www.elsevier.com/books-and-journals

Publisher: Zoe Kruze
Acquisitions Editor: Ashlie M. Jackman
Developmental Editor: Jhon Michael Peñano
Production Project Manager: James Selvam
Cover Designer: Matthew Limbert

Typeset by STRAIVE, India

Contents

Contributors

Taruna Ahrodia
Translational Health Science and Technology Institute, NCR Biotech Science Cluster, Faridabad, Haryana, India

Md Jahangir Alam
Department of Biotechnology, National Institute of Pharmaceutical Education and Research (NIPER), Guwahati, India

Susmita Bakshi
Translational Health Science and Technology Institute, NCR Biotech Science Cluster, Faridabad, Haryana, India

Purbita Bandopadhyay
IICB-Translational Research Unit of Excellence, CSIR-Indian Institute of Chemical Biology, Kolkata; Academy of Scientific and Innovative Research (AcSIR), Ghaziabad, India

Sanjay K. Banerjee
Department of Biotechnology, National Institute of Pharmaceutical Education and Research (NIPER), Guwahati, India

Gargi Bhattacharjee
Department of Biosciences, School of Science, Indrashil University, Rajpur, Mehsana, India

Dinh-Toi Chu
Center for Biomedicine and Community Health; Faculty of Applied Sciences, International School, Vietnam National University, Hanoi, Vietnam

Sarat Dalai
Institute of Science, Nirma University, Ahmedabad, Gujarat, India

Bhabatosh Das
Molecular Genetics Laboratory, Centre for Human Microbial Ecology, Translational Health Science and Technology Institute, NCR Biotech Science Cluster; Molecular Genetics Laboratory, Infection and Immunology Division, Translational Health Science and Technology Institute, Faridabad, Haryana, India

Mrinmoy Das
International Centre for Genetic Engineering and Biotechnology, New Delhi, India

Santanu Das
Translational Health Science and Technology Institute, NCR Biotech Science Cluster, Faridabad, Haryana, India

Dipyaman Ganguly
IICB-Translational Research Unit of Excellence, CSIR-Indian Institute of Chemical Biology, Kolkata; Academy of Scientific and Innovative Research (AcSIR), Ghaziabad, India

Hiral Gautam
Department of Biosciences, School of Science, Indrashil University, Rajpur, Mehsana, India

Tarini Shankar Ghosh
APC Microbiome Ireland, University College Cork, Cork, Ireland

Jigresh Gohil
Department of Biosciences, School of Science, Indrashil University, Rajpur, Mehsana, India

Nisarg Gohil
Department of Biosciences, School of Science, Indrashil University, Rajpur, Mehsana, India

Bharti Kandiyal
Molecular Genetics Laboratory, Centre for Human Microbial Ecology, Translational Health Science and Technology Institute, NCR Biotech Science Cluster, Faridabad, Haryana, India

Khushal Khambhati
Department of Biosciences, School of Science, Indrashil University, Rajpur, Mehsana, India

Soumendu Mahapatra
Chromatin and Epigenetic Unit, Institute of Life Sciences; Kalinga Institute of Industrial Technology (KIIT), School of Biotechnology, Bhubaneswar, Odisha, India

Subir Kumar Maulik
Indian Council Of Medical Research, New Delhi, India

Rupesh Maurya
Department of Biosciences, School of Science, Indrashil University, Rajpur, Mehsana, India

Tanshi Mehrotra
Translational Health Science and Technology Institute, Faridabad, Haryana, India

Rasmita Mishra
Chromatin and Epigenetic Unit, Institute of Life Sciences, Bhubaneswar, Odisha, India

Smrutishree Mohanty
Chromatin and Epigenetic Unit, Institute of Life Sciences, Bhubaneswar, Odisha, India

Archana Pant
Molecular Genetics Lab, National Institute of Immunology, New Delhi, India

Sweta Patel
Institute of Science, Nirma University, Ahmedabad, Gujarat, India

Punit Prasad
Chromatin and Epigenetic Unit, Institute of Life Sciences, Bhubaneswar, Odisha, India

Ayushi Purohit
Molecular Genetics Laboratory, Centre for Human Microbial Ecology, Translational Health Science and Technology Institute, NCR Biotech Science Cluster, Faridabad, Haryana, India

Suresh Ramakrishna
College of Medicine; Graduate School of Biomedical Science and Engineering, Hanyang University, Seoul, South Korea

Sriram Seshadri
Institute of Science, Nirma University, Ahmedabad, Gujarat, India

Shalimar
Gastroenterology and Human Nutrition, All India Institute of Medical Sciences, New Delhi, India

Priyanka Singh
Department of Biosciences, School of Science, Indrashil University, Rajpur, Mehsana, India

Vijai Singh
Department of Biosciences, School of Science, Indrashil University, Rajpur, Mehsana, India

Preface

Among biological entities studied by microbiologists, microorganisms, the unicellular microscopic living entities, displayed incredible diversity both in terms of their structure and functional potency. Based on the nucleotide sequences of ribosomal RNA gene and unique enzymatic functions, the microscopic organisms are classified into three domains: Bacteria, Archaea, and Eukarya. They are ubiquitously colonized on the surfaces of living and non-living substances distributed across the planet. The term "Microbiome" refers to the assemblage of a complex community of microbiota and their biomolecules occupying a well-defined habitat in or on living or non-living objects. Like other living organisms, the environmentally exposed surfaces of the human body habitat are also colonized with thousands of microbial taxa from all three domains of life. Multitudes of symbiotics, commensals, and pathogenic microbes are isolated and identified from the skin, mucous membranes, gastrointestinal tracts, and reproductive tracts. The rapid development of nucleic acids sequencing technology and computational biology tremendously helped to identify and characterize complex microbial communities associated with any specific ecosystem without cultivating the microorganisms. In the last 20 years, microbiologists, clinicians, mathematicians, and computational biologists have demonstrated how crucial these microbes are for our health, metabolism, development, behavior, adaptation, and evolution.

We conceptualized this volume as providing concise and updated information on human microbiome structures, diversity, functions, and their association with communicable and noncommunicable diseases. In this volume, we included cancer, metabolic diseases, and non-alcoholic fatty liver disease. We have also included two chapters covering emerging tools for decoding the human microbiome and the role of human microbiome in drugs metabolism. The concluding chapter covers the recent discovery of microbiome-based therapeutics.

We believe that this volume will enjoy a wide readership from research scientists, clinicians, students, and researchers. The updated information and knowledge will help in mechanistic understanding of microbiome research and the discovery of biomarkers and microbiome-based therapeutics. We also believe that this volume will be an excellent primer in which scientific knowledge would grow and widen in the field of microbiome biology in health and disease.

Bhabatosh Das
Vijai Singh

CHAPTER ONE

An introduction to human microbiome

Bhabatosh Das*
Molecular Genetics Laboratory, Centre for Human Microbial Ecology, Translational Health Science and Technology Institute, NCR Biotech Science Cluster, Faridabad, Haryana, India
*Corresponding author: e-mail address: bhabatosh@thsti.res.in

Contents

Progress in Molecular Biology and Translational Science, Volume 191
ISSN 1877-1173
https://doi.org/10.1016/bs.pmbts.2022.06.026

Abstract

The microbiome is an assemblage of a complex community of microbes (bacteria, archaea, fungi, algae, protists, and viruses) and their biomolecules occupying a well-defined habitat in or on a living or non-living object. All the environmentally exposed surfaces of the human body are colonized with trillions of microbes from all three major domains of life, including bacteria, archea, and microscopic eukarya. However, the richness, abundance, and functional potency of microbial taxa living in different parts of the human body are distinct. The Presence of common microbial taxa in different body habitats is also very rare. With the recent development of next generation sequencing technologies, it has been established that the indigenous microbial community in the human body and their functional attributes within a given body habitat vary over time, between ethnic groups and health status of the host. Perturbation of homeostasis in community structures or functions due to any extrinsic factors can alter mutualistic host–microbe interactions and may lead to disease. In addition, the dysbiotic state of the microbiome can also affect the efficacy of therapeutics, prolong treatment duration and lead to undesired treatment outcomes. In this chapter, structure, functions, diversity and dynamics of human microbiome in health and diseases, factors that alter microbial composition, interactions between microbial taxa and xenobiotics, and therapeutic efficacy of drugs in dysbiotic conditions are highlighted.

1. Introduction

Human physiology is profoundly influenced by the microbial communities and microbial metabolites produced by resident taxa in different body habitats.[1] Although the presence of microbes in the human body has been evident for a long time, until recently very little was known about (i) who they are? (ii) where are they? (iii) where do they come from? (iv) how are they selected? (v) what do they contribute to our health? Recent developments in nucleic acid sequencing technology in decoding orders of nucleotides in the complex mixture of nucleic acids have helped researchers in deciphering composition, diversity, and functional potency of complex microbial ecosystems associated with each and every environmentally exposed surface of the human body.[2] Metagenome-based analyses also reduced the culture bias- and helped in exploring the nearly identical structure of microbial ecosystems and the genetic potential of microbial taxa. This was achieved by directly analyzing mixtures of DNA fragments obtained from biological or environmental samples. It is well evident that microbes from all three domains of life including bacteria, archaea and eukarya are native members of the human microbiome. Several studies have established that the microbial content and composition of the microbial communities of healthy individuals differ significantly in different body habitats such as gastrointestinal tract, respiratory tract, skin, reproductive and urinary tract.[3] The highest microbial diversity in terms of community membership exists in the gastrointestinal tract and minimal compositional diversity at the species level is observed in the vaginal milieu. However, the richness of indigenous microbial communities and functional potency of microbiome in any body sites are not static. Over time, the microbial community composition varies. The highest heterogeneity is observed in the oral cavity, whereas the indigenous microbial taxa stay stable in the vagina and gastrointestinal tract.[4] The structural and functional changes in the microbial community in different body sites are modulated by diet, environmental exposure, host genetics, age, gender, health status and lifestyle.[5–7] The diversity of microbiomes of any ecosystem, including a given body habitat of a healthy human can be defined by the total number of specific microbes and their relative abundance distribution compared to the other microbes living in the same or a distinct habitat. Microbial diversity, functional potency of microbiota, complex interactions between microbiota and humans is directly linked with health and human diseases.[8] For example, reduced microbial diversity in the digestive tract can lead to a lower abundance of metabolic functions, an increased level of pro-inflammatory molecules and a lower capacity for xenobiotic degradations. Such alterations can cause malnutrition, inflammatory bowel disease, metabolic diseases and reduce the efficacy of several therapeutics. However, the diversity index and its association with health status are not uniform for all the body habitats and increased microbial diversity can also lead to a disease such as bacterial vaginosis. In a given body site, the microbial taxa or its genome (genes), proteins and metabolites may potentially serve as biomarkers for early prediction of health disorders and also in forecasting therapeutic potency of a drug commonly used to treat infectious or metabolic diseases.

2. Microbiome in different sites of human body

In a healthy adult human body there are around 39 trillion bacterial cells which means, we house 1.3 bacterial cells for every cell.[9] However, these numbers represent only one component of the microbiome (bacteria) and don't take into consideration other microbial cells including viruses, fungi and protists. The microbial communities that colonize different parts of the human body are very distinct and represent the major source of extraneous genetic heterogeneity and metabolic differences between individuals (Fig. 1). An estimated 1000–1500 species of microbes live in different parts of the human body at any one time with a body site-specific biogeography. The number of subspecies with unique genomic content could be 10 times higher than the estimated species numbers. In addition, the diversity of total genetic content of the microbiome associated with our body is almost 500 times higher than our estimated 21,306 protein-coding genes and 21,856 non-coding genes. However, due to their microscopic size, microbial biomass represents only 1–3% of our total body mass.

When the composition and functions of microbiomes at a given body site are compared, individuals fit into a specific clade based on their microbial diversity within people of the same ethnicity rather than clustering into random groups.[10] The accuracy of such clustering based on microbiome diversity is robust but varies depending on the selected body habitats and confounding factors excluded during the study. Ethnicity specific microbial signatures at the genus level may represent from 17% to 84% in different body habitats. The signature taxa may not be present in some communities (0% at the detection level) but are widely distributed in the entire population (100%) of other groups.

2.1 Gastrointestinal tract microbiome

The environmentally exposed surface of the human gastrointestinal tract is one of the most favorable niche for anaerobic and aerobic microbial taxa to colonize and live. Microbial colonization in the gut of neonates begins immediately after delivery, although the presence of microbes in an infant's meconium is also reported. In the early-life, gut microbiome composition is heterogeneous, mostly dominated by proteobacteria, actinobacteria and firmicutes.[11] Several longitudinal studies have reported that the microbial community in the gut of adult human remains relatively stable and contrasts with the rapid changes over the first 2 years of life. At the age of 3 years the microbiome of the gut takes a stable shape and represents composition and diversity very similar to an adult. Gut microbiota of a healthy adult is mainly dominated by four bacterial phyla *Firmicutes, Bacteroidetes, Actinobacteria* and *Proteobacteria* which represent ~98% of the total microbiota.[12] Remaining 2% of bacterial taxa represent 6–8 rare bacterial phyla

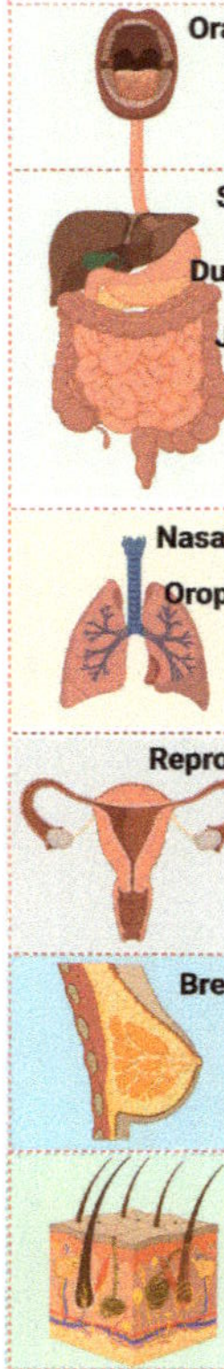

Body Site	Microbial load	Key microbial taxa	Important functions
Oral cavity	10^{12} (T)	*Porphyromonas, Streptococcus, Prevotella, Propionibacterium*	Reduce the risk of caries, periodontitis, endodontic infections, alveolar osteitis, and tonsillitis cardiovascular disease, stroke, preterm birth, diabetes and pneumonia
Stomach	10^2/ml	*Helicobacter pylori, Lactobacillus, Veillonella*	CAZymes production and energy harvest, Immune protection, Cell siganaling molecules production, Lantibiotics production, Vitamin biosynthesis, SCFAs synthesis, Xenobiotics degradation, Drugs metabolism, Resistance to colonization by enteric pathogens
Duodenum	10^4/ml	*Faecalibacterium, Clostridium, Lactobacillus, Ruminococcus, Blautia, Dorea, Megasphaera, Enterobacter, Streptococcus, Prevotella, Bacteroides, Bifidobacterium, Collinsella, Escherichia, Klebsiella, Fusobacterium, Brachyspira, Methylobacterium, Akkermansia Saccharomyces, Candida, Methanobrevibacter*	
Jejunum	10^5/ml		
Ileum	10^8/ml		
Colon	10^{11}/ml		
Nasal cavity	10^3/ml	*Streptococcus, Veillonella, Rothia, Haemophilus, Moraxella, Leptotrichia, Tropheryma, Corynebacterium, Staphylococcus, Prevotella, Dolosigranulum*	Mucosal immunity, Decrease lung inflammation, Neutrophil recruitment, AMPs production, Resistance to colonization by respiratory pathogens
Oropharynx	10^6/ml		
Lungs	10^2/ml		
Reproductive Tract	10^9/ml	*Lactobacillus crispatus, L. iners L. gasseri, L. jensenii, L. fermentum, L. coleohominis, L. johnsonii, Halomonas, Gardnerella vaginalis, Sneathia sanguinigens, Atopibium*	Resistance to colonization of pathogens, Antimicrobial peptides production, Lactic acid production, Reduce inflammation, Reduce preterm birth, Reduce HIV infection,
Breast Milk	10^9/L	*Bifidobacterium, Pseudomonas, Lactobacillus, Propionibacterium, Streptococcus*	Establish the infant gut microbiota, Reduce allergy, asthma, and obesity in breastfed children
Skin	10^{11} (T)	*Corynebacterium, Cutibacterium, Haemophilus, Streptococcus, Micrococcus, Branhamella, Pseudomonas, Streptophyta, Micrococcineae, Staphylococcus, Veillonella*	Modulation of expression of cytokines, Components of complement, AMPs production, Resistance to colonization of skin and respiratory tract pathogens

Fig. 1 Microbial load, key microbial taxa and functions associated with different parts of human body. Among all the environmentally exposed surfaces of the human body the highest microbial density and diversity is found in the colon. The functional potential of the microbial community in each body site is very distinct and linked with the physiological characteristics of the individual. T = Total.

including *Verrucomicrobia, Fusobacteria, Tenericutes, Spirochetes, TM7, Synergistetes* and *Euryarchaeota*. Each bacterial phylum is represented by specific core and variable bacterial taxa across the globe (Fig. 1). The microbiota living in the gastrointestinal tract is mostly commensal or symbiotic in nature and native abundance of these autochthonous bacterial taxa is crucial for health. However, dysbiosis in the composition or functional potency in the gut microbiota, both from the dominant and rare bacterial phyla, may act as strong risk factors for developing several communicable and metabolic diseases (Fig. 2). In addition, such dysbiosis may also affect the efficacy of several oral drugs by modifying the structure, permeability, bioavailability and stability. Thus, the understanding of baseline microbial composition, function, diversity and dynamics in the human gastrointestinal tract helped immensely in identifying key microbial taxa associated with health and diseases and developing personal medicine.

2.2 Respiratory tract microbiome

The human respiratory tract is a highly aerobic complex organ system that is divided into mainly two parts, upper respiratory tract and lower respiratory tract. Both the parts are colonized with a set of diverse region specific bacterial communities. Highest microbial load is reported in the upper respiratory tract (oropharynx region) and minimal microbial load is observed in the lungs (Fig. 1). Bacteria that are most commonly found in the upper respiratory tract are *Staphylococcus* spp., *Propionibacterium* spp. and *Corynebacterium* spp. *Moraxella* spp., *Dolosigranulum* spp. and *Streptococcus*. In the early age, lower respiratory tract microbiome is dominated by Firmicutes (*Staphylococcus* spp. and *Streptococcus* spp.) and Proteobacteria (*Moraxella* spp., *Haemophilus* spp.), while in the adult it is dominated by Firmicutes (*Streptococcus* spp., *Veillonella* spp.), Bacteroidetes (*Prevotella* spp.) and Actinobacteria (*Tropheryma whipplei*).[13] In addition to bacterial members, the healthy lungs of adult humans also harbor several members of fungi including *Eremothecium, Malassezia* and *Systenostrema*. The autochthonous microbial community in the respiratory tract provides protective functions and inhibits the colonization of pathogenic microbiota that causes respiratory infection.

2.3 Reproductive tract microbiome

The reproductive tract of the women in the reproductive age harbors stable but complex assemblages of discrete microbial communities, which plays an

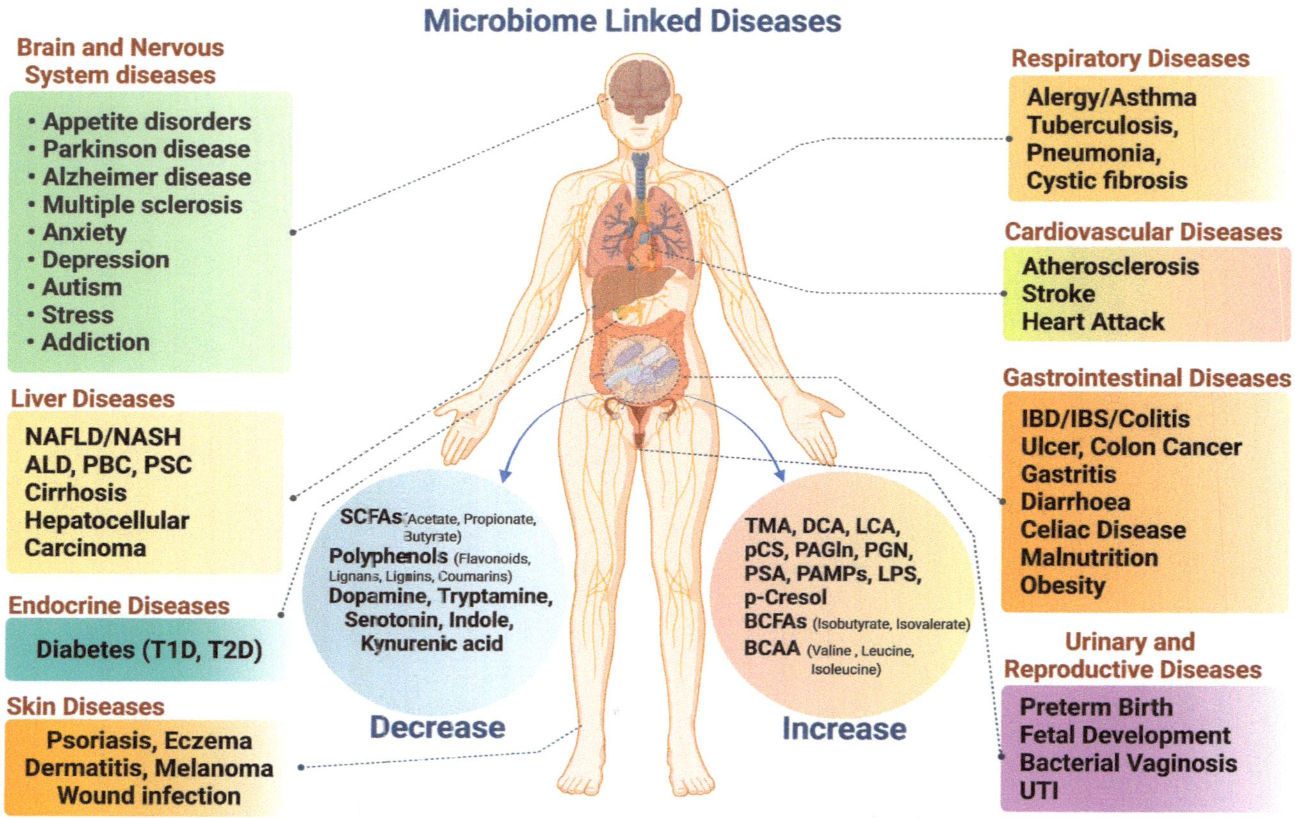

Fig. 2 Different diseases associated with the dysbiosis of human microbiome. Hundreds of microbial taxa reside in the human body providing important metabolic functions and several signaling molecules to the host. Optimal level of these functions is maintained in their homeostasis condition. However, abrupt changes (increase or decrease) often interfere with host physiology and lead to health disorders. TMA, Trimethylamine; DCA, Deoxycholic acid; LCA, Lithocholic acid; pCA, para-Cresol; PAGln, Phenylacetylglutamine; PGN, *Bacterial* peptidoglycan; PSA, Polysaccharide A; PAMPs, Pathogen-associated molecular patterns; LPS, Lipopolysaccharides; SCFAs, Short-chain fatty acids; BCFA, Branched- chain fatty acids; BCAA, branched-chain amino acids.

important role in the health and contribute substantially in the birth outcomes.[14] The biomolecules produced and secreted by the vaginal microbiota inhibit the colonization of allochthonous microbial taxa, reduce inflammation and maintain an acidic environment. In addition, the reproductive tract microbial communities of the biological mother act as the primary source of initial gut microbiota of a vaginally delivered baby. *Lactobacillus,* a non-spore forming microaerophilic, lactic acid producing Firmicutes, is the most dominant genus in the vaginal bacterial communities across the globe.[15] More than 35 different *Lactobacillus* species are reported in the vaginal bacterial communities and most common species are *L. iners, L. crispatus, L. gasseri, L. jensenii* and *L. johnsonii.*[16] Several metagenomic-based analyses established that three clades including (i) clade 1, *Lactobacillus* species exclusive (ii) clade 2, mixture of *Lactobacillus* and other bacterial species and (iii) clade 3, non-*Lactobacillus* bacterial taxa comprise the vaginal microbiome. The prevalence of each clade varies depending on the different ethnic groups (white, black, Hispanic, and Asian). Clade 1 and Clade 2 are common in American white and Asian women whereas Clade 3 is common in the vaginal microbial communities of the American black, Hispanic and African women. The common bacterial taxa in the Clade 3 are *Cryptobacterium, Prevotella, Sneathia, Dialister, Megasphaera, Atopobium, Gardnerella, Peptoniphilus, Atopobium, Halomonas, Eggerthella, Aerococcus, Finegoldia, and Mobiluncus Gardnerella, Prevotella, Porphyromonas, Peptoniphilus Atopobium, Halomonas* and others (Fig. 1). The higher pH values in the black (~pH 4.7) and Hispanic (~pH 5.0) women favor the higher prevalence of non-Lactobacillus dominated communities (Clade 3). Composition and functional attributes of the vaginal microbiome of all three types of clade play an important role in birth outcomes. Increased diversity of non-Lactobacillus vaginal microbiota in the reproductive tract cause a dysbiotic state known as bacterial vaginosis (BV) which is strongly linked with the higher risk of preterm birth (PTB) delivery.[17] The elevated level of microbial pro-inflammatory molecules, a molecular signature of BV, may contribute to the pathophysiology of PTB delivery by increasing the level of systemic and local inflammatory cytokines and interstitial collagenase production. In contrast, *Lactobacillus* dominated vaginal bacterial communities keep the vaginal milieu more acidic (~pH 4.3), low diverse and in non-inflammatory state that favors term birth delivery.

2.4 Skin microbiome

Skin is the largest organ in our body where hundreds of microbial species from bacteria and fungi live together mutually over time. The microbes residing the skin play important roles in the host resistance functions against colonization of pathogens and educating the immune system for differentiation of commensal and pathogenic strains in the microbial community either by recognizing specific microbial antigens or through a microbial signaling molecule. The composition of skin microbiome is highly heterogeneous and mostly influenced by the physiology of the skin including moist, dry and sebaceous microenvironments.[18] Lipophilic *Propionibacterium* species dominate sebaceous sites, whereas *Staphylococcus* and *Corynebacterium* species are most abundant in humid skin environments (on the feet and bends of the elbows). *Propionibacterium acnes* and *Staphylococcus epidermidis* strains are very common and dominant bacterial species in the human skin (Fig. 1). Longitudinal studies of the human skin microbiome revealed that though exposed to constant environmental changes, the microbial composition in the human skin stayed largely stable over time.[19]

3. Microbiome diversity and dynamics

Advancement and availability of NGS technology tremendously helped us understand the diversity, stability, plasticity and dynamics of human microbiome overtime and in different disease contexts. With current understanding, we believe that 10^{14} number of bacterial cells are associated with each human body while approximately 10^{30} number of bacterial and archeal cells inhabit the earth. The sequencing advancement and availability also help to study the species richness in the complex microbial communities and their stability upon exposure to xenobiotics and other environmental factors (Fig. 3). It is well evident that the stability vs plasticity dichotomy of the microbial community structure in different body habitats is distinct and modulated by various factors.[20] Three key components, environment, host genetics and microbiome interact with each other and maintain the stability and homeostasis of the gastrointestinal tract and other body habitats (Fig. 4).

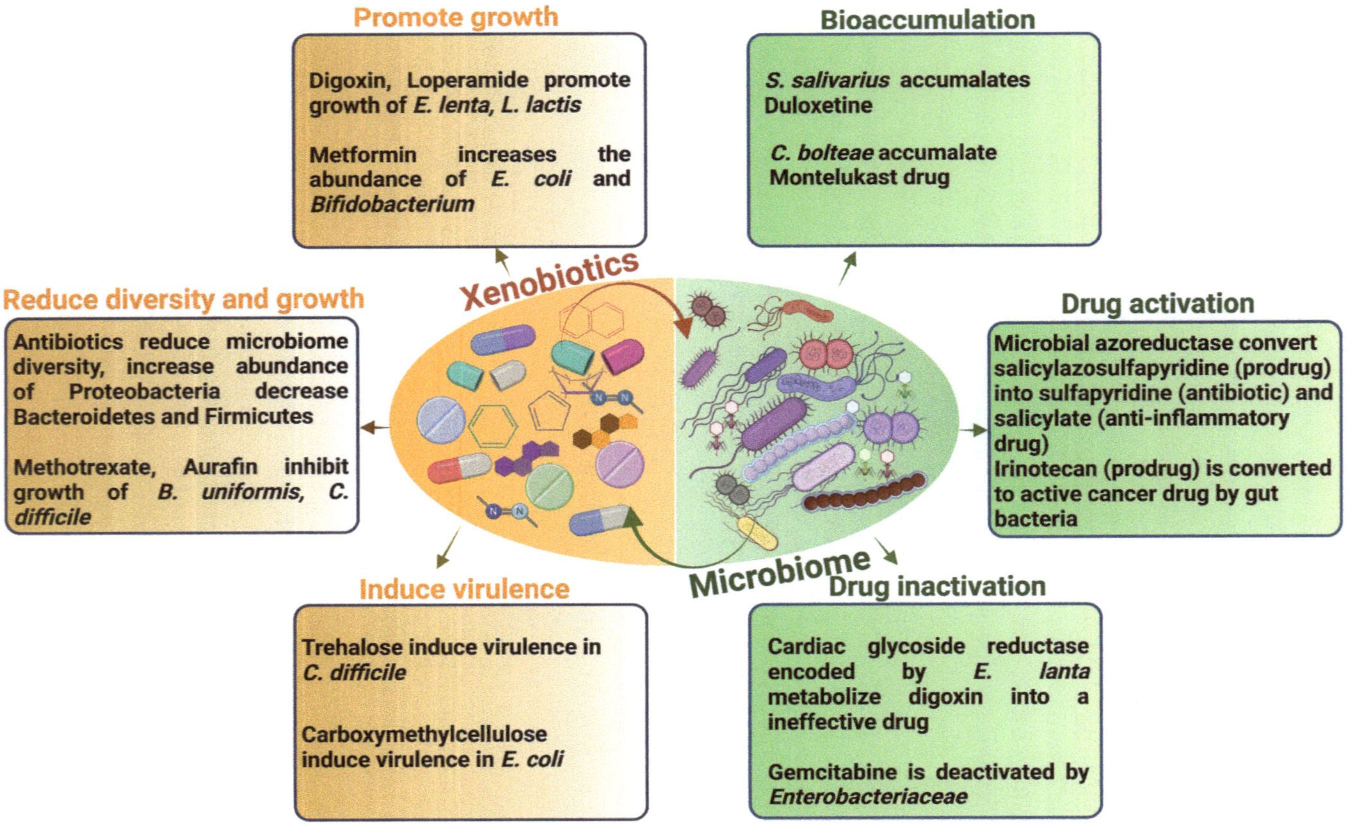

Fig. 3 Human microbiome and xenobiotics metabolisms. Microbes can activate, deactivate or accumulate xenobiotics including drugs. Prior to the entry in the systemic circulation such modifications affects toxicity, bioavailability, absorption and excretion of xenobiotics. Different xenobiotics including antimicrobials, pesticides, food additives, dietary bioactive compounds, industrial chemicals and environmental pollutants can also modulate (induce or repress) microbial growth in the human body and lead to the microbial dysbiosis in specific body habitats.

3.1 Geography

Geographic, socio-economic and cultural settings are important factors that modulate microbial community composition, diversity and genomic repertoires in the different sites of the human body. Highest differences in the taxonomic composition of microbial communities are found in the gastrointestinal tract and skin of children and adult individuals living in different countries.[21] It is reported that the ratios of Firmicutes to Bacteroidetes in individuals living in India is markedly different than those living in Japan.[22] Analyses with higher taxonomic resolution at the genus level showed that *Prevotella* and *Bacteroides* are the most dominant bacterial genera in the gastrointestinal tract of healthy adult Indian and Japanese population, respectively. Further analysis at the species level showed that *P. copri* is the highly abundant bacterial species followed by the *P. stercorea* and *Prevotella* sp. However, in the Japanese individuals these species are rare with minimal representation in a subset of subjects. While considering the genus *Bacteroides*, it is well established that *B. uniformis, Bacteroides* sp., *B. ovatus*, and *B. fragilis* are well conserved in the gut microbiome of Japanese individuals. In the fungal community, Ascomycota and Basidiomycota are the dominant phyla in both Indian and Japanese populations. Compared to healthy Indians abundance of *Basidiomycota* is very high in the gut of healthy adult Japanese individuals. Genus level analysis revealed that *Candida* and *Saccharomyces* are the two major fungal genera in the gut of Indian and Japanese individuals, respectively. *C. albicans, C. tropicalis* and *C. glabrata* are prevalent species of the genus *Candida* in the gut of Indian individuals while *Saccharomyces cerevisiae* mostly dominates the gut of Japanese individuals.

Studies also showed that the composition of gut microbial communities between the Danish and Indian individuals is significantly different.[23,24] The microbial richness in the Danish population is higher than the Indian population and a total of 12% variation is country specific. At genus level, the gut microbiota of the Indian individuals is enriched with *Prevotella* group 9, *Megasphaera* and *Lactobacillus*. While the gut microbiota of the Danish individuals harbors a higher abundance of *Bacteroides, Alistipes* and *Akkermansia*. Some bacterial genera are very country specific. For example, Achromobacter is present in the gut of Indian individuals (47%) and completely absent in the Danish population ($n = 500$). In contrast, *Anaerotruncus* is highly prevalent in the Danish population (56%) and absent in the Indians population (97%).

3.2 Host genetics

Although environmental factors have ascendancy over host genetics in shaping the structure of gut microbial ecology, a significant association between the host genotype and gut microbial composition has been well established.[25] Comparative analysis of the gut microbiome composition between monozygotic and dizygotic twins revealed the microbiota of monozygotic twins is more similar compared to the dizygotic twins. Genome wide association study showed that at least 9 loci of the human genome are associated with altered gut microbial diversity. Several analyses established the involvement of these loci in complex diseases, cell signaling, inflammation, prevention of microbial invasion in the systemic circulation and immune response. The effector molecules of these loci are known to interact with gut microbiota and modulate the production of pro-inflammatory cytokines when activated. There is a direct association of functional LCT (Lactase persistence gene) single-nucleotide polymorphism (SNP) with the relative abundance of *Bifidobacterium*. Similarly, SNP in the *Nod2* gene significantly correlated with the abundance of *Dorea* and *Subdoligranulum* spp. *CARD9* genes are associated with inflammatory bowel disease (IBD) and increase abundance of pro-inflammatory bacteria.[26] Riboflavin is a redox mediator and induces the growth of a widely distributed gut commensal *Faecalibacterium prauznitsii*. SNPs at the *CLEC4A–FAM90A1* locus affect the riboflavin biosynthetic process and reduce the abundance of *F. prauznitsii*. SNP at the MEFV locus, which encodes pyrin, has been associated with lower bacterial richness in the gut, especially in patients with familial Mediterranean fever. Deficiency of secretory immunoglobulin IgA in humans may lead to IBD. IgA deficient mice shed more *Salmonella Typhimurium* compared to the wild-type mice.

3.3 Age

Culture independent DNA-based studies showed evidence of bacterial presence in the placenta of biological mothers, in amniotic fluid, and in the meconium of term and preterm babies.[27] In parturition, the initial colonization and diversity of microbial community in the gut of new-borns depends on the mode of delivery, physiology and health status of mothers, antibiotic intake before delivery and the environment of the birth place.[28] Gastrointestinal tract of the new-born are primarily dominated by the facultative anaerobic members of Firmicutes and Proteobacteria, which play important role in reducing oxygen level in the gut and prepare the environment for stable colonization of strict anaerobic members of Bacteroidetes, Firmicutes and Actinobacteria. Several species of *Bifidobacterium* including *B. longum B. bifidum, B. breve, B. pseudocatenulatum* and *B. adolescentis* dominate the new-born gut microbiota based on their feeding patterns. *B. longum* ssp. *infantis* is frequently observed in the gut of breast fed infants, whereas *B. longum* ssp. *longum* dominates the gut microbiome of bottle-fed infants. However, interpersonal variation both in terms of richness and diversity is significantly higher among children than among adult individuals.[29] During the first year of life, the taxonomic composition of gut microbiota is less complex but highly heterogeneous between individuals and over time. Microbial composition matures in the gut of children at the age of 2 years and achieves similar diversity and richness like adults at the age of 3 years. In the elderly population relative abundance of some bacterial genera including *Bifidobacterium* is significantly reduced. However, currently there is no consensus definition of old-age-specific gastrointestinal microbiota. Like the gastrointestinal tract, composition and diversity of the vaginal microbiome also change with age. While the vagina of reproductive age women is dominated by different *lactobacilli*, the post-menopausal vaginal microbiota has significant depletion of *Lactobacillus* and also the total microbial load.

3.4 Diet

Diet is the most crucial factor in shaping microbial composition in the gastrointestinal tract and selection of gut microbiota.[30] Studies showed that vegans, unique plant species consumption, gluten free special diet, no dairy vs milk/cheese intake, fiber rich vegetables, whole grains, antibiotic-treated animal products, poultry, red meat, high-fat red meat, eggs, salted snacks, sugary sweets, ready to eat meals, sugar sweetened drink, and regular consumption of seafood significantly changes gut microbial community compositions and functional potency of the bacterial genome.[31–33] Regular intake of the animal-based diet increased the relative abundance of *Alistipes, Bilophila* and *Bacteroides* (bile-tolerant microorganisms) and decreased the levels of *Roseburia, Eubacterium rectale* and *Ruminococcus bromii* (Firmicutes) which encode large number of Carbohydrate-Active enZymes (CAZymes) and metabolize dietary plant polysaccharides efficiently.[34] Relative abundance of sulfite-reducing pathobiont, *Bilophila wadsworthia* is significantly increased on the animal-based diet. Consumption of milk derived saturated fat, but not plant derived polyunsaturated fat, also increases the relative abundance of *B. wadsworthia*. Dietary habits of host also facilitate horizontal gene transfer in the microbial communities and increase the functional potency of microbial genome. The gut of Japanese individuals frequently harbors *Bacteroides plebeius*, a bacteroidetes able to metabolize sulphated polysaccharide porphyran of marine red algae. The porphyranases encoding genes (*bp1689*, *porA*, *porB*, *porC*, *porD* and *porE*) and associated functions are transferred to the *B. plebeius* from a marine bacterium *Zobellia galactanivorans*. However, *B. plebeius* is absent in the gut microbiome of North American individuals.

3.5 Therapeutics

Different therapeutics used for prophylaxis or treatment of infectious and metabolic diseases cause short-term disturbance in the richness, diversity and functional potency and long-term ecological impacts on the human microbiome.[35] Among therapeutics, the highest impact on the microbiome is observed because of antimicrobial administration.[4] Although antibiotics are used to kill pathogenic microbes, most antibiotics are non-specific and kill both pathogenic and commensal bacteria in the absence of resistance functions. Antibiotic administration causes acute perturbations to the microbiome and leads to reduce richness and diversity of the gut microbiota. Broad-spectrum antibiotics including doxycycline, aminoglycosides, carbapenems, azithromycin can reduce the abundances of more than 30% of the gut bacteria and cause acute and huge drops in taxonomic and functional richness and diversity. The short-term effects of antibiotic administration include diarrhea, *C. difficile* infection, fungal colonization, whereas the long-term consequences of some antibiotics include the development of allergic conditions (asthma or food allergies), malnutrition and obesity. The degree of interruptions depends on the class of antibiotic, dosage and duration of treatment, route of administration, time for elimination and mechanism of action. For example, 7 days clindamycin administration drastically reduced richness and diversity of the human gut microbiota. A sharp reduction in the clonal diversity of Bacteroides persisted for more than 2 years. Oral administration of macrolides reduces bacterial diversity in the intestinal microbiota and leads to a shift in the abundance of Bifidobacterium and Bacteroides. Antibiotic administration in tuberculosis patients can vary from 6 months (newly infected) to 9 months (MDR-TB) with a combination of four antibiotics including isoniazid (INH), rifampicin (RIF), pyrazinamide (PZA), and ethambutol (EMB). Except, rifampicin, all three antibiotics are narrow spectrum and work mainly on the Mycobacterium species. However, such prolonged treatment profoundly affects the microbial richness and diversity in the gastrointestinal tract. Abundance of different Firmicutes (*Ruminococcus, Eubacterium, Blautia, Clostridium, Roseburia*) are significantly reduced while the relative abundance of *Prevotella* (Bacteroidetes) is significantly increased. Although the change in the microbiome composition due to antibiotic administration is mostly reversible, the recovery time may vary depending on the antibiotic regimen. In case of tuberculosis treatment the dysbiosis in the human gut microbiome persists for a long-term even after completion of treatment.

Recent findings demonstrated that more than 240 non-antibiotic drugs with human targets used in the clinics also modulate (reduce or increase) the richness, abundance, diversity and functional potency of human gut bacteria[35] including commensal *Clostridia* species (*C. ramosum* and *C. saccharolyticum*), pathobionts (*C. perfringens, C. difficile, Fusobacterium nucleatum,* enterotoxigenic *B. fragilis*), probiotic (*Lactobacillus paracasei*) (Fig. 3). Level of susceptibility to drugs with human targets for different bacterial species showed significant differences, with some members of Firmicutes (*Roseburia intestinalis, Eubacterium rectale*) and Bacteroidetes (*Bacteroides vulgatus*) being most sensitive, and bacterial species from γ-proteobacteria showed the most resistance. Several commensal bacteria which contribute significantly in the human health including butyrate producing *Enterobacter rectale, Coprococcus comes, Ruminococcus intestinalis* and propionate producing, *Prevotella copri, Bacteroides vulgatus, Blautia obeum* are relatively more susceptible to the non-antibiotic drugs. It is quite evident that in contrast to most of the routinely used antibiotics, human-targeted drugs have long-term impact in the ecology of gut microbiota.

3.6 Lifestyle

Alcohol consumption, smoking, exercise, vitamin supplementation and probiotics significantly modulate the taxonomic and functional composition and diversity of the gut microbial community.[7] Regular consumption of alcohol can robustly segregate microbiota profiles from non-drinker individuals. Alpha diversity of the gut microbiota significantly increased in the gut of frequent alcohol consumers. At the genus level, relative abundance of *Bifidobacterium* ASV is significantly increased in the gut of frequent alcohol consumers. Further analysis showed that increased microbial diversity and richness in the gut of red wine, beer/cider and white wine

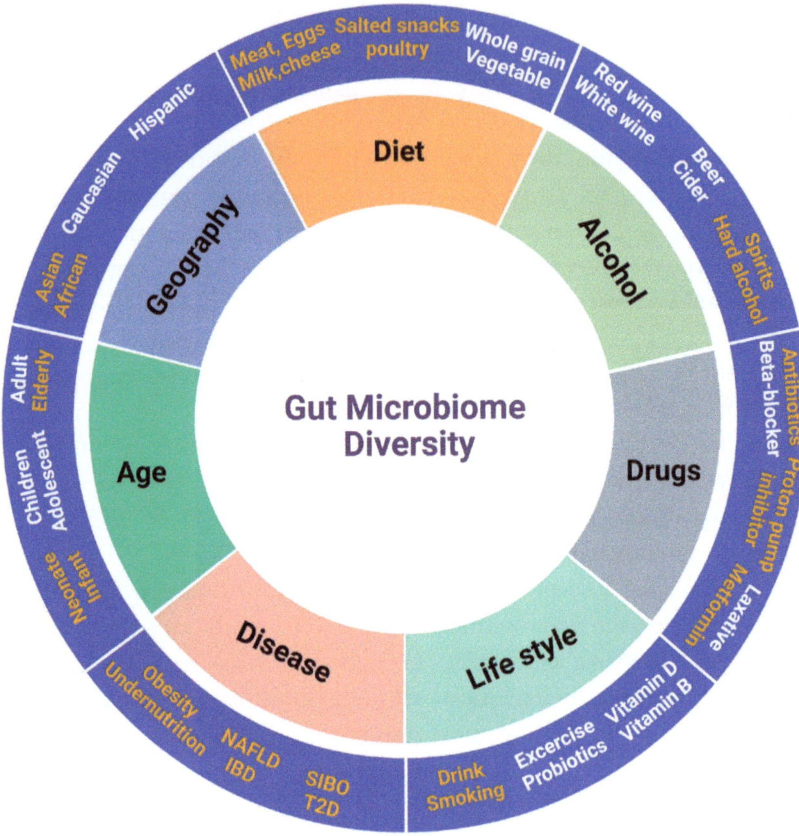

Fig. 4 Factors that modulate gut microbial diversity. Although gut microbial composition is quite robust, overtime in individuals, several factors can increase or reduce gut microbial diversity depending on the nature of the chemical compounds, quantity, and exposure frequency. In addition to diet and xenobiotics, the most important factors in shaping gut microbial composition, geography, age, host physiology, and lifestyle can also contribute to gut microbiome diversity. Outer ring letter colors; White, increases diversity; Panton, reduces diversity.

consumers is consistent and robust. Daily use of alcohol has the maximum effect on the increased microbial diversity index compared to never, rare and occasional consumers (Fig. 4). However, the difference of diversity between spirits/hard alcohol consumers and non-drinkers is not very high. Microbial diversity is also increased upon vitamins and probiotic supplementations.

4. Key functions of microbiome in human physiology

Composition of microbial communities and their functional potency are key for the sustainability of a microbial ecosystem. Microbes associated with different parts of the human body use their cell surface receptors and cytoplasmic sensory systems to monitor their surrounding environment in order to assimilate nutrients and energy for their multiplication and adaptation in stressful conditions. In the homeostatic condition, microbial functions and the end products of microbial metabolic reactions profoundly contribute to human physiology and reduce risk of different metabolic and infectious diseases.[36] In addition, microbial derived functions directly interact with different therapeutic agents and modify their structure, permeability, bioavailability and stability, which may increase or decrease its efficacy.

4.1 Metabolic functions

Microbes inhabiting the human body provide a broad range of metabolic functions for the fermentation of non-digestible complex plant polysaccharides and endogenous mucus. Human genome is not equipped with the enzymes for several metabolic functions including digestion of complex dietary macronutrients, biosynthesis of essential amino acids and vitamins and production of subset of cellular signaling molecules. Thousands of microbes reside in our body providing all such important metabolic functions and maintaining the healthy host physiology. Digestion of dietary fibers, mostly plant polysaccharides, needs a number of CAZymes. Gut bacterial taxa, mostly Firmicutes and Bacteroidetes, are the major source of hundreds of different CAZymes like glycoside hydrolases (GHs), glycosyl transferases (GTs), carbohydrate esterases (CEs), polysaccharide lyases (PLs), and auxiliary activities (AAs).[33,37] A healthy human passes 20–60 g carbohydrates and 5–20 g protein per day in the colon and cecum. Indigestible dietary fibers and some proteins, which escape the digestion in the stomach and small intestine, are fermented to short-chain fatty acids (acetate, propionate, and butyrate) and branched-chain fatty acids (isobutyrate, isovalerate and 2-methylbutyrate) in the cecum and colon.[38] Short-chain fatty acids (SCFAs) bind to the GPR41, GPR43, GPR109A and modulates IL18 production, FoxP3 functions, differentiation of colonic T_{regs}, and HDAC9 functions.[39] Other than providing digestive enzymes, the gut microbial community is also involved in the biosynthesis of essential amino acids and vitamins. Different *Lactobacillus* species living in the vaginal milieu use oligosaccharides derived from vaginal epithelial glycogen and produce D- and L-lactic acids. *Lactobacillus iners* produces only L-lactic acids, whereas *L. crispatus produces both* D- and L-lactic acids. Such metabolites play a crucial role in preventing dysbiosis.[40] However, their protective efficacy differs depending on their structure. For example, D-lactic acid is more protective against colonization of allochthonous bacterial taxa in the vaginal milieu. Metabolites derived from microbes compete with drugs for host xenobiotic metabolic functions. The tyrosine derived microbial product 4-methylphenol (p-cresol) and acetaminophenare recognized by the host encoded sulfotransferase SULT1A1. Increased levels of p-cresol reduce the accessibility of SULT1A1 to acetaminophen (active form). Therefore, increased levels of p-cresol reduce the biotransformation of the acetaminophen to acetaminophen sulfate (the inactive form) and drastically reduce its therapeutic dose in a subset of individuals enriched with the p-cresol producing gut bacteria.[41]

4.2 Xenobiotic degradation

Xenobiotics, compounds that are non-endogenous origin in a biological system including drugs, pesticides, food additives, dietary bioactive compounds, industrial chemicals and environmental pollutants are often metabolized by human gut microbiota to modulate their toxicity, bioavailability, absorption and excretion through urinary system.[42] Most commonly observed modifications are reduction and hydrolysis. Several pro-drugs that contain azo bonds in their scaffold require reduction for its therapeutic effects. The sulfonamide antibiotic prontosil is not active in vitro against bacterial species. However, the prontosil inhibits the growth of gut bacteria in vivo. The difference of antibiotic activity of prontosil between in vivo and in vitro conditions is due to its biotransformation (prontosil to triaminobenzene and sulfanilamide) in the gut by a microbiota encoded azoreductase, enzyme belonging to the family of oxidoreductases which use 2 NAD(*P*)H to reduce one azo compound substrate. Azoreductase enzymes are widespread across different members of the gut microbial communities and possess broad substrate compatibility. The bioavailability of more than 50 different oral drugs in the systemic circulation depends on the extent of their interactions with gut microbial and hepatic enzymes. Gut bacteria derived β-glucuronidases also modulate the biological activity and toxicity of different oral drugs used to treat cancer and inflammation. Bile acid metabolizing bacteria *Clostridium scindens* generates the secondary bile acids deoxycholate and lithocholate, which has the unique ability to reduce the growth of *C. difficile* and prevent its colonization in the gut.

Enzymes produced by the different members of Firmicutes, Bacteroidetes, Actinobacteria, and proteobacteria can reduce the bioavailability of active drugs (Fig. 3). For example, patients suffering from cardiac arrhythmia and heart failure are often treated with the drug cardiac glycoside digoxin. Therapeutic range of digoxin is very small, 0.5–2 ng/mL. It was demonstrated that a subset of autochthonous gut bacterial species, *Eggerthella lenta*, reduces digoxin to dihydrodigoxin, and reduces the bioavailability of the drugs.[43] In approximately 10% of patients, more than 50% of the administered digoxin is inactivated to dihydrodigoxin very rapidly. Dihydrodigoxin was identified to be excreted rapidly through the urine which decreased systemic digoxin concentration. Biotransformation of digoxin to dihydrodigoxin is mediated by plasmid encoded cardiac glycoside reductases Cgr1 and Cgr2. Some strains of *E. lenta* harbor *cgr* positive plasmid and some are not. Only the *cgr* positive *E. lenta* has the ability to reduce digoxin. However, the expression level and activity of Cgr1 and Cgr2 are sensitive to the protein rich diets and some other gut bacterial species. *Enterococcus faecalis*, a well conserved Firmicutes in the human colon, encodes pyridoxal phosphate-dependent tyrosine decarboxylase (TyrDC), which decarboxylates L-dopa to dopamine.[44] Decarboxylation of L-dopa limits its bio-availability and also contributes to drug toxicity. Molybdenum-dependent dehydroxylase, encoded by the *E. lenta* converts dopamine to m-tyramine. Heterogenous response of Parkinson's patients to dopamine could be due to difference of their gut microbiome.

4.3 Immunity

Commensal microbiota associated with different parts of the human body play an important role in modulating functions of localized and systemic immune systems by directly interacting with the effector cells or their metabolites.[45] The conserved molecular structures of the Pattern-recognition receptors (PRRs) in the innate immune system interact with microbial antigens or metabolic products to sense the environment and translate the signals into host physiological responses. Toll-like receptors (TLRs), the C-type lectin receptors, the RIG-I-like receptors, the nucleotide-binding oligomerization (NOD)-like receptors (NLRs), the absence of melanoma 2 (AIM2)-like receptors and the OAS-like receptors are well characterized PRRs expressed by different cells located across the body like gastrointestinal tract, oral cavity, lung, urogenital tract, bone marrow and skin.[46] Toll-like receptors (TLRs) are located at the cell surface of the host cell. Its external leucine-rich repeat (LRR) domain recognizes microbial antigens and intracellular interleukin-1 receptor (IL-1R)–like signaling domain activates transcription factor NF-κB, leading to the induction of expression of pro-inflammatory genes. The Nod receptors (Nod1 and Nod2) are intracellular and widely distributed in the cytosol of immune cells and epithelial cells. Activation of TLRs and NOD-like receptors lead to the production of mucus and antimicrobial peptides (RegIIIγ, RegIIIβ, Ang4 and Itln1) in the gut and restrict the growth of allochthonous microbiota. Cell surface receptors and cytoplasmic receptors of the epithelial cells recognize microbial metabolites like SCFAs (acetate, butyrate, propionate), polyamines (spermine), taurine, histamine, indole and modulate the activity of inflammasome and promotes the barrier function of the epithelium.[47] Commensal microorganisms positively modulate the maturation and tissue-specific functions of innate lymphoid cells (natural killer cells, and non-cytotoxic subsets ILC1, ILC2 and ILC3). For example, polysaccharide A (PSA) produced by the commensal gut bacteria *B. fragilis* interact with TLR2 and modulate the balance of Th1/Th2 cell and suppresses the production of pro-inflammatory IL-17 and promotes expression of IL-10.[48] Cells associated with innate immunity express Pathogen-associated molecular patterns (PAMP) recognition receptors in their surface membrane for sensing pathogen antigens. Lipopolysaccharide (LPS), an abundant molecule in the outer membrane of Proteobacteria and Bacteroidetes, is recognized by Toll-like receptor 4 (TLR4) and MD-2 type PAMPs. Depending on the structure of lipid A molecule of LPS, TLR4 activates the transcription factor NF-κB and increases the level of pro-inflammatory cytokines leading to inflammation and pathogen clearance.

4.4 Resistance to infection

Autochthonous microbiota inhabiting different parts of the human body provide a crucial line of resistance to colonization by allochthonous microbiota and are very important in prevention of microbial invasion to systemic circulation, tissues and organs by different pathogens. Germ-free animals are highly susceptible to microbial infections. Depletion of endogenous bacteria due to therapeutic exposure also makes the host susceptible to bacterial and fungal infections.[49] For example, *C. difficile, Shigella flexneri, Citrobacter rodentium, Listeria monocytogenes* and *S. enterica* serovar Typhimurium infections are often linked with oral antibiotic use. Native microbiota can also over-compete the non-endogenous microbiota in their natural habitats. Such as, *C. difficile* infection can efficiently (~90%)

be cured with fecal microbial transplantation.[50] Higher microbial diversity in the gut with increased abundance of Dorea and Coprococcus is significantly resistant to *Campylobacter jejuni* infection compared with people who have low-diversity gut microbiota. Autochthonous microbiota inhibit the colonization of pathogens and infection by producing antimicrobial peptides (bacteriocin), limiting space and nutrients and also by producing signaling molecules that repress the expression of colonization factor in pathogenic bacteria. Several gut bacterial species including *L. salivarius, L. gasseri, L. acidophilus, L. crispatus* and *Enterococcus* spp. produced different types of bacteriocin like ABP-118, enterocin B, salivaricin B, gassericin T, lactacin B and gassericin A. *Ruminococcus obeum*, a member of the Firmicutes and native member of gut microbiota, can prevent the intestinal colonization of the cholera causing pathogen *Vibrio cholerae* by producing autoinducer-2 (furanone signal), which reduces the expression of colonization factors (TCP) of the pathogen, an essential virulence factor for the development of disease.[51]

5. Microbiome dysbiosis and associated diseases

Compositional and/or functional imbalance of resident microbes in an environment to an extent that surpasses its native functional capacity and resilience proficiencies is commonly known as microbiome dysbiosis. It is subject and context specific. Several factors including environment, availability of nutrients, abundance and nature of xenobiotic, host genetics may lead to dysbiosis. Microbial dysbiosis is a potential risk factor for several communicable and non-communicable diseases (Fig. 2). However, mechanistic insights are evolving to explain the causal link between dysbiosis and disease susceptibility and progression.

5.1 Infectious diseases

Several allochthonous gut bacteria including *Escherichia coli, Shigella, Salmonella, Campylobacter, Aeromonas, V. cholerae* and *C. difficile* are mostly associated with bacterial diarrhea. Autochthonous members of the human gut microbiota reduce the susceptibility and also help in quick recovery from diarrhea. *R. obeum* is involved in the quick recovery from *V. cholerae* infection.[51]

Dysbiosis of gut microbiome is also associated with Crohn's disease (CD) and ulcerative colitis (UC), together known as inflammatory bowel diseases (IBDs), a chronic condition of the gastrointestinal tract linked with multiple host and environmental factors.[52] Both CD and UC inter-individual variation of the gut microbiomes is high, showing decreases in taxonomic diversity compared to the gut microbiome of healthy individuals. At the phylum-level, relative abundance of Firmicutes is decreased, while the relative abundance of Proteobacteria significantly increased. In CD at the family level resolution relative abundance of Lachnospiraceae and Ruminococcaceae is decreased, while at the genera level proportion of *Roseburia* and *Faecalibacterium* are decreased, whereas abundance of *Ruminococcus gnavus* increases.

Severe acute respiratory syndrome coronavirus 2 (SARS-CoV-2), the causative agent of Coronavirus disease 2019 (COVID-19), infects primarily the upper and lower respiratory tracts but also use other environmentally exposed mucosal lining of the human body for entry to the host body. For the same virus, susceptibility and disease severity widely vary between individuals. Several studies showed that the disease risk and heterogeneity in the pathophysiology mostly depend on the host immunity and also possibly gut and respiratory tract microbiome diversity.[53] Several bacterial species in the gut microbiota modulate host innate immunity and provide protection against bacterial and viral infections. It was observed that the relative abundance of bacterial taxa belonging to family Lachnospiraceae and Bifidobacteria, were significantly low in patients with acute respiratory distress syndrome (ARDS) compared to the COVID-19 patients with mild syndrome. In addition, increased relative abundance of *Staphylococcus* and *Enterococcus* were also observed in ARDS patients. Different members of the human microbiome positively modulate blood plasma cytokine and thus play an important role determining disease susceptibility, pathophysiology and recovery.[54]

5.2 Metabolic diseases

Most metabolic diseases are multifactorial and a combination of host genetics and environmental factors lead toward the development of the disease. Host-associated microbial taxa and microbial metabolites have recently been recognized as crucial factors for chronic metabolic diseases.[55] Functional potency and metabolic capacity of human-associated microbiota are far more versatile than those encoded by the human genome. A large number of microbial metabolic functions are crucial for human physiology and digestion functions. However, over-abundance of such metabolic functions in the autochthonous bacteria and inclusion of spurious metabolites or enzymes through allochthonous bacteria are linked with multiple metabolic disorders in humans (Fig. 2). Microbiome-wide association studies have reported associations between the gut microbiome and metabolic diseases, including obesity, malnutrition, type 2 diabetes (T2D)[56] and non-alcoholic fatty liver disease (NAFLD).[57] A dozen of Firmicutes (*Faecalibacterium prausnitzii, Clostridium leptum, family Ruminococcaceae, Eubacterium rectale,* and *Roseburia* spp.) and few Bacteroidetes (*Alistipes* and *Butyricimonas*) have been implicated in the fermentation of dietary fibers that leads to the production of SCFAs, a group of carboxylic acids with aliphatic tails smaller than six carbons principally acetate, propionate, and butyrate. SCFAs enter into host cells by both via mono-carboxylic acid transporters and passive diffusion. SCFAs play an important role as signaling molecules. Propionate, acetate, and butyrate (lesser extent) are ligands for G protein-coupled receptor (GPCRs), Gpr41 which stimulates production of leptin, a polypeptide hormone associated with appetite and energy metabolism. In the homeostasis condition normal gut micro- biome produce 50–100 mmol/L SCFAs per day. Optimal level of SCFA (butyrate) in the gut is associated with reduced insulin resistance, whereas reduced production or absorption of propionate is linked with increased risk of T2D.

Metagenomic studies have also revealed that dysbiosis of gut microbiota at different taxonomic levels affects host metabolism and energy assimilation. Different genes coding for proteins with metabolic, fermentation and transport functions including glycoside hydrolases, β-fructosidase, phosphotransferases, fermentation enzymes and transport proteins are enriched in the gut microbiome of obese individuals. Such enrichment is observed because of the increase in Firmicutes/Bacteroidetes ratio in the gut microbiome of obese individuals. However, selected members of Firmicutes can also reduce obesity by producing SCFA (Butyrate). Fecal transplantation from lean donors into obese individuals increases the relative abundance of butyrate producing *Roseburia intestinalis,* which helps in reducing obesity.

Dysbiosis of the gut microbiome also contributes significantly to the pathogenesis of NAFLD, a broad spectrum of liver abnormalities, ranging from uncomplicated hepatic steatosis to different degrees of inflammation and fibrosis in the hepatic cells (NASH). Microbial dysbiosis with increase Firmicutes, proinflammatory bacteria including *Collinsella aerofaciens* and *Clostridium coccoides*, and reduced abundance of Bacteroidetes affects the tight junctions of gut epithelial cells, increases gut permeability to microbial products and increases hepatic exposure to inflammatory and toxic molecules that induce inflammation and fibrosis in the hepatic cells.[58] In addition, the optimal level of SCFAs produced in the gut also lowers activity of the hepatic fatty acid synthetase (FAS) and reduced lipid accumulation. GPR43 is also present on intestinal neutrophils and might get activated due to increased levels of butyrate, therefore contributing to NASH pathogenesis during dysbiosis.

Gut bacterial metabolic pathways can efficiently use choline, phosphatidylcholine and L-carnitine and produce trimethylamine (TMA) in the gastrointestinal tract. Through portal circulation, TMA enters into the liver and hepatic flavin monooxygenase 3 (FMO3) oxidizes it into TMA-N-oxide (TMAO). Increased level of TMAO in the systemic circulation is a strong risk factor for atherosclerosis in humans.[59] In mice models, oral application of 3,3-dimethyl-1-butanol (analogue of choline) reduces the gut microbiota mediated production of TMA and prevents atherosclerosis by lowering systemic TMAO level.

6. Perspectives

Human microbiome studies were initiated more than a hundred years ago. Initially, the human-associated microbial taxa were primarily studied through culturing single species. However, culture-independent study of human microbiomes showed previously unknown tremendous microbial richness, functional potency, health and age associated dynamics across the body habitats. The dysbiotic state linked with gastrointestinal, respiratory, uro-reproductive, metabolic, hepatic, psychiatric, neurologic, oncologic, metabolic and cardiovascular illness opens new avenues for non-invasive biomarker discovery by targeting signature microbial taxa or their genomic contents, proteins and metabolites. Deeper understanding in the functional potency of human associated microbiota, association of microbial metabolites with host physiology will help us in decoding causality and disease mechanisms and developing microbiome based therapeutics like Fecal microbiome transplantation (FMT), vaginal microbiome transplantation and skin microbiome transplantation. Understanding of metabolic capacity of disease preventing microbes will help in formulation of prebiotics and synbiotics. Several clinical and pre-clinical studies, including (i) gut microbiome-based therapy like FMT to treat *C. difficile* infection, (ii) inhibition of trimethylamine synthesis in gut bacteria for atherosclerosis, (iii) *B. fragilis* for ameliorated intestinal dysbiosis and reducing anxiety, (iv) *Lactobacillus johnsonii* mediated airway protection against both allergen challenge and respiratory viral infection and (v) engineered *E. coli* for reducing adiposity provided the tremendous potential of microbiome-based therapeutics for improving public health.

7. Conclusion

Microbes are found associated with virtually any multicellular organism and have developed mutualistic relations with the host to assist their host metabolisms, provide resistance against infections, and sustain their relationship with the host. Different body sites of the human being are colonized by taxonomically and functionally distinct microbiota. Acquisitions of the microbial taxa in early life happen vertically from the mother and laterally from the environment. The microbial load, composition, richness, dynamics and functions are context and anatomical site specific. However, the discoveries of molecular links between identities of specific microbial taxa, microbial derived molecules linked with host physiology, signaling pathways, and protection against infections and disease susceptibility remain to be elucidated to a large extent. This is a frontier for human microbiome research to identify preventive medicine and contribute to the medical management of infectious and chronic metabolic diseases.

Acknowledgment

B.D. acknowledges Lekshmi N. for critical reading of the chapter. Images are created using BioRender.

Author contributions

B.D. designed, wrote, reviewed and edited the chapter.

Funding

This study received financial support from the Department of Biotechnology (DBT), Govt. of India (Grant No. BT/PR38173/MED/97/474/2020) and Translational Health Science and Technology Institute (Intramural, 2021–2022).

Conflict of interest

Author declares that there is no conflict of interest.

References

1. Lynch SV, Pedersen O. The human intestinal microbiome in health and disease. *N Engl J Med*. 2016;375(24):2369–2379.
2. Franzosa EA, Hsu T, Sirota-Madi A, et al. Sequencing and beyond: Integrating molecular 'omics' for microbial community profiling. *Nat Rev Microbiol*. 2015;13(6):360–372.
3. Lloyd-Price J, Mahurkar A, Rahnavard G, et al. Strains, functions and dynamics in the expanded human microbiome project. *Nature*. 2017;550(7674):61–66.

4. Cho I, Yamanishi S, Cox L, et al. Antibiotics in early life alter the murine colonic microbiome and adiposity. *Nature*. 2012;488(7413):621–626.
5. Gacesa R, Kurilshikov A, Vich Vila A, et al. Environmental factors shaping the gut microbiome in a Dutch population. *Nature*. 2022;604(7907):732–739. https://doi.org/10.1038/s41586-022-04567-7.
6. Sanna S, Kurilshikov A, van der Graaf A, Fu J, Zhernakova A. Challenges and future directions for studying effects of host genetics on the gut microbiome. *Nat Genet*. 2022;54(2):100–106. https://doi.org/10.1038/s41588-021-00983-z.
7. Vujkovic-Cvijin I, Sklar J, Jiang L, Natarajan L, Knight R, Belkaid Y. Host variables confound gut microbiota studies of human disease. *Nature*. 2020;587:448–454.
8. Das B, Nair GB. Homeostasis and dysbiosis of the gut microbiome in health and disease. *J Biosci*. 2019;44(5).
9. Sender R, Fuchs S, Milo R. Are we really vastly outnumbered? Revisiting the ratio of bacterial to host cells in humans. *Cell*. 2016;164(3):337–340.
10. Human Microbiome Project C. Structure function and diversity of the healthy human microbiome. *Nature*. 2012;486(7402):207–214.
11. Gensollen T, Iyer SS, Kasper DL, Blumberg RS. How colonization by microbiota in early life shapes the immune system. *Science*. 2016;352(6285):539–544.
12. Das B, Ghosh TS, Kedia S, et al. Analysis of the gut microbiome of rural and urban healthy Indians living in sea level and high altitude areas. *Sci Rep*. 2018;8(1):10104.
13. Ho Man W, Wouter AA, Piters S, Bogaert D. The microbiota of the respiratory tract: Gatekeeper to respiratory health. *Nat Revw Microb*. 2017;15:259–270.
14. Kumar S, Kumari N, Talukdar D, et al. GARBH-Ini study group. The vaginal microbial signatures of preterm birth delivery in Indian women. *Front Cell Infect Microbiol*. 2021;11 [622474].
15. Ravel J, Gajer P, Abdo Z, et al. Vaginal microbiome of reproductive-age women. *Proc Natl Acad of Sci USA*. 2011;106:8.
16. Mehta O, Ghosh TS, Kothidar A, et al. Vaginal microbiome of pregnant Indian women: Insights into the genome of dominant lactobacillus species. *Microb Ecol*. 2020;80(2):487–499.
17. Callahan BJ, DiGiulio DB, Goltsman DSA, et al. Replication and refinement of a vaginal microbial signature of preterm birth in two racially distinct cohorts of US women. *Proc Natl Acad of Sci USA*. 2017;114(37):9966–9971.
18. Byrd AL, Belkaid Y, Segre JA. The human skin microbiome. *Nat Rev Microbiol*. 2018;16 (3):143–155.
19. Grice EA, Segre JA. The skin microbiome. *Nat Rev Microbiol*. 2011;9(4):244–253.
20. Lozupone CA, Stombaugh JI, Gordon JI, Jansson JK, Knight R. Diversity, stability and resilience of the human gut microbiota. *Nature*. 2012;489(7415):220–230.
21. Yatsunenko T, Rey FE, Manary MJ, et al. Human gut microbiome viewed across age and geography. *Nature*. 2012;486(7402):222–227.
22. Pareek S, Kurakawa T, Das B, et al. Comparison of Japanese and Indian intestinal microbiota shows diet-dependent interaction between bacteria and fungi. *NPJ Biofilms Microbiomes*. 2019;5(1):37.
23. Alvarez-Silva C, Kashani A, Hansen TH, et al. Trans-ethnic gut microbiota signatures of type 2 diabetes in Denmark and India. *Genome Med*. 2021;13(1):37.
24. Pinna NK, Anjana RM, Saxena S, et al. Trans-ethnic gut microbial signatures of prediabetic subjects from India and Denmark. *Genome Med*. 2021;13(1):36.
25. Bonder MJ, Kurilshikov A, Tigchelaar EF, et al. The effect of host genetics on the gut microbiome. *Nat Genet*. 2016;48(11):1407–1412.
26. Knights D, Silverberg MS, Weersma RK, et al. Complex host genetics influence the microbiome in inflammatory bowel disease. *Genome Med*. 2014;6(12):107.

27. Aagaard K, Ma J, Antony KM, Ganu R, Petrosino J, Versalovic J. The placenta harbors a unique microbiome. *Sci Transl Med.* 2014;6(237) [237ra65].
28. Dominguez-Bello MG, Costello EK, Contreras M, et al. Delivery mode shapes the acquisition and structure of the initial microbiota across multiple body habitats in newborns. *Proc Natl Acad Sci U S A.* 2010;107(26):11971–11975.
29. Sommer F, Backhed F. The gut microbiota masters of host development and physiology. *Nat Rev Microbiol.* 2013;11(4):227–238.
30. De Filippo C, Cavalieri D, Di Paola M, et al. Impact of diet in shaping gut microbiota revealed by a comparative study in children from Europe and rural Africa. *Proc Natl Acad Sci U S A.* 2010;107(33):14691–14696.
31. Claesson MJ, Jeffery IB, Conde S, et al. Gut microbiota composition correlates with diet and health in the elderly. *Nature.* 2012;488(7410):178–184.
32. Muegge BD, Kuczynski J, Knights D, et al. Diet drives convergence in gut microbiome functions across mammalian phylogeny and within humans. *Science.* 2011;332 (6032):970–974.
33. Wardman JF, Bains RK, Rahfeld P, Withers SG. Carbohydrate-active enzymes (CAZymes) in the gut microbiome. *Nat Rev Microbiol.* 2022. https://doi.org/10.1038/s41579-022-00712-1.
34. Hehemann JH, Correc G, Barbeyron T, Helbert W, Czjzek M, Michel G. Transfer of carbohydrate-active enzymes from marine bacteria to Japanese gut microbiota. *Nature.* 2010;464(7290):908–912.
35. Maier L, Pruteanu M, Kuhn M, et al. Extensive impact of non-antibiotic drugs on human gut bacteria. *Nature.* 2018;555(7698):623–628.
36. Fan Y, Pedersen O. Gut microbiota in human metabolic health and disease. *Nat Rev Microbiol.* 2021;19(1):55–71.
37. El Kaoutari A, Armougom F, Gordon JI, Raoult D, Henrissat B. The abundance and variety of carbohydrate-active enzymes in the human gut microbiota. *Nat Rev Microbiol.* 2013;11(7):497–504.
38. Sanna S, van Zuydam NR, Mahajan A, et al. Causal relationships among the gut microbiome, short-chain fatty acids and metabolic diseases. *Nat Genet.* 2019;51 (4):600–605.
39. Koh A, De Vadder F, Kovatcheva-Datchary P, Backhed F. From dietary fiber to host physiology: Short-chain fatty acids as key bacterial metabolites. *Cell.* 2016;165 (6):1332–1345.
40. Witkin SS, Mendes-Soares H, Linhares IM, Jayaram A, Ledger WJ, Forney LJ. Influence of vaginal bacteria and D- and L-lactic acid isomers on vaginal extracellular matrix metalloproteinase inducer: Implications for protection against upper genital tract infections. *MBio.* 2013;4(4):[e00460-13].
41. Spanogiannopoulos P, Bess EN, Carmody RN, Turnbaugh PJ. The microbial pharmacists within us: A metagenomic view of xenobiotic metabolism. *Nat Rev Microbiol.* 2016;14(5):273–287.
42. Koppel N, Maini Rekdal V, Balskus EP. Chemical transformation of xenobiotics by the human gut microbiota. *Science.* 2017;356(6344).
43. Haiser HJ, Gootenberg DB, Chatman K, Sirasani G, Balskus EP, Turnbaugh PJ. Predicting and manipulating cardiac drug inactivation by the human gut bacterium Eggerthella lenta. *Science.* 2013;341(6143):295–298.
44. Maini Rekdal V, Bess EN, Bisanz JE, Turnbaugh PJ, Balskus EP. Discovery and inhibition of an interspecies gut bacterial pathway for levodopa metabolism. *Science.* 2019;364(6445):[eaau6323].
45. Macdonald TT, Monteleone G. Immunity, inflammation, and allergy in the gut. *Science.* 2005;307(5717):1920–1925.

46. Belkaid Y, Naik S. Compartmentalized and systemic control of tissue immunity by commensals. *Nat Immunol.* 2013;14(7):646–653.
47. Thaiss CA, Zmora N, Levy M, Elinav E. The microbiome and innate immunity. *Nature.* 2016;535(7610):65–74.
48. Levy M, Blacher E, Elinav E. Microbiome, metabolites and host immunity. *Curr Opin Microbiol.* 2016;35:8–15.
49. Smits WK, Lyras D, Lacy DB, Wilcox MH, Kuijper EJ. Clostridium difficile infection. *Nat Rev Dis Primers.* 2016;2:16020.
50. van Nood E, Dijkgraaf MG, Keller JJ. Duodenal infusion of feces for recurrent Clostridium difficile. *N Engl J Med.* 2013;368(22):2145.
51. Hsiao A, Ahmed AM, Subramanian S, et al. Members of the human gut microbiota involved in recovery from vibrio cholerae infection. *Nature.* 2014;515(7527):423–426.
52. Maloy KJ, Powrie F. Intestinal homeostasis and its breakdown in inflammatory bowel disease. *Nature.* 2011;474(7351):298–306.
53. Beresford-Jones BS, Shao Y. Through the gut, down the drain. *Nat Rev Microbiol.* 2021;19:345.
54. Yeoh YK, et al. Gut microbiota composition reflects disease severity and dysfunctional immune responses in patients with COVID-19. *Gut.* 2021;2021(70):698–706.
55. Musso G, Gambino R, Cassader M. Interactions between gut microbiota and host metabolism predisposing to obesity and diabetes. *Annu Rev Med.* 2011;62:361–380.
56. Forslund K, Hildebrand F, Nielsen T, et al. Disentangling type 2 diabetes and metformin treatment signatures in the human gut microbiota. *Nature.* 2015;528(7581):262–266.
57. Hindson J. Distinct faecal mycobiome in patients with NAFLD. *Nat Rev Gastroenterol Hepatol.* 2022;19(2):79.
58. Brunt EM, Wong VW, Nobili V, et al. Nonalcoholic fatty liver disease. *Nat Rev Dis Primers.* 2015;17(1):15080.
59. Falony G, Vieira-Silva S, Raes J. Microbiology meets big data: The case of gut microbiota-derived trimethylamine. *Annu Rev Microbiol.* 2015;69:305–321.

Further reading

60. Cho I, Blaser MJ. The human microbiome: At the interface of health and disease. *Nat Rev Genet.* 2012;13(4):260–270.
61. David LA, Maurice CF, Carmody RN, et al. Diet rapidly and reproducibly alters the human gut microbiome. *Nature.* 2014;505(7484):559–563.

CHAPTER TWO

Emerging tools for understanding the human microbiome

Tarini Shankar Ghosh[a,]* and Mrinmoy Das[b]
[a]APC Microbiome Ireland, University College Cork, Cork, Ireland
[b]International Centre for Genetic Engineering and Biotechnology, New Delhi, India
*Corresponding author: e-mail addresses: tarini.ghosh@ucc.ie; tarini.ghosh@gmail.com

Contents

Abstract

Recent advances in sequencing technologies, experimental protocols and approaches in data generation and analysis have enabled us to investigate the human microbiome at an unprecedented level of resolution. The current chapter aims to provide an understanding of the different computational and bioinformatic strategies adopted to answer the different questions of a typical microbiome investigation and how the upstream DNA sequencing methodologies can affect this. The chapter enlist the state-of-the-art in metagenomic data analysis along with the available strategies to perform an integrated investigation of the human microbiome along with other data layers.

1. Introduction

Rapid advances in sequencing technologies, molecular microbiological techniques and approaches in data generation and analysis have enabled us to investigate the human microbiome at an unprecedented level of resolution. Human microbiome research has now transformed from being a field primarily reliant on microscopy and laboratory-based culturing and genetic engineering of specific culture-amenable microbes. It has now become a paradigm of research that aims to characterize and comprehensively investigate the hitherto unexplored diversity of culture-resistant host-associated organisms on a routine-basis. The questions addressed as part

Progress in Molecular Biology and Translational Science, Volume 191
ISSN 1877-1173
https://doi.org/10.1016/bs.pmbts.2022.06.027

of these investigations as well as the population cohort studied are versatile, and so are experimental and computational data analytic techniques. However, the insights obtained from these multitude of individual studies can potentially act as the cumulative stepping-stones toward answering fundamental questions that can revolutionize personalized health-care. Some of these questions include: What is the compositional landscape of a "healthy" microbiome? How do host-specific factors like age, socio-economic status, geography, diet, medications and inter-individual contacts shape the architecture of the human microbiome? How is the microbiome altered during specific disease? Do we observe microbiome alterations that are shared across diseases along with those that are disease-specific? How does the human population-strata (host demographics, host anthropometric and host life-style) influence these alterations?

The current chapter aims to provide an understanding of the different computational and bioinformatic strategies adopted to answer the different questions of a typical microbiome investigation (summarized in Fig. 1). Next, we provide examples of how state-of-the-art data analytic techniques have enabled provided vital clues into the role of the microbiome as a robust diagnostic marker and a potential therapeutic target for various human disorders as well as a modulator for detecting personalized host responses to diet and other therapeutic interventions. Finally, we cover

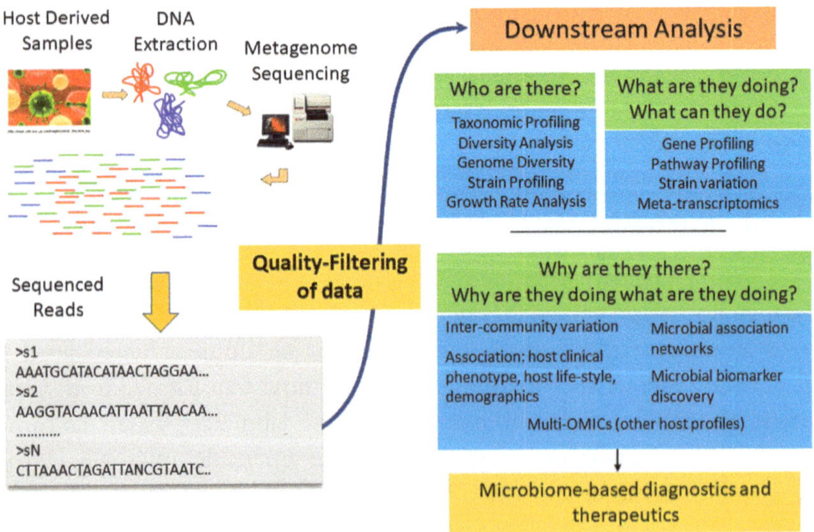

Fig. 1 Schematic overview of a microbiome investigation workflow.

the limitations of the different methodologies, the caution that needs to be exercised while using some of these methodologies and the questions that remain to be answered.

2. Questions addressed in a microbiome investigation

A typical human microbiome investigation aims to answer three different questions:

(a) Who are there? These include identifying what microbes a given body site of an individual. In most microbiome investigations, obtaining accurate and high-resolution snapshots of the composition of the underlying host-associated communities is the key to almost all objectives of the concerned study. Accurate estimation of the taxonomic composition of the gut microbiome is not only the key to understanding and comparing the diversity of host samples from different clinical or other phenotypes, but also facilitate insights into all downstream objectives. These include identifying the microbial markers associated with a given trait, understanding the interaction patterns within a microbiome, and how they are affected across different host phenotypes, as well as making fairly reliable predictions on the underlying metabolic capabilities of the microbiome.

Microbiome sequencing is typically performed by either sequencing the entire genomic content of all the microbes residing in a given environment or by sequencing the hypervariable regions of specific taxonomic marker genes like 16S rRNA gene. Subsequently, the taxonomic composition of the given sample is obtained by bioinformatically classifying each read to a taxa in the microbial tree of life. However, the taxonomic resolution (phylum or class or order or family or genus or species or strain) and the accuracy of this classification depends on a range of factors (Fig. 2). These include the approach used to sequence the microbiome (16S v/s shotgun), the accuracy and the fidelity of the sequencing approach,[1] the depth at which the sample is sequenced as well as the down-stream taxonomic classification approach used to obtain the taxonomic classification of the sample (summarized in detail in[2]). Obtaining taxonomic composition profiles at the taxonomic levels of genus or higher can be efficiently achieved using amplicon sequencing (e.g., like 16S rRNA gene sequencing) using high-throughput sequencing approaches like Illumina (discussed later). For obtaining deeper resolutions like that of species, the efficiency of 16S rRNA gene sequencing approaches may be limited and is dependent upon the bioinformatic method, the quality of the DNA sequence, the divergent

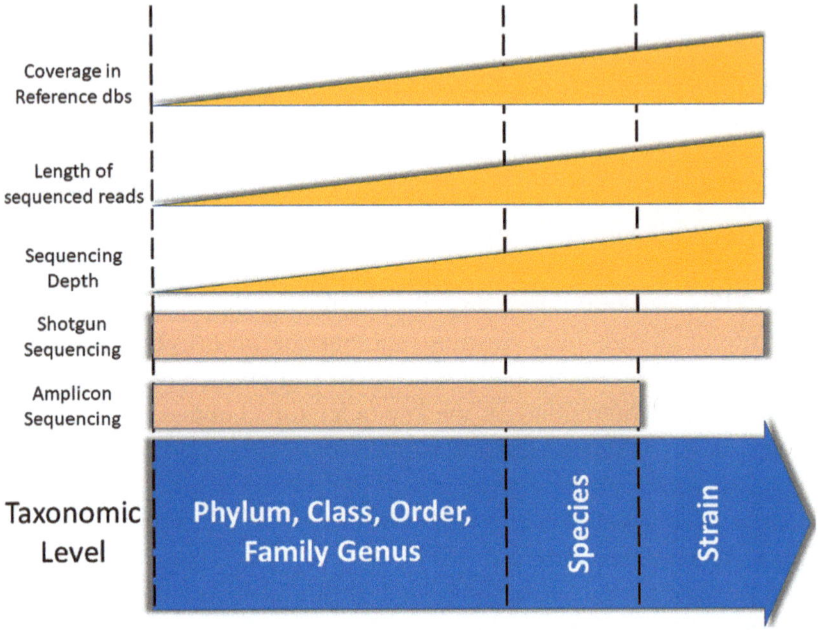

Fig. 2 Factors dictating the taxonomic resolution and accuracy of classification in typical microbiome investigation pipelines.

nature of the species being detected and the host environment that is being investigated. To further elaborate on this, in terms of the bioinformatic methods, while methods like SPINGO are customized for and perform reasonably well for classifying reads at the level of species,[3] other approaches like the RDP Classifier can only perform the classification at the level of genus.[4] There are other approaches like DADA2 (discussed later) and other approaches using the SILVA database which can classify reads at any taxonomic level, but the classification accuracy is dependent on a multitude of factors, including the length and the quality of the sequenced reads, the enrichment of the reference databases toward species from specific environments. Classification of reads from well-characterized environments like the human gut (and especially of individuals from western countries) can be achieved with reasonably accuracy and reliability, those from other human body sites and even for the gut environments from non-westernized individuals may have lower accuracy and larger number of reads being unclassified. These limitations with respect to reference databases has been recently addressed to a certain extent by increased sequencing of metagenomes and metagenome-assembled genomes from individuals of non-industrialized and from other relatively uncharacterized human body sites.[5,6] However, for a large number of such metagenome-assembled genomes, the corresponding taxonomic annotations still remains to be achieved at finer resolutions of species. Another factor that determines how well we are able to characterize the microbiome is the depth of metagenomic sequencing (both shotgun v/s 16S). Having a reasonable depth of sequencing is important for two reasons. First, it enables the isolation of sequence reads and the subsequent characterization of rarer taxa in the microbiome. This is especially important in scenarios where the microbiome composition of an environment adopts an especially long-tailed distribution, for example, an *Helicobacter pylori* dominated stomach environment or a *Lactobacillus* dominated vaginal environment.[7,8] In such cases, it is important to characterize and investigate the co-occurrence of the other rarer taxa that may co-occur with dominant members but also have consequences for host susceptibility to different diseases and clinical compositions like pre-term birth.[9,10] Secondly, higher sequencing depth is also important for studies that aim to profile strain-wise differences even among the dominant taxa resident in a microbiome, including assembling of metagenome assembled genomes.[5] Higher depth of sequencing that enables recovery of genomes with better resolution can facilitate accurate inference of associated metabolite production/consumption capabilities (even before the functional classification or gene-finding approaches are used).[11–13]

Another aspect that plays in the resolution and depth of classification is the length of sequenced reads, especially in amplicon sequencing and for strain-tracking studies. While shorter reads obtained using high-throughput sequencers like Illumina or Ion-Torrent are reasonably good for shotgun metagenomic sequencing, the resolution of assignments obtained using such sequencing techniques for amplicon-based datasets are often lower. For such scenario, long read sequencing is better as they facilitate recovery of the full amplicon sequence. Recently state-of-the-art hybrid sequencing approaches are also being used, where the same sample is sequenced using both long-read and short-read sequencing approaches.[14] While long read sequencing, while having relatively higher error rates,[15] can still act as scaffolds for sequence assembly, the high-quality Illumina while being smaller in length can help to refine the errors resulting in long high quality reads. The major objective behind the advent of such long-read or hybrid sequencing technologies is the desire to enable investigating the microbiome at the level of individual strains. There are a multitude of bioinformatic approaches developed for this purpose (comprehensively summarized in Ref. [16]).

(b) What are they doing? What can they do? Answering these questions involves identifying what functionalities or genes or metabolic functionalities are encoded by the genomes of the resident microbes; investigating or predicting how fast they are growing in the given environment and; investigating the transcription rates of the different genes encoded by the different members of the microbiome and; identifying the highly growing and metabolically/functionally active (or dormant) members of the microbiome. There are different available approaches to answer these and that depends on the nature of data we have to investigate. For Amplicon-based sequencing data sets, we only have reads originating from the taxonomic marker regions of the genome. So, identification of other genes that are encoded by the genomes of the resident microbes is not possible. However, in these cases, a functional profile can still be inferred or predicted if the gene presence/absence information in the already sequenced genomes of the resident species (or their close phylogenetic relatives) or the experimentally validated metabolite production/consumption of the species detected in the microbiome are previously known. These classes of methods can be described as Reference-Information-based functional prediction methods. The former information is utilized by methods like PICRUSt (Phylogenetic Investigation of Communities by Reconstruction of Unobserved States),[17] PICRUSt-2,[18] Tax4Fun,[19] Piphillin,[20] Faprotax[21] and Vikodak.[22] The second kind of information has been utilized by previous studies by our group.[11–13] A third kind of prior-knowledge-based prediction methods that rely on a two-step strategy to predict the complete metabolic potential of a given microbial community.[23,24] In general, the first step of these methods involves the species-specific prediction of specific gene families. This is generally performed utilizing an approach as performed by the PICRUSt-class of methods. Once this information is obtained, the second step is performed which utilizes a different mapping that lists the gene families or enzymes that constitute the pathways corresponding to the production or consumption of various metabolites. Thus, a complete picture of the metabolic potential can be inferred based on the species composition obtained using any of the available 16S rRNA gene amplicon-based or shotgun-based taxonomic classification approaches.

The above class of methods have been very effective in providing unique insights into a range of diverse environments, ranging from intestinal environments of various species[25,26] to earth's diverse geochemical niches.[27] However, these methods have two primary limitations. First, their applicability is limited by the availability of the reference genomes corresponding to the resident species in the corresponding environments. Second, while the methods can, in principle, be able to predict the core gene components or pathways corresponding to a given species, their applicability with respect to the prediction of accessory genes (and pathways), is limited. This is especially important when the environment is characterized by hitherto unknown sub-strains of known species containing unique functional traits.

To address the first kind of limitations for the methods described above, a second category of methods have been developed that utilize pre-defined models.[23,28,29] Ideally, these pre-defined models are generated separately for microbiomes from different environments (that is different models for different environments). These model-creation or training phase in the development of these methods aim to answer the following question: Given a specific species profile from a microbiome from an environment X, what is the functional or metabolic profile that is expected with the highest likelihood. To answer this question, this phase generally involves paired analysis of metagenomic species profile and metagenomic functional profile and/or the analysis of metagenomic species profile and the paired metabolomic profile, to generate a set of reference or training models that link species-composition pattern to functional gene composition pattern and/or metabolomic composition pattern. This is typically done in an environment-specific manner. In the next phase, which can be described as the "application" phase, given a new microbiome species composition profile, these models are then utilized to infer the corresponding functional/metabolomic profile utilizing the above models. These methods do not utilize reference databases but direct relationships between microbiome species composition and microbiome functional/metabolite composition in previously characterized microbiomes from the same environment. Consequently, they should be better equipped to handle scenarios accounting for both "known" (or genomically/functionally characterized) or hitherto unknown taxa as well as scenarios pertaining to the presence of hitherto uncharacterized functional variants of known species that are unique to a given environment. However, the efficacy of these methods is again strongly influenced by how comprehensible the training models are and may drastically limit their applicability for low microbial biomass host environments, where in generation of shotgun sequencing-based metagenomic gene information may be extremely difficult.

Shotgun-sequencing approach attempts to sequence and profile the entire genomic content of all microbes residing in a given environment. Consequently, the reads data contains sequence information pertaining to the gene regions of the genomes of the resident microbes. Consequently, gene profiling of the microbiome can be performed by performing a nucleotide-to-nucleotide or translated nucleotide-to-protein sequence alignment searches of the individual sequence reads against specific reference databases[30,31]. Alternatively, individual reads from the metagenomic datasets can be assembled into metagenome-assembled genomes.[5] These metagenome-assembled genomes then be investigated further to not only investigate the presence of specific genes or gene families, but also performing investigations at the level or gene synteny or gene-pair co-occurrence as well as investigations at operonic levels. However, functional annotations within metagenomes has it's own challenges.[32] A typical gut metagenome sequencing projects generally has less than 15% of reads being assigned to a defined function or pathway.[33] A primary reason for this is the current bias in the reference databases toward genomes from model organisms and culture-friendly members of the gut microbiome. This gap needs to be substantially by large-scale culturomics-based initiatives that focus on isolating and sequencing microbial genomes from human-associated environments like the gut,[6,34,35] especially from hitherto uncharacterized human populations. High throughput and extremely accurate bioinformatics analysis pipelines need to be developed that facilitate the assembly and subsequent functional characterization of metagenome assembled genomes directly from metagenomic reads datasets.[5] Finally, improved metagenomic sequencing strategies need to be developed, especially those utilizing hybrid approaches like combining high throughput short read Illumina and long read Oxford Nanopore Technologies (ONT), that facilitate accurate recovery of near full-length genomes directly from a host-derived sample.[14] A culmination of these efforts will pave the way for extensive characterization of the current reference databases resulting in positive feedback cycle, where in improved recovery and characterization of isolate genomes (directly from host-derived samples) further improves the functional annotation of metagenomes and subsequent improvements in reference databases. And better annotated reference databases facilitating further improvements in

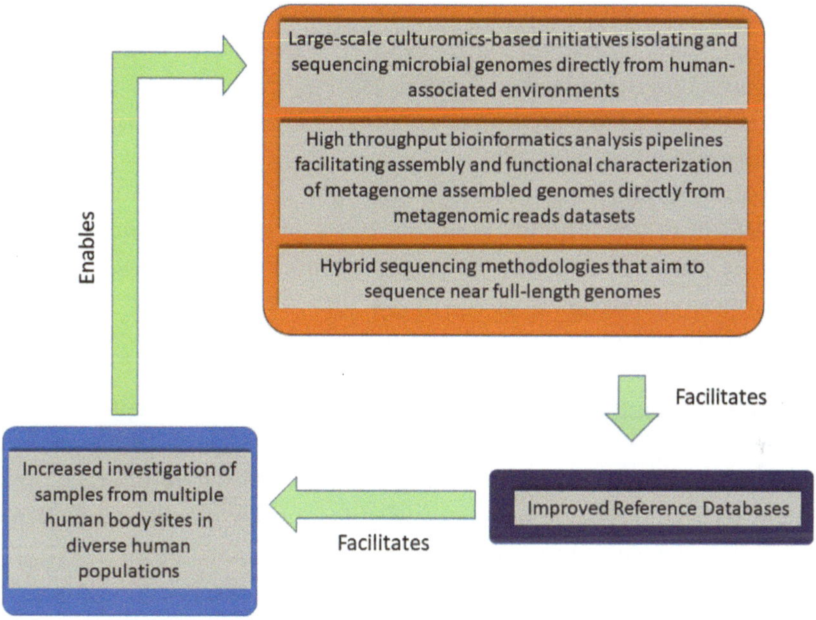

Fig. 3 Positive feedback loop showing that: 1. Improvements in culturing, bioinformatic and sequencing strategies can help to improve the reference functional annotation databases. 2. These improvements can help to increase the accuracy and sensitivity of functional annotations from hitherto uncharacterized environments within the human body. 3. Increased characterization of samples from these human body sites can help further enable the application of the strategies described in the Step 1.

annotations of newly sequenced host-derived metagenomes (Fig. 3). Functional annotation of the metagenomic reads can provide a fair information of what the majority of the microbiome can do. To know what they are actually doing at a given point of time, we need to investigate what genes are transcribed (using meta-transcriptomics) and what proteins are being produced (meta-proteomics). Meta-transcriptomics and meta-proteomic investigations are fairly recent in the area of human microbiome and recent studies and improved approaches are coming up that can potentially open a whole new avenue in this area.[36,37] One of the pioneering studies in this regard is the Health Professionals Follow-Up Study by Zhang *et al.*, which showed that the microbiome of individuals is characterized by a "meta-transcriptomic" core set of genes that are stable across individuals and stable across times, but transcribed by different sets of micro-organisms.[36] Similarly, one of the recent meta-proteomic investigations is the recent study by Mills *et al.* which identified the production of specific proteases from gut *Bacteroides vulgatus* to be associated with ulcerative colitis severity.[38] These later two studies can be regarded as benchmarks of multi-OMICs investigations that can be taken as references to plan the analysis protocol for any human microbiome investigations.

Besides investigating what activities the gut microbiome members can perform or are currently performing at a given point of time, another aspect of the microbiome that can now be computationally (or bio-informatically) inferred is how fast are they doing? Recent times have witnessed an emergence of methods that aim to predict the growth rates of individual species directly from the metagenomic reads sequenced from a given sample.[39–41] The premise is the fact that the replication of a bacterial/archaeal genomes typically starts (by the creation of a replication for) from the origin (*ori*) sequence (Fig. 4). This replication fork then opens up further as the replication proceeds bi-directionally toward a single terminus (*ter*) sequence. Thus, for a single replicating cell, regions that have already been replicated will have two copies of the genome while the

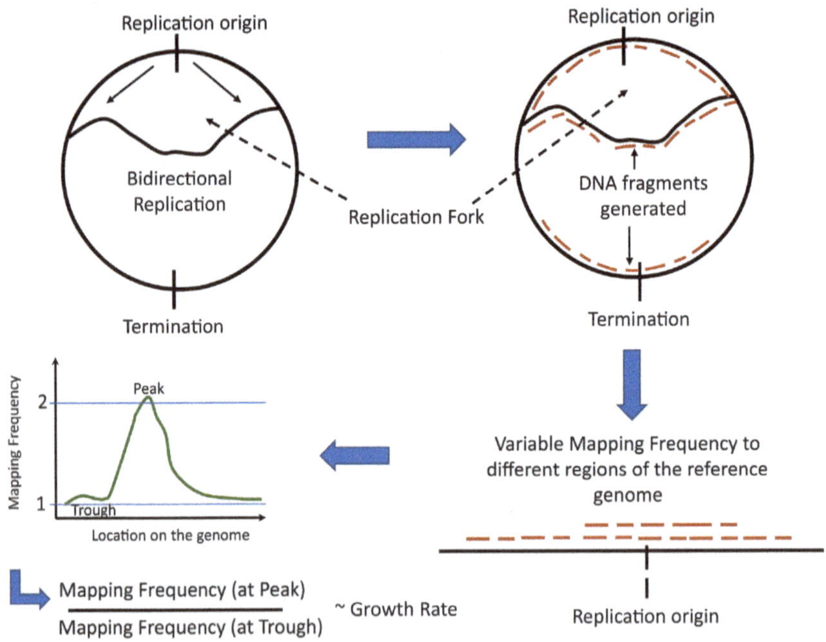

Fig. 4 Premise the bioinformatic approaches developed to compute growth rates of individual species based on the ratio of the peak-to-trough mapping frequency of reads in the corresponding reference genomes.

unreplicated regions will have a single copy. At a population level, where in many cells of the same bacteria, different cells may be at different stages. Nevertheless, at cumulative level, highly replicating bacterial species will have much higher number of replicating forks and thus a much higher frequency of reads that map to the origin region (peak mapping frequency) as compared to the terminus region (trough mapping frequency). In contrast, non-replicating species will have a ratio closer to 1. Thus, calculating the peak-to-trough ratio of the mapping frequencies is likely to give an indication of how fast the given species is growing. This concept is utilized in different but related manners by the above three methods to infer the species-specific growth rates in metagenomic samples.

Why are they there? Why are they doing whatever are they doing? These are the two most important questions in a microbiome investigation, when it comes to understanding the role of the human microbiome as a diagnostic marker or a therapeutic target. Answering these questions involves investigating multiple microbiomic profiles and identify traits that account for the variations in host phenotype (host disease, host transcriptome, host metabolome and host physiological function); that associate with specific host behavior and life-style like diet and medications as well as; identifying groups of microbes that tend to co-occur and co-inhibit each other. Understanding these involves the use of efficient state-of-the-art data analytic strategies and integrated OMICs strategies. These two aspects will be covered in the subsequent sections. However, prior to that, we will discuss a key development in recent times that has potentially revolutionized this research area. This is the advent long-read and hybrid NGS.

3. The advent of long-read and hybrid NGS

Beyond doubts, the NGS technology also known as second-generation sequencing has enabled tremendous advances in many aspects of biological studies. However, second-generation technologies had certain drawbacks which included short read length and limitation in discovery of sequence variants in repeat genomic locations or elements.[42] This resulted in developments of methods that enabled generation of longer reads at a faster rate and could also be used for epigenetic studies. The term third generation sequencing was first used in 2009 when newer technologies were developed and are currently still in development.[43] These sequencing approaches are considered a game changer and a complete overhaul from its predecessors. Two of the most commonly used long-read approaches are

SMRT (single-molecule, real time) sequencing from Pacific Biosciences (PacBio) and Oxford Nanopore Technologies.[44,45] An important advantage of these methodologies is that the equipment is smaller, portable and requires minimal sample processing enabling real time data generation in settings like extreme environments where there is difficulty in traveling, bringing samples and hospitals which requires results in very short time.[46,47]

Pacific Biosciences or PacBio sequencing: SMRT sequencing approach used by PacBio works by converting electromagnetic waves into light spectra for different nucletotides.[48] In this methodology, a circular DNA molecule is used as a template. This structure is known as SMRTbell and it contains the DNA insert which is the region to be sequenced and can be over 100 kb in length. The structure also contains hairpin adapters at both ends. A polymerase molecule is then attached to the SMRTbell structure along with the DNA insert. A SMRT Cell is an object which contains the SMRTbell along with millions of zero-mode waveguides (ZMW). ZMW objects enables conversion of electromagnetic wave into light by allowing one molecule of nucleotide base to be accessible to the polymerase enzyme at a time. Nucleotide base contains a fluorescent dye which gets cleaved off once the complementary base is added by the polymerase enzyme into the nascent strand. This emission and release of the fluorescent dye is recorded by the detector at the time of each incorporation. Base calling is done based on the type of dye released which is repeated multiple times to identify each base present in the SMRTbell structure.

Reads generated by PacBio falls under two categories: Continuous long reads (CLR) and High-fidelity (HiFi) reads. CLR are the most common read types of PacBio sequencing. As large inserts of DNA is used, the polymerase usually passes once or only a few times through the template. Despite generating longer reads, it has been observed that the reads accuracy typically ranges between 85% and 92% which is lower than Illumina. Homopolymer regions also resulted in generation of more error-prone reads. HiFi reads are the most recently developed read type of PacBio. In this case, usually smaller fragments of the DNA is used which allows the polymerase enzyme to pass through the SMRTbell template multiple times. This approach to sequencing is termed circular consensus sequencing (CCS). Due to multiple passes by the polymerase, reads generated are much longer containing the multiple subreads. The subreads allows generation of very long consensus reads using the CCS formula which are termed as HiFi consensus reads with extremely high accuracy (>99%). The formation of multiple subreads and the generation of consensus reads using the CCS algorithm results in very long processing time.[49]

Oxford Nanopore Technologies (ONTs): Oxford Nanopore sequencing comprises of sequencing adaptor, a motor protein and thousands of nanopore structures loaded into a flow cell.[45] The system contains electrolyte solution and constant electric current. Once the motor protein unwinds the DNA, the nanopore structure allows only single stranded DNA to pass through it one nucleotide at a time. As the nucleotide pass through the nanopore, it causes a characteristic change in the current which is dependent on the type of nucleotide molecule. The electric flux is recorded in real time as the nucleotide passes through thereby determining the sequence of the base in the sequencing strand. This approach has been used to successfully generate reads over 1 Megabases (Mb) in length.

ONTs generate standard ONT long reads and ONT ultra-long reads. Standard ONT long reads are the most commonly used with an accuracy between 87% and 98% and usually several kb in length. DNA extraction and prepartion have been a major limiting factor for ONT read lengths. Loman's lab was the first to generate ultra-long reads using ONT which were over 100 kb and up to several Mb in length with similar accuracy as standard ONT long reads. Despite generating extremely long reads, the overall throughput is much lower compared to standard ONT and the generation of ultra-long reads can tak several weeks if high fold coverage is required.[49]

To improve read accuracy various approaches have been developed. This includes tools like Quiver for PacBio sequencing reads.[50] It utilizes the raw pulse data and CLR alignments to generate true sequence bases on consensus. Tools like INC-Seq can be applied to ONT sequences which improves the consensus read accuracy.[51] These two methods have made tremendous impact in several areas which includes recovery of near full genome from metagenomics, improved characterization of strains and species, detection of sequence variants, and improved capture of gene isoforms.[52,53]

Despite the tremendous improvement in long-reads technologies which offers numerous advantages over second-generation sequencing, the cost of long-read sequencing is still high and has stringent DNA extraction and processing requirement. To establish feasibility in using third-generation technologies, various hybrid approaches have been developed that utilizes both third-generation sequencing reads and NGS reads to generate highly accurate and near complete genomes assemblies down to subspecies levels. An example of such tool is OPERA-MS which has utilizes both Illumina reads and ONT reads.[14] In brief, OPERA-MS builds contigs from the short reads and these contigs are compared against the long-reads bundlers and a mapper tool to generate contig coverage and an assembly graph. This graph contains contiguity information of the metagenomes can be resolved in clusters and subclusters based on genomic distance, which results in generation of subspecies level genome contigs. OPERA-MS not only provided higher base quality than individual short reads or long reads tools but also required less fold coverage compared to short read tools. It also generated better contiguity with less coverage and resulted near completeness of the assembled genomes.

4. State-of-the-art data-analysis methodologies

The subsequent stage of the microbiome investigation (that is, the stage succeeding the generation of sequence data) is the downstream data analysis. Microbiome research is a booming field and is highly reliant on new approaches in every aspect, from sampling, processing, sequencing to generation of data and its analyses. Various methods, tools, etc. have been developed to capture the accurate nature of the microbiome niche and obtain a holistic view on their functions and activities. As discussed earlier, newer sequencing approaches have allowed us to obtain subspecies level information on the microbiome. However, these approaches are not without their limitations in terms of accuracy, computational time requirement, and other technical biases. The first and foremost requirement for obtaining good quality data is extraction of good quality DNA with high yield. DNA extraction and processing protocols severely affects all subsequent steps as they can result in unwanted contamination, low yield, sheared low quality DNA. A key requirement in microbiome research is the generation of accurate biological data in terms of taxonomy, population, functionality etc. Among the most popularly used set of tools is the QIIME2 repository.[54] One of the key reasons for using QIIME2 is that it is collaborative effect of various researchers across the world to create a hub of tools which can be uniformly accessed by all researchers and it allows addition of newer tools as plugins. This is huge repository of various tools available to all types of researchers along with multiple choices. Another widely used set of tools known as Microbiome Analysis Tools have been developed by Bork and Zeller Groups at EMBL (https://microbiome-tools.embl.de/). In brief, data generation involves quality filtering of sequencing reads, assembly/mapping to references (genomes, genes, marker regions), annotation of the mapped reads, tabulation and finally downstream exploratory and statistical analyses and visualizations.

To resolve issues with contamination and low-quality sequences, various new tools have been developed. Among them is SourceTracker; a tool that allows filtering out contaminant reads from the sequencing runs based on a reference set of contaminants.[55] Quality filtering involves various steps which includes removing low quality reads, potential contaminants or chimeras, and denoising the dataset. It is crucial to generate high quality error free reads as certain taxa can show very high levels of similarity (e.g., 16S region of *Enterobacteriaceae* family).[56] In such cases, strain specific resolution can be identified through site specific polymorphisms which requires extremely high-quality error corrected data. This issue is common with the use of 16S region for sequencing as the region can be highly conserved at lower taxonomic level (sub-genus). One of the recently developed tools that allows discrimination between closely related taxa in 16S data is Divisive Amplicon Denoising Algorithm (DADA2) which resolves sequences down to single nucleotide changes and terms each representative sequences as sequence variants (SVs).[57]

Alternative to 16S data, metagenomic whole genome sequencing (mWGS) is considered and attempts to resolve whole genome of the various taxa present in the samples. However, issues with low quality, generating genomes of closely related species is a persistent problem in this case also. Assembly based methods applied in mWGS were unable to distinguish very closely species or strain as it generated a consensus genome by collapsing strain level changes into single genome.[58] As strain specific differences can occur extremely small section of the genome or in plasmids, genomic assembly-based methods required various modification to incorporate these factors. Well known tools like Velvet, and IDBA developed MetaVelvet-SL, and Meta-IDBA for mWGS.[59,60] To overcome the limitation of assembly-based approach, specific or strain specific markers regions were considered as reference and tools that directly mapped reads to these regions were developed. Other examples in this group of tools include MetaPhlAn3, StrainPhlAn and K-SLAM.[61–64] However, the use of reference database is a major limitation as taxa that are not present in the reference database or any new novel taxa will not be detected. Currently the use of third generation sequencing along with tools like OPERA-MS (discussed earlier) serves as the next step in microbiome data generation.

Microbiome research has gained a lot of momentum is the last two decades. Advancement in various aspects from sampling, sequencing, to computational infrastructure support has allowed researchers to investigate large scale high-throughput microbiome data rapidly and study complex interactions and patterns. Various downstream analyses methodologies applied to microbiome research are adopted from ecological or transcriptomic studies. These methodologies have performed reasonably well in discerning meaningful biological conclusion but have also highlighted the inherent differences in the nature of microbiome dataset. To address the nature of microbiome data, various statistical approaches are developed which would not only address the nature of the microbiome data but also be sensitive in understanding the biological relationship. Examples of these include is the use of rarefaction method for normalizing the microbiome data or proportional normalization.[65] Although these are widely used to normalize sequencing depth issue, it has drawn criticism from world experts as they may be statistically inaccurate and does not consider heteroscedasticity.[66] The use of other methods like cumulative sum scaling, variance stabilization or upper quantile scaling may serve as a better alternative[67,68] (https://www.cbcb.umd.edu/software/metagenomeSeq). Libraries such as phyloseq, MetagenomeSeq and ALDEx2 have been developed in R statistical tool specifically to analyze microbiome data[69,70] (https://www.cbcb.umd.edu/software/metagenomeSeq). These methods implement various strategies to address the sparseness, non-normal distribution and compositional nature of microbiome data. Independent testing of various methods and normalizing techniques describes the strengths and weakness of these approaches. These notable points include agreement with community structure, sensitivity toward biologically meaningful context.[71,72]

UniFrac is a distance measurement method developed specifically for microbiome studies.[73] It measures the dissimilarity between samples based not only on the abundance of the taxa but also considers the phylogenetic distance between the taxa. Two variations of this method are implemented: Weighted UniFrac which considers the abundance of taxa in the samples and Unweighted UniFrac where the presence/absence of the taxa is accounted. Both variations build a phylogenetic relationship between the taxa under investigation.

As more microbiome data becomes available across the world, more complex methodologies can be applied. Machine learning is a supervised approach which requires large amount of data to uncover complex interaction and features. Machine learning in simple terms can study the features of microbiome data in a more robust, holistic manner and identify key determinants/highly discriminating features corroborated through various cross-validation, boot-strapping, and test datasets. Two of the most commonly used machine learning approaches are XGBoost (eXtreme Gradient Boosting) and Random Forest (RF) which have shown remarkable performance against various benchmark datasets. Both of these methods implement decisions trees to 'learn' from the features and returns a set of most important/highly discriminatory features.

5. Integrated OMICs for the microbiome

The final goal of any human microbiome investigation is to understand the interaction between the microbiome and the host, that is to understand the functional associations of the microbiome alterations with the host phenotype, behavior and life-style. A majority of alterations in the gut microbiome of individuals may be stochastic. However, those alterations that are associated with a given disease in a given human population cohort as well as linked to specific changes in host clinical profiles at the level of host metabolome, host transcriptome and host proteome or host life-style at the level of diet, socio-economic status and medications, are likely to be indicative of the underlying functional mechanism. These insights can be subsequently investigated for causality in pre-clinical models or can be utilized to develop microbiome-based diagnostics for an underlying clinical complication.

The basic premise of performing multi-OMICs investigations is to understand the "how" along with the "what" (Fig. 5). These investigations can be strictly multi-OMICs, that is performed between host-associated OMICs profiles, like host genomics, bulk transcriptomics, single-cell host transcriptomics, proteomics, serum and/or fecal metabolomics and epigenomics with microbiome-associated OMICs profiles like metagenomics, meta-transcriptomics, microbiome proteomics and 16S rRNA gene amplicon profiles. Alternatively, they can also incorporate additional host functional and life-style profiles, like age, BMI, gender, habitual diet, co-morbidities, medications, etc.

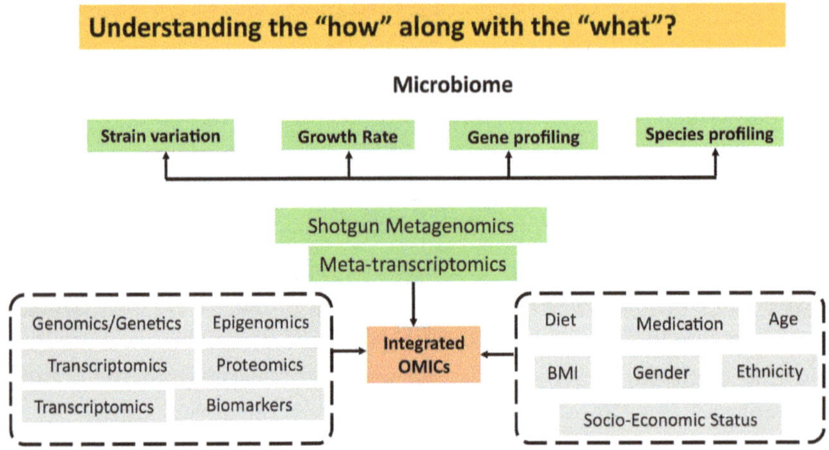

Fig. 5 Overview of multiple data layers in typical host-microbiome-interaction investigation studies.

There are a multitude of analytic protocols that are available for performing such kind of investigations (Fig. 6). These methods can range from correlations and regression models that can investigate pairwise-relationships between variables across multiple OMICs layers to approaches like Procustes or Co-inertia that can profile the association or correspondence between two data layers as a whole. In between these two extremes are a range of more pertinent approaches that can either identify clusters or groups of either features that tend to show similar pattern of variations across the studied environments or identify groups of environments that show similar levels of the same cluster of features. These include methods like WGCNA (Weighted Gene Co-expression Network Analysis), Matrix-based NMF (Non-negative Matrix Factorization), Machine-learning-based multi-view factorization auto encoder. Alternatively, there are other methods like iBBiG (Iterative Binary Biclustering of Genesets) that can facilitate both (feature-clustering as well as site-clustering) simultaneously.

The primary objective of these approaches is to identify modules of strongly features across multiple data-layers. Once modules, a module abundance (or module rank) is then computed across each site, which is then associated with clinical traits. Identification of such modules not only reduces the features for association testing thereby increasing the sensitivity of the statistical tests, but also provide a list of biological entities like important metabolites, bacteria and host gene and/or protein levels that can then be investigated further.

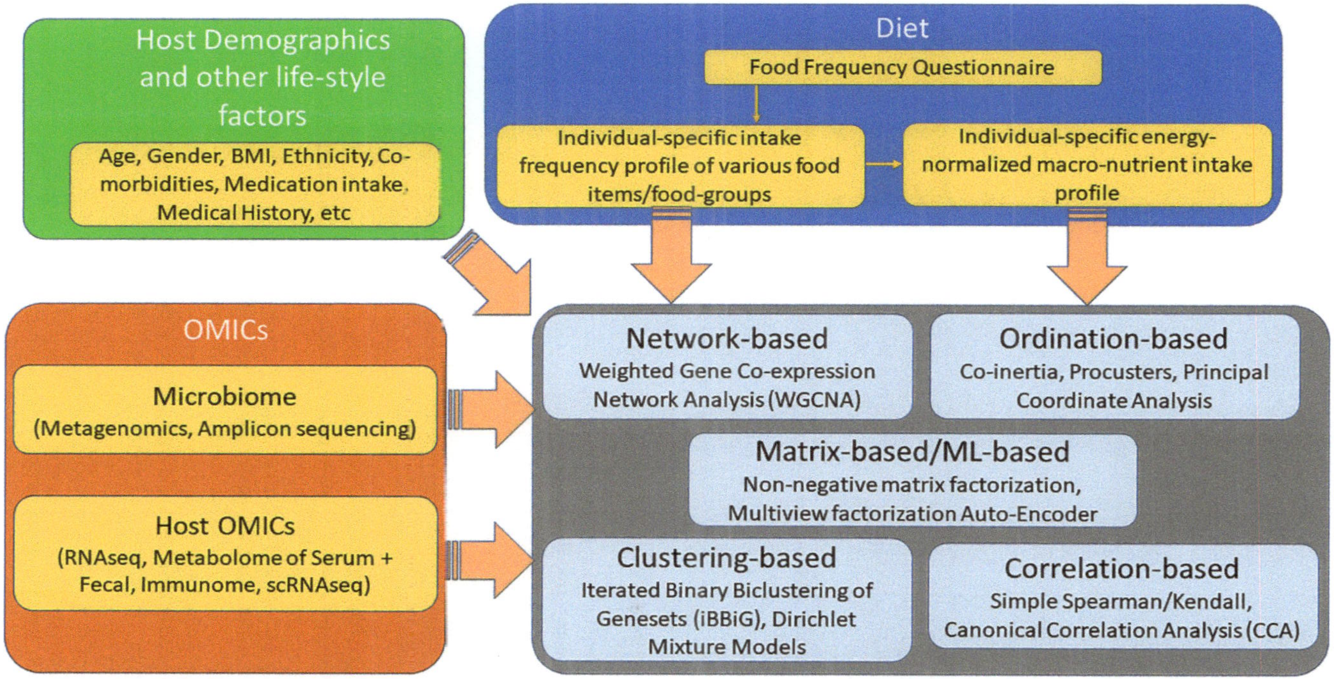

Fig. 6 Availability of different classes for performing multi-data-layer integrative studies.

References

1. Bag S, et al. An improved method for high quality metagenomics DNA extraction from human and environmental samples. *Sci Rep*. 2016;6:26775.
2. Mande SS, Mohammed MH, Ghosh TS. Classification of metagenomic sequences: methods and challenges. *Brief Bioinform*. 2012;13:669–681.
3. Allard G, Ryan FJ, Jeffery IB, Claesson MJ. SPINGO: a rapid species-classifier for microbial amplicon sequences. *BMC Bioinformatics*. 2015;16:324.
4. Wang Q, Garrity GM, Tiedje JM, Cole JR. Naive Bayesian classifier for rapid assignment of rRNA sequences into the new bacterial taxonomy. *Appl Environ Microbiol*. 2007;73:5261–5267.
5. Pasolli E, et al. Extensive unexplored human microbiome diversity revealed by over 150,000 genomes from metagenomes spanning age, geography, and lifestyle. *Cell*. 2019;176:649–662. e620.
6. Almeida A, et al. A unified catalog of 204,938 reference genomes from the human gut microbiome. *Nat Biotechnol*. 2021;39:105–114.
7. Mehta O, et al. Vaginal microbiome of pregnant Indian women: insights into the genome of dominant Lactobacillus species. *Microb Ecol*. 2020;80:487–499.
8. Das A, et al. Gastric microbiome of Indian patients with Helicobacter pylori infection, and their interaction networks. *Sci Rep*. 2017;7:15438.
9. Haque MM, Merchant M, Kumar PN, Dutta A, Mande SS. First-trimester vaginal microbiome diversity: a potential indicator of preterm delivery risk. *Sci Rep*. 2017;7:16145.
10. Kumar S, et al. The vaginal microbial signatures of preterm birth delivery in Indian women. *Front Cell Infect Microbiol*. 2021;11:622474.
11. Ghosh TS, Das M, Jeffery IB, O'Toole PW. Adjusting for age improves identification of gut microbiome alterations in multiple diseases. *elife*. 2020;9(e50240). https://doi.org/10.7554/eLife.50240.
12. Ghosh TS, et al. Mediterranean diet intervention alters the gut microbiome in older people reducing frailty and improving health status: the NU-AGE 1-year dietary intervention across five European countries. *Gut*. 2020;69:1218–1228.
13. Keohane DM, et al. Microbiome and health implications for ethnic minorities after enforced lifestyle changes. *Nat Med*. 2020;26(7):1089–1095. https://doi.org/10.1038/s41591-020-0963-8.
14. Bertrand D, et al. Hybrid metagenomic assembly enables high-resolution analysis of resistance determinants and mobile elements in human microbiomes. *Nat Biotechnol*. 2019;37:937–944.
15. Delahaye C, Nicolas J. Sequencing DNA with nanopores: troubles and biases. *PLoS One*. 2021;16, e0257521.
16. Van Rossum T, Ferretti P, Maistrenko OM, Bork P. Diversity within species: interpreting strains in microbiomes. *Nat Rev Microbiol*. 2020;18:491–506.
17. Langille MG, et al. Predictive functional profiling of microbial communities using 16S rRNA marker gene sequences. *Nat Biotechnol*. 2013;31:814–821.
18. Douglas GM, et al. PICRUSt2 for prediction of metagenome functions. *Nat Biotechnol*. 2020;38:685–688.
19. Aßhauer KP, Wemheuer B, Daniel R, Meinicke P. Tax4Fun: predicting functional profiles from metagenomic 16S rRNA data. *Bioinformatics (Oxford, England)*. 2015;31: 2882–2884.
20. Iwai S, et al. Piphillin: improved prediction of metagenomic content by direct inference from human microbiomes. *PLoS One*. 2016;11, e0166104.
21. Louca S, Parfrey LW, Doebeli M. Decoupling function and taxonomy in the global ocean microbiome. *Science (New York, NY)*. 2016;353:1272–1277.

22. Nagpal S, Haque MM, Mande SS. Vikodak—a modular framework for inferring functional potential of microbial communities from 16S metagenomic datasets. *PLoS One.* 2016;11, e0148347.
23. Noecker C, et al. Metabolic model-based integration of microbiome taxonomic and metabolomic profiles elucidates mechanistic links between ecological and metabolic variation. *mSystems.* 2016;1(1):e00013–e00015. https://doi.org/10.1128/mSystems.00013-15.
24. Noronha A, et al. The virtual metabolic human database: integrating human and gut microbiome metabolism with nutrition and disease. *Nucleic Acids Res.* 2018;47(D1): D614–D624. https://doi.org/10.1093/nar/gky992.
25. Ma C, Chen C, Jia L, He X, Zhang B. Comparison of the intestinal microbiota composition and function in healthy and diseased Yunlong Grouper. *AMB Express.* 2019;9:187.
26. Ishii C, et al. A metabologenomic approach reveals changes in the intestinal environment of mice fed on American diet. *Int J Mol Sci.* 2018;19.
27. Falkowski PG, Fenchel T, Delong EF. The microbial engines that drive Earth's biogeochemical cycles. *Science (New York, NY).* 2008;320:1034–1039.
28. Das M, Ghosh TS, Jeffery IB. IPCO: inference of pathways from co-variance analysis. *BMC Bioinformatics.* 2020;21:62.
29. Mallick H, et al. Predictive metabolomic profiling of microbial communities using amplicon or metagenomic sequences. *Nat Commun.* 2019;10:3136.
30. Franzosa EA, et al. Species-level functional profiling of metagenomes and metatranscriptomes. *Nat Methods.* 2018;15:962–968.
31. Bose T, Haque MM, Reddy C, Mande SS. COGNIZER: a framework for functional annotation of metagenomic datasets. *PLoS One.* 2015;10, e0142102.
32. New FN, Brito IL. What is metagenomics teaching us, and what is missed? *Annu Rev Microbiol.* 2020;74:117–135.
33. Ghosh TS, Keohane D, Jeffery IB, Molloy M, O'Toole PW, Shanahan F. Microbiome and health consequences for ethnic minorities after enforced lifestyle changes. *Nat Med.* 2020;26:1089–1095. Under revision.
34. Almeida A, et al. A new genomic blueprint of the human gut microbiota. *Nature.* 2019;568:499–504.
35. Perez M, et al. A synthetic consortium of 100 gut commensals modulates the composition and function in a colon model of the microbiome of elderly subjects. *Gut Microbes.* 2021;13:1–19.
36. Zhang Y, et al. Metatranscriptomics for the human microbiome and microbial community functional profiling. *Annu Rev Biomed Data Sci.* 2021;4:279–311.
37. Heyer R, et al. A robust and universal metaproteomics workflow for research studies and routine diagnostics within 24 h using phenol extraction, FASP digest, and the MetaProteomeAnalyzer. *Front Microbiol.* 2019;10:1883.
38. Mills RH, et al. Multi-omics analyses of the ulcerative colitis gut microbiome link Bacteroides vulgatus proteases with disease severity. *Nat Microbiol.* 2022;7:262–276.
39. Korem T, et al. Growth dynamics of gut microbiota in health and disease inferred from single metagenomic samples. *Science (New York, NY).* 2015;349:1101–1106.
40. Emiola A, Oh J. High throughput in situ metagenomic measurement of bacterial replication at ultra-low sequencing coverage. *Nat Commun.* 2018;9:4956.
41. Brown CT, Olm MR, Thomas BC, Banfield JF. Measurement of bacterial replication rates in microbial communities. *Nat Biotechnol.* 2016;34:1256–1263.
42. Tedersoo L, Albertsen M, Anslan S, Callahan B. Perspectives and benefits of high-throughput long-read sequencing in microbial ecology. *Appl Environ Microbiol.* 2021;87, e0062621.

43. Check Hayden E. Genome sequencing: the third generation. *Nature.* 2009;457 (7231):768–769. https://doi.org/10.1038/news.2009.86.
44. Levene MJ, et al. Zero-mode waveguides for single-molecule analysis at high concentrations. *Science (New York, NY).* 2003;299:682–686.
45. Branton D, et al. The potential and challenges of nanopore sequencing. *Nat Biotechnol.* 2008;26:1146–1153.
46. de Siqueira GMV, Pereira-Dos-Santos FM, Silva-Rocha R, Guazzaroni ME. Nanopore sequencing provides rapid and reliable insight into microbial profiles of intensive care units. *Front Public Health.* 2021;9:710985.
47. Latorre-Pérez A, et al. A round trip to the desert: in situ nanopore sequencing informs targeted bioprospecting. *Front Microbiol.* 2021;12:768240.
48. Rhoads A, Au KF. PacBio sequencing and its applications. *Genomics Proteomics Bioinformatics.* 2015;13:278–289.
49. Logsdon GA, Vollger MR, Eichler EE. Long-read human genome sequencing and its applications. *Nat Rev Genet.* 2020;21:597–614.
50. Chin CS, et al. Nonhybrid, finished microbial genome assemblies from long-read SMRT sequencing data. *Nat Methods.* 2013;10:563–569.
51. Li C, et al. INC-Seq: accurate single molecule reads using nanopore sequencing. *Gigascience.* 2016;5:34.
52. Ciuffreda L, Rodríguez-Pérez H, Flores C. Nanopore sequencing and its application to the study of microbial communities. *Comput Struct Biotechnol J.* 2021;19:1497–1511.
53. Wang X, et al. Full-length transcriptome reconstruction reveals a large diversity of RNA and protein isoforms in rat hippocampus. *Nat Commun.* 2019;10:5009.
54. Bolyen E, et al. Reproducible, interactive, scalable and extensible microbiome data science using QIIME 2. *Nat Biotechnol.* 2019;37:852–857.
55. Knights D, et al. Bayesian community-wide culture-independent microbial source tracking. *Nat Methods.* 2011;8:761–763.
56. Jovel J, et al. Characterization of the gut microbiome using 16S or shotgun metagenomics. *Front Microbiol.* 2016;7:459.
57. Callahan BJ, et al. DADA2: high-resolution sample inference from Illumina amplicon data. *Nat Methods.* 2016;13:581–583.
58. Breitwieser FP, Lu J, Salzberg SL. A review of methods and databases for metagenomic classification and assembly. *Brief Bioinform.* 2019;20:1125–1136.
59. Afiahayati, Sato K, Sakakibara Y. MetaVelvet-SL: an extension of the Velvet assembler to a de novo metagenomic assembler utilizing supervised learning. *DNA Res.* 2015;22:69–77.
60. Peng Y, Leung HC, Yiu SM, Chin FY. Meta-IDBA: a de novo assembler for metagenomic data. *Bioinformatics (Oxford, England).* 2011;27:i94–i101.
61. Beghini F, et al. Integrating taxonomic, functional, and strain-level profiling of diverse microbial communities with bioBakery 3. *elife.* 2021;10.
62. Truong DT, Tett A, Pasolli E, Huttenhower C, Segata N. Microbial strain-level population structure and genetic diversity from metagenomes. *Genome Res.* 2017;27:626–638.
63. Truong DT, et al. MetaPhlAn2 for enhanced metagenomic taxonomic profiling. *Nat Methods.* 2015;12:902–903.
64. Ainsworth D, Sternberg MJE, Raczy C, Butcher SA. k-SLAM: accurate and ultra-fast taxonomic classification and gene identification for large metagenomic data sets. *Nucleic Acids Res.* 2017;45:1649–1656.
65. Hughes JB, Hellmann JJ. The application of rarefaction techniques to molecular inventories of microbial diversity. *Methods Enzymol.* 2005;397:292–308.
66. McMurdie PJ, Holmes S. Waste not, want not: why rarefying microbiome data is inadmissible. *PLoS Comput Biol.* 2014;10, e1003531.

67. Love MI, Huber W, Anders S. Moderated estimation of fold change and dispersion for RNA-seq data with DESeq2. *Genome Biol.* 2014;15:550.
68. Robinson MD, McCarthy DJ, Smyth GK. edgeR: a bioconductor package for differential expression analysis of digital gene expression data. *Bioinformatics (Oxford, England).* 2010;26:139–140.
69. McMurdie PJ, Holmes S. phyloseq: an R package for reproducible interactive analysis and graphics of microbiome census data. *PLoS One.* 2013;8:e61217.
70. Fernandes AD, et al. Unifying the analysis of high-throughput sequencing datasets: characterizing RNA-seq, 16S rRNA gene sequencing and selective growth experiments by compositional data analysis. *Microbiome.* 2014;2:15.
71. McKnight DT, et al. Methods for normalizing microbiome data: an ecological perspective. *Methods Ecol Evol.* 2019;10:389–400.
72. Weiss S, et al. Normalization and microbial differential abundance strategies depend upon data characteristics. *Microbiome.* 2017;5:27.
73. Lozupone C, Knight R. UniFrac: a new phylogenetic method for comparing microbial communities. *Appl Environ Microbiol.* 2005;71:8228–8235.

CHAPTER THREE

Structure, functions, and diversity of the healthy human microbiome

Taruna Ahrodia[a], Santanu Das[a], Susmita Bakshi[a], and Bhabatosh Das[b,*]
[a]Translational Health Science and Technology Institute, NCR Biotech Science Cluster, Faridabad, Haryana, India
[b]Molecular Genetics Laboratory, Centre for Human Microbial Ecology, Translational Health Science and Technology Institute, NCR Biotech Science Cluster, Faridabad, Haryana, India
*Corresponding author: e-mail address: bhabatosh@thsti.res.in

Contents

Abstract

Taxonomic composition and functional potency of microbes associated with different parts of the human body have largely been explored by culture-independent metagenome sequencing. The diverse microbiota living throughout the human body is made up of thousands of microbial taxa from all three domains of life: Archaea, Bacteria, and Eukarya. Microbial load and functional potency in different body sites

Progress in Molecular Biology and Translational Science, Volume 191
ISSN 1877-1173
https://doi.org/10.1016/bs.pmbts.2022.07.003

are well distinct and have minimal resemblance at higher taxonomic levels between the two habitats. The highest microbial load, diversity, and functional potency including biosynthesis of essential nutrients, chemical modifications of dietary components, and sources of immunomodulatory molecules, are found in the gut microbiome. However, the inter-individual diversity and dynamics of the human microbiome in a given body habitat vary greatly over time. Both environmental factors and host genetics contribute significantly to shaping microbial community structure and its stability. A basic understanding of native microbial compositions and their functional potency and stability in different parts of healthy humans living across geography will help us to identify disease-specific microbiota and develop potential microbiome-based therapeutics. Here, we updated our current understanding of the diversity, dynamics, and functional potency of microbiomes associated with different parts of the human body.

Abbreviations

CAZyme	carbohydrate-active enzymes
CSTs	community state types
GPCRs	G protein-coupled receptors
rRNA	ribosomal RNA
SCFAs	short-chain fatty acids
TMAO	trimethylamine N-oxide

1. Introduction

Genetic variations within and between human populations are very few and any two individuals are likely 99.9% genetically identical.[1] The human microbiome, or the overall genetic content and functional potency of bacteria that live in our bodies, is significantly varied, with just a third of its constituent genes present in the majority of healthy people.[2,3] Even the microbial genetic contents in different body parts of the healthy individuals including gut, skin, reproductive tract, mouth, and lungs vary significantly. While food, climate, host biology, lifestyle, and early microbial exposure have all been implicated, much of this diversity remains unknown. In response to host and environmental factors, the human microbiome is constantly changing. Age, genes, lifestyle, nutrition, hormonal swings, and underlying sickness are all important determinants of the makeup and functional potency of the human microbiome at any particular moment. However, abrupt changes in the human microbiota's makeup (dysbiosis) can result in metabolic and communicable illnesses.[4] Other factors that have been found to influence the richness and diversity of the human microbiota include the mode of delivery at birth (cesarean or vaginal), diet during childhood (breast milk or formula), and adulthood (meat-dependent or vegan/vegetarian), as well as exposure to antibiotics and other therapeutics originating from the atmosphere or resident microbial taxa.[5–7] The microbiota works in tandem with the host cellular machinery for various vital activities, and thereby maintains a homeostatic environment. However, sudden changes in the composition of the human microbiota termed dysbiosis can lead to metabolic and communicable diseases.[4] Previous studies on the human microbiome have reported that subtle changes in the immunological environment can lead to dysbiotic gut flora. As dysbiotic conditions alter the microbial composition and, lead to the proliferation of pathogens, it has also been linked to life-threatening diseases such as cancer, cardiovascular disease, bowel inflammatory disease, and antibiotic resistance.[8–10] To modulate the dysbiotic microbiome, various approaches are being currently used, such as the use of probiotics, prebiotics, and synbiotics is being a trend across the world. However, the use of prebiotics in diseased states is showing significant benefits, but their role in healthy states required more studies. Furthermore, recent studies show the role of phage-based tools to manipulate the human microbiome and restore the dysbiotic states.[11]

Due to advancements in the next-generation high-throughput sequencing, and mass spectrometry platform, the diversity, the number of metabolites produced, and the function potency of these microorganisms are well documented. Biomarker gene sequencing, such as the sequencing of the 16S ribosomal RNA (16S rRNA) gene to determine the identity and abundance of bacteria or archaea, allows us the evaluation of native microbial composition in any body site. Identification of individual taxa in a microbial community can lead to the discovery of signature microbiota or its functional potency linked with a specific health condition. Shotgun sequencing of isolated community DNA (metagenomics) or RNA (metatranscriptomics/RNA Seq) has advanced our efforts to explore the fitness associated with candidate microbial genes, and adaptive potency of microbes in different environmental conditions, regulation of microbial metabolic pathways upon exposure to any xenobiotics. Parallel to this, targeted and untargeted mass spectrometry has aided the detection and characterization of microbiome-associated products, such as metabolites and proteins. Due to advances in high-throughput technology, the understanding of human-associated microbiomes and their metabolically derived products showed their aid in the interactions of the microbiota within the host.[12]

2. Compositions and diversity of human microbiome

Trillions of microbes living throughout the human body with distinct body site-specific community structures profoundly influence host physiology, provide protection against colonization of allochthonous microbes and modulate the risk of chronic and acute diseases depending on their metabolic potency. The skin, mucosa, gastrointestinal tract (GIT), respiratory tract, urogenital tract, and mammary gland are all colonized by microbial taxa (Fig. 1). They develop a specialized and intricate ecosystem that responds to the niche's unique environmental circumstances.[4] The interactions between the host cells and endogenous microbiota start with birth. These relationships are crucial in maintaining homeostasis thereby promoting overall wellness and harmony. Coevolution has allowed microbiota organisms to actively adapt to their diverse habitats and exist in their respective niches within the human body.[13,14]

2.1 Microbiome of the gastrointestinal tract (GIT)

One of our body's most densely inhabited microbial habitats is the gastrointestinal tract. To better understand the role of the gut microbiome in health and illness, researchers have studied the diversity, richness, dynamics, stability, and sustainability of microbial ecology in the GIT of healthy individuals.[15] The gut microbiome comprises roughly 70% of all microorganisms residing in the human body[16] with an average bacterial cell count of 10^{11}–10^{12} per gram of fecal material in the colon.[17] Compared to other body habitats, the gut microbiota contains the greatest number of bacterial species and exhibits the widest diversity of organisms. Most of the studies to profile the gut microbiota makes use of fecal samples, owing to the availability of the sample and non-invasive procedures during collection. However, certain studies to encompass the gut microbiota through intestinal tissue and biopsy samples also have been undertaken. The constituents of the gut microbiota differ significantly between healthy individuals (Table 1), partly due to species and strain-level diversity, as well as differences in microbial growth rates and structural variants within microbial genes, and partly due to inter-individual variability such as diet, genetic makeup, lifestyle, and age, among other factors.[32] Although no higher taxonomic level has been found for universal healthy gut microbiota, the innate ecology of healthy human gastrointestinal tracts is dominated (>99%) by Firmicutes, Bacteroidetes, Actinobacteria, and Proteobacteria.[33] The prominent bacterial families are *Lachnospiraceae, Ruminococcaceae, Prevotellaceae, Bacteroidaceae, and Rikenellaceae.*[34] Mostly the members of the Lactobacillaceae and Enterobacteriaceae families, on the other hand, colonize the small intestine. *Faecalibacterium prausnitzii, Roseburia intestinalis, Ruminococcus intestinalis, Blautiaobeum, Dorea formicigenerans,*

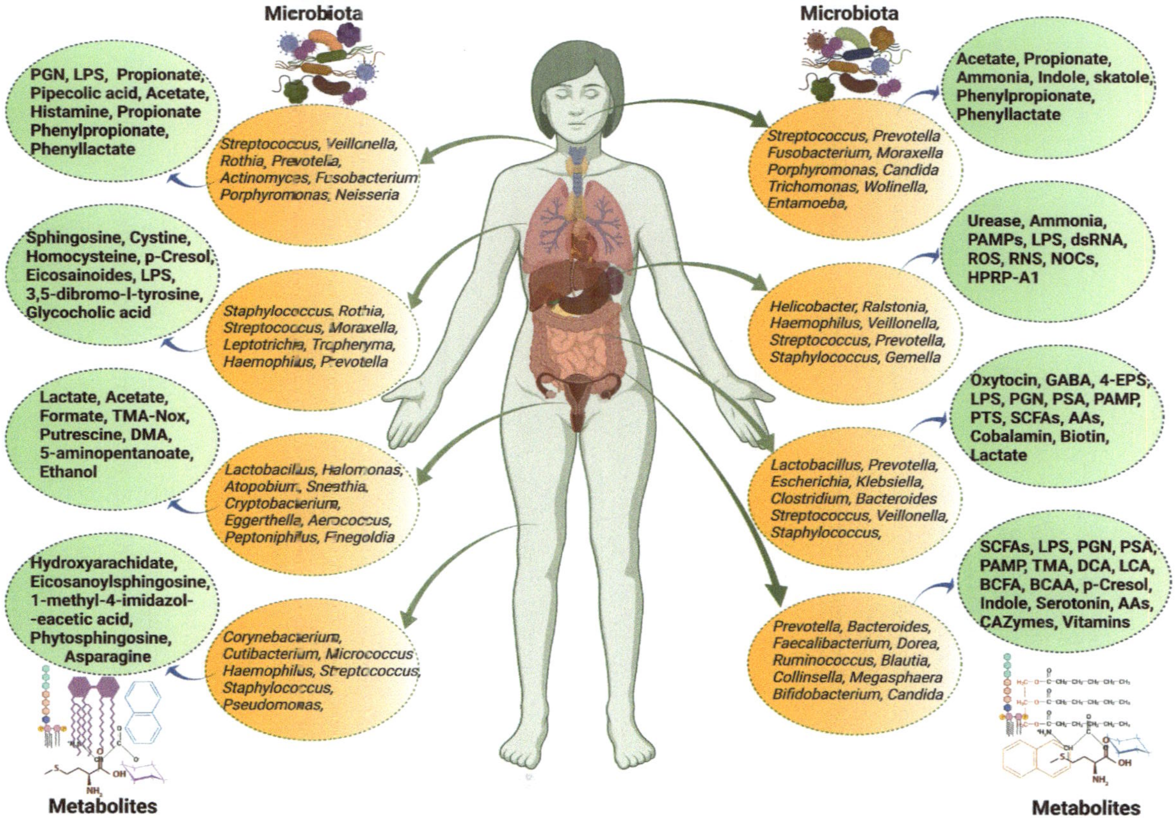

Fig. 1 Commonly found abundant microbial taxa and metabolites in different body-sites that have been identified using multi-omics approaches. Different members of Firmicutes mostly dominate throughout the healthy human. Maximum metabolic potency has been reported in the microbial community living in the gastrointestinal tract. SCFA, short-chain fatty acids; LPS, lipopolysaccharides; PGN, peptidoglycan; PSA, polysaccharide A; PAMP, pathogen-associated molecular patterns; TMA, trimethylamine; DMA, dimethylamine; DCA, deoxycholic acid; LCA, lithocholic acid; BCFA, branched chain fatty acids; BCAA, branched chain amino acids; GABA, gamma-aminobutyric acid; 4-EPS, 4-ethylphenylsulfate; PTS, PEP-dependent phosphotransferase system; AA, amino acids; ROS, reactive oxygen species; RNS, reactive nitrogen species; NOC, nucleoid occlusion protein; HPRP-A1, amphipathic α-helical peptide.

Table 1 Geo-graphical variation within the different body sites.

Body sites	African's core genus	Indian's core genus	Danish's core genus	American's core genus
Gut	*Treponema*, Bacteriodetes and *Prevotella*[18]	*Faecalibacterium*, *Roseburia*, *Dorea*, Agathobacter, *Collinsella*, *Lachnoclostridium*, *Lachnospira*, *Blautia*, *Megasphera*, *Lactobacillus*[19]	*Parabacteroides* and *Alistipes*[19]	*Bacteroides*[20]
Oral	*Porphyromonas*[21]	*Neisseria*, *Streptococcus*, *Prevotella*, *Porphyromonas*, *Haemophilus*[22]	*Streptococcus*, *Veillonella*[23]	*Streptococcus*, *Prevotella_7*, *Haemophilus*, *Veillonella*[24]
Skin	*Staphylococcus*, *Micrococcus*, *Kocuria*[25]	*Corynebacterium*, *Alloiococcus*, *Staphylococcus*, *Streptococcus*, *Anaerococcus*[22]	*Corynebacterium*, *Acinetobacter*, *Moraxella*, *S epidermidis*, *P acnes*[26]	*Cutibacterium*, *Corynebacterium*, *Staphylococcus*[27]
Vagina	*Prevotella*, *Sneathia*[28]	*Lactobacillus iners*, *L. crispatus*, *L. gasseri*, *L. jensenii*, *L. johnsonii*[29]	*Veillonela spp.*, *Prevotella spp*[30]	*Lactobacillus* spp.[31]

Clostridium cluster XIVa, *Eubacterium hallii*, *Prevotella copri*, *Bacteroides uniformis*, *Collinsella aerofaciens*, *Bifidobacterium bifidum*, and *Escherichia coli* are most common bacteria found in healthy people.[35] The gut microbiome includes methanogenic archaea (*Methanobrevibactor smithii*), eukaryotes (*Saccharomyces* and *Candida*), and viruses (bacteriophages) in healthy individuals.[36]

Despite the core microbiota's homogeneity, the biodiversity and richness of bacterial species present in healthy people's GI tracts vary substantially between individuals and between geographical locations (Table 1). In a study conducted among Indians by[33] it was found that *Firmicutes* (62%), was the most common phylum among the healthy Indian vegetarians followed by *Bacteroidetes* (24%). Whereas, in the case of the Indian balanced non-vegetarians, the abundance of Bacteriodetes is increased. Several comparative studies were undertaken to understand the impact of geographical variation among gut microbiota. Such a study, carried out at our laboratory identified the gut microbial signatures between the Indian and Danish populations.[19] We observed that the Danish cohort exhibited a more diverse microbiota as compared to that of the Indian cohort. Firmicutes, followed by Bacteroidetes, were the dominant phyla in both populations across all samples. While *Actinobacteria*, *Proteobacteria*, and *Elusimicrobia* were found in greater abundance in the Indian cohort; *Bacteroidetes*, *Tenericutes*, *Verrucomicrobia*, and *Synergistetes* were shown to be considerably richer in the Danish cohort.[19] Furthermore, it was identified that *Dorea*, *Agathobacter*, *Collinsella*, *Lachnoclostridium*, *Lachnospira*, *Blautia*, *Faecalibacterium*, *Roseburia*, and *Subdoligranulum* as core genera among both the groups. *Megasphera* and *Lactobacillus* were also identified as core microbiota in the Indian population, however, *Parabacteroides* and *Alistipes* were only identified as core taxa in the Danish Cohort. In another study, a comparison of Indian and Japanese gut microbiota shows similar microbial diversity and richness in both populations.[37] However, at the genus level, *Prevotella* and *Candida* are the most dominant microbiota in the Indian gut, whereas *Bacteroides* and *Saccharomyces* are the most dominant microbes in the Japanese gut. A comparative study on the composition and diversity of gut microbiome between African and European children showed the population-specific microbial signatures.[38] The diversity of the gut microbiota of rural African populations is particularly striking, as evident from numerous studies. In general, the African population exhibits a higher level of microbial richness and diversity.[38] The representative major taxa include *Prevotella*, *Treponema*, *Suturella*, *Anaerovibrio*, members of the Clostridiaceae family, *Faecalibacterium*, and *Dialister*; the majority of which are enriched with CAZyme functions and are fibrinolytic.[38,39] In addition, *Bifidobacteria* and *Lactobacillus* species are rarely detected in the gut microbial communities of African children. In a study carried out on the African population by[18] it was observed that the gut parasite *Entamoeba* is substantially connected with microbiome composition and diversity, indicating cross-kingdom microbial interactions in the gut environment.[18] The influence of those who migrate from Latin America to the United States at an early age has been the subject of fascinating research. They have found that the immigrants have lower *Prevotella* to *Bacteroides* ratios that last throughout their lives.

In early life, intrapersonal variation of the gut microbiome compositions and functions are higher than in adulthood. Within 3 years after birth, the diversity and abundance of the gut microbiota developed into an adult-like structure, regardless of geography.[35] Table 1 depicts the geographical variance within different microbiota among several body locations. Although there are few genetic differences between human groups, the genetic and metabolic diversity of gut microbiota varies greatly. In adulthood, the gut microbiota helps to maintain human health by aiding in metabolic functions, releasing nutrients, encouraging the isolation of certain host tissues, reducing the likelihood of pathogen invasion of the intestine, and modulating the immune system, among other things.[40]

Gut microbial diversity represents the abundance and the distribution of various taxa residing in an individual's gastrointestinal tract. Loss or gain in the gut microbial diversity disturbs the gut homeostasis, a condition often termed dysbiosis.[41] The dysbiotic condition leads to various gastrointestinal problems; For example, obesity and inflammatory bowel disease (IBD) are linked to reduced diversity of gut commensals.[41,42] whereas pathogenic taxa increases incase of liver disease, malnutrition. Patients with colon cancer and IBD had higher *Bacteroidetes* (55%) than *Firmicutes* (26%) but lower *Bacteroidetes: Firmicutes* ratios (3.8:1 and 2.4:1, respectively). Verrucomicrobiota and Actinobacteria phyla's were found in 23% and 5% of IBD and colon patients, respectively.[32] Several bacterial species from less common phyla, such as *Verrucomicrobia, Fusobacteria, Tenericutes,* and *Spirochetes,* are also present.[43]

2.2 Microbiome of the oral cavity

The oral cavity of humans houses over 700 bacterial species, making it the second-largest site of the human body with the most bacterial diversity. The oral microbiota is essential for sustaining oral and systemic health in addition to being the starting point for digestion of nutrients. It is one of the most well-studied habitats in microbiome research due to the simplicity of obtaining samples. Most of the oral microbiome studies used saliva or oral swabs as biological samples to explore microbial diversity and dynamics. Bacteria begin populating the mouth cavity shortly after a person is born. The oral microbiota, like the gut microbiota, is impacted by the manner of delivery and feeding strategy.[44] There are 185 genera and 12 phyla among 700 species residing orally, with 54% being formally named, 14% unknown (but culturable), and 32% solely acknowledged as uncultivated phylotypes.[45] The 12 phyla are Firmicutes, Proteobacteria, Actinobacteria, Bacteroidetes, Chlamydiae, Chloroflexi, Spirochaetes, Synergistetes, Saccharobacteria, and Gracilibacteria.[46] The mouth cavity is also home to bacteria such as protozoa, fungus, and viruses. *Entamoeba gingivalis* and *Trichomonas tenax* are the most common protozoa, both of which are saprophytic. The Candida genus is the most often encountered fungus in the mouth.[47] Oral bacteria exhibit both pro and anti-inflammatory actions, which are important for maintaining homeostasis in strongly populated areas like the mouth. Despite intense microbial colonization, acute infections of the oral mucosa are uncommon due to interactions between the host's immune system and its microbial symbionts.[48] Observations in immunocompromised individuals, who can develop life-threatening viral and fungal infections of the mucous membranes, as well as oral infections by non-oral species, underscore the relevance of these host-microbe interactions[49]

2.3 Microbiome of the genitourinary tract

The vaginal microbiota is a dynamic system that changes throughout a woman's life, including throughout her menstrual cycle. The most dominant species of the human vaginal flora is *Lactobacillus*, which uses glycogen which is regulated by estrogens, resulting in unique vaginal conditions (Fig. 1). *Lactobacilli* and other fermentative bacteria, in collaboration with vaginal epithelial cells, synthesize lactic acid, which acidifies the vaginal environment. This acidic environment provides stability to the vaginal microbiota and prevents foreign bacteria from colonizing, thus preventing infection. The major population state forms in the vagina are characterized by individual lactobacilli species, primarily *Lactobacillus crispatus*. In addition to lactobacilli, the vaginal microbiota of healthy women is also colonized by non-lactobacillary microbiota, the rate of which is determined by ethnicity (white, black, Hispanic, and Asian)[29] for example, different species of *Lactobacilli* dominate the vaginal microbiota of Asian and white American women, whereas different species of anaerobic bacteria such as *Cryptobacterium*, *Gardnerella*, *Prevotella*, *Porphyromonas*, *Atopobium*, *Peptoniphilus* dominate the vaginal microbiota of a large percentage of Hispanic and black American women.[50] The vaginal microbiota of reproductive-age women constantly exposes physiological changes such as hormonal changes and thus is dynamic.

In terms of the vaginal microbiota, however, there are significant variations between non-pregnant and pregnant women. The richness and diversity of the vaginal microbiome of pregnant women are declining in previous research. Pregnant women have a higher prevalence of *Lactobacillus* spp., *Actinomycetales*, *Clostridiales*, and *Bacteroidales* whereas *Lactobacillus* spp., *Actinobacteria*, *Prevotella*, *Veillonellaceae*, *Streptococcus*, *Proteobacteria*, *Bifidobacteriaceae*, *Bacteroides*, and *Burkholderiales* predominate in non-pregnant women.[51] As a result, the vaginal microbiota varies throughout time in a single person's lifespan. Furthermore, the vaginal microbiota varies greatly across people, and these changes are caused by disparities in sexual activity,[52] douching,[53] chronic stress,[54] geographical disparity,[55] race,[56] and other variables.[57] According to the research of 396 asymptomatic women from four ethnic groups in North America, single or multiple *Lactobacillus* species dominate vaginal microbiomes, which are divided into five community state categories (CSTs). *L. crispatus*, *L. gasseri*, *L. iners*, and *L. jensenii* dominate CSTs I, II, III, and V, respectively, whereas CST IV refers to a diverse microbial community dominated by obligate anaerobic bacteria. High Nugent scores are most commonly associated with CST IV, but they can also be seen in other CSTs. CSTs I, II, III, and V are present in 89.7% of white women and 80.2% of Asian women, respectively, although they are 61.9% and 59.6% in black and Hispanic women, respectively.[56] The immune system, ligands on the surface of epithelial cells, and the quantity and components of vaginal discharge may all play a role in the variability in vaginal microbiota by race of women. Host factors may be more important in developing the vaginal microbiota among races than behavioral and cultural difference.

2.4 Microbiome of the skin

Skin is considered the largest organ of a human, which is colonized with beneficial microbes and functions as a physical barrier against disease invasion. This barrier maintains the balance between the commensals and pathogens and during the diseased condition this barrier is destroyed. Also, they play an important role in protecting us from invading pathogens, educating our immune system, and breaking down natural products. The physiology of the skin site was revealed to be the most important factor in bacterial colonization and abundance, with differences in bacterial taxonomy linked with humid, dry, and sebaceous microenvironments.[58–61] Lipophilic *Propionibacterium* species are known to dominate sebaceous locales, whereas bacteria such as *Staphylococcus* and *Corynebacterium* are known to dominate humid places such as the bends of the elbows and the foot (Fig. 1). The skin microbiota, in comparison to other places, retains the makeup of their microbial communities for at least 2 years; yet, the factors that form and sustain these complex communities are little understood.[62] Microbial colonization of the skin is dependent on the delivery method, as it is with any other area; neonates delivered vaginally get microorganisms that are residing in the vagina, whereas neonates born by Cesarean section acquire microorganisms linked with the skin.[63,64] However, due to hormonal changes, an individual's skin microbiota changes dramatically during puberty. As a result, the skin microbiota of postpubescent people promotes the growth of lipophilic bacteria like *Propionibacterium* spp. and *Corynebacterium spp.*, as well as the fungus *Malassezia* spp. Firmicutes (*Streptococcaceae* etc.), Bacteroidetes, and Proteobacteria (*beta-proteobacteria* and *gammaproteobacteria*) are more abundant in prepubescent individuals, and the fungus community is more diversified.[62] In a process known as 'colonization resistance,' interactions among members of the skin microbiota modify the resident microbial community and prevent harmful bacteria from colonizing the skin. Bacteria that are normally helpful to their hosts might become pathogens under specific circumstances. Dysbiosis, or a shift in the microbial balance, is linked to a variety of common skin illnesses. Changes in the abundance of commensal organisms are frequently the cause of dysbiosis. *Propionibacterium acne*, *Staphylococcus aureus*, and *Staphylococcus epidermidis*, for example, cause acne, eczema, and chronic wounds, respectively. Individual species and changes in the microbial population are hypothesized to have underlying contributions to both unusual and common skin problems. Skin health depends on the delicate balance between our cells and the microorganisms that live on its surface. When *Staphylococcus epidermidis*, one of the most frequent bacteria on human skin, was administered to germ-free mice, it was shown that it reduced sensitivity to allochthonous bacterial infections. The addition of a single species of bacteria to the skin of mice boosted immunological function. Mice with *S. epidermidis* were able to defend themselves against a parasite, whereas mice without the bacterium were unable to do so.[65] The bacteria on the scalp surface are provided with a distinct habitat by the host physiological variables, which include sebum material, moisture, pH, and topography.[66] A dysbiosis in the cutaneous microbiome has been found in episodes of dandruff, seborrheic dermatitis, and atopic dermatitis.[67,68] The previous study on the scalp microbiome, on the other hand, has primarily looked at the taxonomic organization and diversity differences among bacterial and fungal species in different populations. This research identified the most frequent fungal and bacterial taxa as *Malassezia*, *Propionibacterium*, and *Staphylococcus*, respectively, and emphasized human or population-specific compositional variations.[67,69]

3. Functional diversity of human microbiome

Variations in microbial ecology are hypothesized to have a role in the development of numerous metabolic and infectious illnesses; microbes in the human body influence metabolic activity and energy balance. The great majority of these metabolites (>95%) are consumed via the gastrointestinal system. Alterations in the gut microbiota, for example, cause long-term changes in inflammation in obese people[70]; bacterial endotoxins like LPS activate pattern recognition receptors like Toll-like receptors, resulting in an innate immune response and insulin resistance.[71] To establish relationships between healthy and pathological phenotypes, it is critical to understand the metabolism and function of the healthy human microbiome. The importance of the gut microbiota in human metabolism and health is shown in Table 2, which has prompted research into the microorganisms involved and their activity in relation to metabolic pathways, especially those involving dietary component metabolism.

3.1 Carbohydrate

Humans' digestive systems only partly break down carbs, leaving the complex undigested to pass onto the colon. To survive, bacteria in the colon rely on undigested food substrates from the upper digestive system. Bacterial fermentation of saccharolytic carbohydrates normally yields helpful metabolites, however, when carbohydrate is low, bacteria switch to alternative

Table 2 Microbe's involved in the metabolism.

Metabolism	Microbes involved in metabolism	References
Short chain fatty acids	*Faecalibacterium prausnitzii, Bifidobacterium longum, R. intestinalis, Lachnospiraceae*	72,73
Hydrogen and carbon dioxide	*C. sporogenes, C. butyricum,* and *C. perfringens*	74
Bile acids	*Rumminococcaceae, Blautia*	75
TMAO	*Bacteroides* and *Bifidobacteria*	76
Vitamin synthesis	*Bacteroidetes, Fusobacteria,* and *Proteobacteria*	74

energy sources, resulting in the production of other metabolites that may be toxic to the host.[77] Following the fermentation of dietary carbohydrates, short-chain fatty acids and gases are the major bacterial fermentation products.

3.1.1 Short-chain fatty acids (SCFA)

The production of SCFAs (linear carboxylic acids with short (1–6 carbons) aliphatic tails, of which acetate (C2), propionate (C3), and butyrate (C4) are the most prevalent, is one of the most important metabolic activities mediated by colonic microbes. SCFAs are the main products of anaerobic fermentation of undigested, microbially available dietary carbohydrates, and they allow the host to get energy from food through absorption and metabolism, as well as modulate host physiology by activating G protein-coupled receptors (GPCRs) like GPR41, GPR43, and GPR109A. Three essential SCFAs serve diverse but equally vital roles in the human body. SCFAs are utilized to generate mucosal lipids and are converted to CO_2 and ketone. Butyrate comes first, followed by propionate, and then acetate. Butyrate is one of the most important SCFAs in human health. It is the principal source of energy for human colonocytes, and its capacity to cause apoptosis in colon cancer cells and modulate gene expression by blocking histone deacetylases suggests that it may have anti-cancer capabilities.[78] Propionate is a source of energy for epithelial cells, but it's also transferred to the liver and helps in gluconeogenesis. It's also getting a lot of attention as a critical molecule in satiety signaling,[79] due to its interactions with the gut receptors GPR41 and GPR43, also known as Free Fatty Acid Receptors FFAR2 and FFAR3, which can cause intestinal gluconeogenesis (IGN).[80,81] Acetate, which is a co-factor/metabolite for the development of other microorganisms, is the most prevalent SCFA. In the human body, acetate is transported to peripheral tissues and employed in cholesterol production and lipogenesis.[82] *Roseburia intestinalis* growth on (fructooligosaccharide) FOS was delayed until enough acetate, generated by *Bifidobacterium longum*, accumulated in the growth medium in co-culture with a different strain of *Bifidobacterium longum*.[73] Also, *Faecalibacterium prausnitzii* will not grow in pure culture without acetate.[72] The most prevalent butyrate makers in the gastrointestinal tract are Firmicutes, such as certain *Lachnospiraceae* and *Faecalibacterium prausnitzii*, while Bacteroides species, Negativicutes, and some *Clostridium* species produce propionate.[83] Some Bacteroidetes and Firmicutes species generate propionate and butyrate as a result of peptide and amino-acid fermentation.[84] Bacteria create intermediate fermentation products including fumarate, succinate, and lactate, but these are often converted to propionate or butyrate by other bacteria, so they're only found in trace amounts in adult feces.[74] Lactate, on the other hand, is detected in slightly greater amounts in ulcerative colitis patients,[85] suggesting that it might be used as a disease predictor.

3.1.2 Hydrogen (H_2) gas and carbon dioxide (CO_2) production

The food components' low molecular mass carbohydrates are rapidly digested by microbial activities and serve as a key source of intestinal H_2 gas. Pyruvate is the most energy-efficient glycolytic intermediary for microbes, hence it's important for H2 production in the microbiota.[86] Colonic anaerobes utilize protons as electron sinks in catabolic events involving carbohydrates and amino acids in the human large intestine. Hydrogen is created by a variety of gut bacteria, the most prevalent of which are *Bacteroides* and *Clostridium*. These genera are well-represented in the microbiota, which means they are the main generators of microbial gas.[71] H_2 may be produced in the large intestine in three ways: by cleaving pyruvate to formate, which is then metabolized to H_2 and CO_2 by formate hydrogen lyase (*Enterobacteria*); by pyruvate: ferredoxin oxidoreductase and hydrogenase, which is present in certain *Clostridia*; and finally, by oxidation of reduced pyridine and flavin nucleotides.[87] In general, if 40–50 g of fermentable carbs are available, daily rates of H_2 generation in humans can approach 1000 mL[87]; this, of course, excludes H_2 created during protein breakdown and amino acid fermentation.

Flatus also emits carbon dioxide, which is a significant gas in terms of volume. Methanogenesis and, to a lesser extent, acetogenesis recycle carbon dioxide, which makes up 5–50% of total flatus volume.[88] Other than bacterial metabolism, a multitude of mechanisms create carbon dioxide. Carbon dioxide can also be produced via the diffusion of carbon dioxide from the blood through the colonic lumen, the acidification of bicarbonate in the upper gastrointestinal tract, and bacterial digestion.[89] Certain clostridia species (e.g., *C. sporogenes*, *C. butyricum*, and *C. perfringens*) release both carbon dioxide and hydrogen during their metabolic activities.[74]

Fiber is the principal source of microbiota-accessible carbohydrates in the human diet. Dietary fibers are scarce in an energy-dense diet. The fiber gap, also known as the low-fiber diet, can cause a considerable loss of gut microbiota diversity and beneficial metabolites, which can contribute to a variety of health issues and illnesses, including metabolic syndrome and obesity-related diabetes, IBD, liver disease, and colon cancer.[90]

3.2 Bile acids

Bile acids (BA) are thought to be important in the control of gut flora. In this context, a study on the gut-liver axis found a link between liver health, fecal bile acid concentrations, and the makeup of the gut microbiota.[91] The researchers looked at the amounts of bile acids in the feces of healthy people and people with early and severe cirrhosis. Low bile acid levels entering the colon as the illness progressed have been linked to cirrhosis-related bacterial dysbiosis.[91,92] In this dysbiosis, the gram-positive components of the normal microbiota, such as *Blautia* and *Ruminococcaceae*, were considerably decreased.[75]

In the hepatocytes of the liver, the traditional or neutral route and the alternative or acidic pathway are employed to produce primary bile acids, cholic acid, and chenodeoxycholic acid. Following a meal, the duodenum secretes cholecystokinin, which induces gallbladder contraction and the release of bile acids into the gut. Primary bile acids (deoxycholic acid and lithocholic acid) are formed in the gut in conjunction with intestinal bacteria, allowing dietary fats and fat-soluble vitamins to be absorbed more easily.[93, 94] By de-conjugating, de-hydroxylating, and re-conjugating these molecules, the gut microbiota has been demonstrated to play a role in the biotransformation of BA.[95] Furthermore, antibacterial properties of BA have been revealed, which can degrade bacterial cell membranes, reducing bacterial expansion. Through nuclear farnesoid X receptor (FXR) and cytoplasmic G protein-coupled receptor (TGR-5) regulation of the expansion and composition of the intestinal microbiota, BA can help protect the liver and gut against inflammation.[96]

3.3 Proteins

According to Macfarlane and Cummings' early research with human gut contents, the colonic microbiota has a large proteolytic capability, converting ingested food protein and endogenous protein from host enzymes, mucus, and sloughed-off intestinal cells into shorter peptides, amino acids and derivatives, short and branched-chain fatty acids, and gases such as ammonia, H2, CO2, and H2S.[97] Protein breakdown activities in the colon, according to Metchnikoff[98] were linked to the onset of senility and the shortening of longevity. He was aware that autointoxication was linked to toxic substances formed during amino acid fermentation, such as NH_3, phenol, and H_2S, and that foetid air, depression, insomnia, fatigue, extreme headaches, elevated body temperature, skin disorders, and other extracolonic physiological symptoms were all influenced by bacteria colonizing the broad intestine.[98] There are various distinct types of protein in the big intestine. They are found in part as nutritional residues (plant proteins, sarcoplasmic and myofibrillar muscle proteins), but they are also found in the host's oral, gastric, pancreatic, and small intestine secretions, which contain enzymes, mucins, and other glycoproteins.

3.4 Trimethylamine N-oxide (TMAO) metabolism

The gut is a circulatory organ that contains around 40% of all human blood. Reduced blood flow to the intestinal epithelial cells promotes cell hypoxia, and anaerobic digestion, and changes the gut microbiota, with a decrease in *Bacteroides* and *Bifidobacteria* and an increase in *Firmicutes* and *Proteobacteria.*[76] Choline, betaine, L-carnitine, croton betaine, trimethyl lysine, butyrobetaine, phosphatidylcholine, glycerophosphocholine, and trimethylamine-N-oxide are examples of trimethylamine (TMA) dietary precursors found in red meat, eggs, fish, and dairy products.[99] These precursors might be converted to TMA by intestinal microbial enzymes. The bulk of TMA is absorbed into the circulatory system, where hepatic flavin-containing monooxygenase (FMO) oxidizes it to TMAO.[100] Based on an untargeted metabolomics analysis, researchers revealed that trimethylamine-N-oxide (TMAO), a chemical molecule originating from the gut microbiota, predicted an increased risk of cardiovascular events.[101] TMAO also impacts cardiac hypertrophy and fibrosis,[99] generates an inflammatory response,[102] and exacerbates mitochondrial dysfunction, according to certain investigations.[103] 3,3-dimethyl-1-butanol and other phytochemicals like berberine,[104] 3,3-diindolylmethane (DIM), and indole-3-carbinol, which target TMAO metabolism, is also predicted as therapeutic targets.[105] Furthermore, regulating gut microbiota composition and focusing on the gut–TMAO axis are also likely to improve HF patient survival.[101]

3.5 Vitamin synthesis

Vitamins, particularly vitamin K and B group vitamins including biotin, cobalamin, folates, nicotinic acid, pantothenic acid, pyridoxine, riboflavin, and thiamine, can be synthesized by gut microorganisms (Table 2). This phenomenon has been studied for more than four decades. However, with the introduction of next-generation sequencing technology,[106] significant insights into the gut microbiota's vitamin production pathways have recently been acquired. J.G. LeBlanc and colleagues investigated the metabolic capabilities of gut microbial genomes from two persons and observed that their microbiomes were rich in Clustered Orthologous Groups (COGs) involved in the manufacture of deoxyxylulose-5-phosphate, a precursor to thiamine and pyridoxal.[107] Magnusdottir et al. also looked at the genomes of 256 common gut bacteria and reported that riboflavin (166 possible producers) and niacin (162 possible producers) were the most often produced vitamins.[108] The researchers discovered that practically all microorganisms from the phyla Bacteroidetes, Fusobacteria, and Proteobacteria possessed the essential pathways for riboflavin and biotin production, but Firmicutes and Actinobacteria had a substantially smaller fraction of the needed pathways.[74]

3.6 Diversity of xenobiotics degradation capacity

The term "xenobiotics" describes the chemical, synthetic, or maybe naturally occurring compounds affecting the normal metabolism of a living organism. The advanced urbanization and industrialization are the major source of introduction of the hazardous compounds. Many of these compounds impact the environment because of their prolonged persistence, high toxicity, and minimal biodegradability.[109] The detrimental effect of xenobiotics on human health is also widely recognized. However, the extent of harm is influenced by inter-and intra-individual genetic variations among humans. Several dietary compounds, environmental pollutants, and drugs are consideredxenobiotics for the human body. Human microbiomes have the potential to metabolize xenobiotics either directly or by influencing the host enzymatic activity. On the contrary, microbiome composition may be altered by xenobiotics leading to dysbiosis. In particular, the microbiome of the human gut has the ability to biotransformation of xenobiotics whether dietary compounds or pharmaceutical products. They usually alter the xenobiotics' half-lives and their effect on the human body. The host's metabolic capability is different from that of the gut microbiota's metabolic capacity. Oxidation and conjugation are majorly carried out by the host enzymes, whereas reduction and hydrolysis are mainly performed by microbial enzymes.[110] Orally administered drug passes from the alimentary canal to upper GIT and is then translocated to the liver through the hepatic portal vein where several modifications occur. Metabolism in the liver occurs in three phases: hydrolysis, oxidation or reduction is referred to as Phase I, conjugation to polar moieties is referred to as Phase II, and phase III mainly involves transport without chemical modification. Host enzymes such as cytochrome P450 monooxygenases or alcohol/aldehyde dehydrogenases plays role in oxidation, cytochrome P450 reductases for reduction and hydrolysis carried out by esterases and epoxide hydrolases play a vital role in transforming the xenobiotics into less harmful and easily excretable. Microbial hydrolytic enzymes such as Sulfatases that breaks sulfur esters of Phase II host metabolism, proteases that hydrolyses peptide chains, Glycosidases that uses water to break the glycosidic bonds, releasing sugars plays a major role in xenobiotic metabolism forms.[110,111]

3.7 Diversity of immunomodulatory capacity

Invading pathogens and endogenous disturbances of homeostasis are both threats to the human immune system, which is made up of a complex network of innate and adaptive components in all tissues.. Host cells and their commensals have co-evolved in maintaining homeostasis. Such a close association necessitates a host immune system that prevents commensals from overusing host resources while preserving immunological tolerance to harmless stimuli.[35] The colonization of the host immune system throughout the first years of growth and development is critical for the maturity of the host immune system. The immune system's priming reaches its pinnacle at this time, and the microbiota composition exhibits the greatest intra- and inter-individual heterogeneity before stabilizing around the age of three years.[35,112] Germ-free (GF) animal models are widely employed in the research of mechanistic causal links between commensal microbiota and host immunity. Experiments on GF mice have shown that the lack of commensal microbiota is linked to significant gut lymphoid tissue architecture and immune function abnormalities.[113] The ability of the gut immune system to establish immune tolerance to an enormous and constantly changing wealth of harmless microorganisms while simultaneously preserving immune responses against pathogenic infection or commensal intrusion into the sterile body milieu is a remarkable phenomenon of the gut immune system.. The microbiota and the agents of the immune system are involved in bi-directional communication, mostly represented by the secretion of antimicrobial peptides (AMP).[114] Paneth cells that secrete the majority of intestinal AMP exhibit manifold interactions with the microbiota and are an essential component in shaping its configuration.[115] In addition, the commensals occupy the space, thereby creating a competition for the invading pathogens. Moreover, the commensals also induce a low level of toxemia in the body, which in turn primes the immune components to readily act against a pathogen.

4. Factors that modulate microbial diversity

As described above, microbial diversity varies among the different body sites between individuals. Moreover, the changes are also evident during the life course of an individual. Several factors such as mode of delivery, lifestyle, disease condition, host genetic makeup, etc. contribute to the establishment and succession of microbial communities.

4.1 Acquiring the microbiota

A fetus is considered to be sterile, or essentially sterile until the rupture of the amniotic sac. After a vaginal delivery, the microbiota of the fetus closely resembles that of the mother's vagina, predominated by *Lactobacillus*. For a breastfed baby, the initial gut microbiota is also predominated by *Lactobacillus* and *Bifidobacteria*, as human milk is enriched with both these bacteria.[116] Modern lifestyles, on the other hand, may impede microbiota transfer from a mother to her child. The influence of medical practice on microbiota composition is clear when a newborn baby is delivered by Cesarean section rather than vaginal birth; with significant changes in the starting microbiota community. Therefore, a child born through cesarean section harbors slightly different microbiota, and the development of a healthy microbiota may take up to months.[117] A fascinating aspect of child microbiota is that it is mostly undifferentiated across many anatomic locations shortly after birth and that delivery style influences which types of bacteria in the infant microbiome colonize first.[63]

4.2 Postnatal influence and development of a stable microbial community

The microbiome of an infant changes with the increase in age and continues to develop a densely populated microbiota. With the development of teeth and the introduction of more complex food materials in the diet of the infant, the oral microbiota starts developing. Bacteria like *Streptococcus oralis*, *Streptococcus mutans*, *Streptococcus salivarius*, *Actinomyces naeslundii*, etc. start colonizing.[118,119] The makeup of the gut microbiota is regulated by environmental variables, geographical location, household exposure, and, most critically, food during the developmental phases.[120] The growing gut microbiota passes through three unique periods: a developmental phase (months 3–14), a transitional phase (months 15–30), and a stable phase (months 31–46), according to a study collecting 12,500 fecal samples from 903 children from three European nations (Germany, Sweden, and Finland), as well as three US states (Colorado, Georgia, and Washington).[120] The development of skin microbiota, depends on numerous intrinsic factors, however hormonal changes during puberty have a major implication for it. *Lactobacillus*, *Prevotella*, and *Sneathia* microorganisms were identified from various parts of the body of newborns delivered through the vaginal canal, including the skin. *Staphylococcus*, *Corynebacterium*, and *Cutibacterium* are among the bacteria found on the skin of children born through cesarean section.[121,122] *Staphylococcus* and *Streptococcus* species are declining, whereas Actinobacteria and Proteobacteria species are increasing. Prepubertal youngsters have a greater variety of fungal species than adolescents and adults.[123,124] In the case of infants up to 6 weeks of age, the vaginal microbiota is represented by *Lactobacilli*, *Streptococci,* and *Enterococci*. However, during the initial years of growth and attaining puberty, the pH of the genitourinary tract remains neutral and harbors a slightly different microbiota.[125] This scenario changes after puberty, due to hormonal changes; specifically due to high estrogen levels the pH of the genitourinary tract turns acidic again. As a result of the vaginal microbiome imbalance, women of reproductive age are more prone to develop urogenital illnesses.[126] *Lactobacilli* are abundant in the vaginal microbiome immediately before menopause and after puberty, when chances of reproduction are high.[127] It has been observed that human microbiota remains dynamic in most of the body sites during developmental phases and attains stability during the middle age. Moreover, loss of microbiota and its function during old age leads to various diseases. This has led to the hypothesis that co-evolved microbiota are adaptive for the human species both by supporting life host activities and in terms of causing host demise later in life.[128]

5. Microbiome diversity-associated health disorders

Currently, numerous studies have focused on exploring the differential microbe populations that occur in a specific disease state, or the temporal microbial changes that are observed throughout a disease. However, the real challenge is to discover if there is a causal relationship between the alternating microbiome and the pathology of a disease.

5.1 Microbial diseases of the skin

Under some situations, common skin commensals such as *Staphylococcus aureus*, *Propionibacterium acnes*, and *Malassezia* spp. can be harmful.

Atopic dermatitis, a chronic and highly inflammatory skin illness that affects both children and adults, is linked to *S. aureus* colonization and infection.[129] Multiple virulence factors encoded by phages, plasmids, and pathogenicity islands have been discovered in the sequenced genome of *S. aureus*. It produces a range of enzymes and toxins to avoid detection by the host's immune system and to effectively establish infection.[130] *Propionibacterium acnes* is also linked to acne vulgaris, a prevalent teenage skin disorder. *Malassezia*, on the other hand, has been implicated as the infectious agent in several skin conditions, including dandruff, atopic dermatitis, tinea versicolor, and, to a lesser extent, psoriasis. *Malassezia* generates lipases, phospholipases, and allergens (*Mala* genes), all of which can harm the skin's integrity by causing inflammation and provoking an immunological reaction.[131]

5.2 Gut microbiota dysbiosis is related to an array of diseases

Colorectal cancers may be caused by the colonic microbiota producing short-chain fatty acids (SCFAs) and other metabolites.[132] Through WNT signaling, SCFAs, particularly butyrate, can cause apoptosis, cell cycle arrest, and differentiation.[133] On the other hand, *Enterococcus faecalis* can induce aneuploidy and tetraploidy which in turn is genotoxic to the epithelial cells.[134] Furthermore, antibiotic therapy alters the expression of host genes involved in cell cycle control, lowering epithelial proliferation and hence the risk of colon cancer.[135] Inflammatory bowel disorder (IBD), which is of major concern nowadays, is linked to dysbiosis in the gut. In IBD patients, the microbiota deviates from normality and is overrepresented by *Enterococcus faecium* and several members of the Proteobacteria phyla. Similarly, in case of liver diseases like alcoholic steatosis, non-alcoholic fatty liver disease the alteration of the normal gut microbiota is observed with overrepresentation of species from the oral microbiota.[136,137] In addition, alteration of the gut microbiota is also observed in lifestyle diseases such as obesity and Type 2 diabetes. In an experiment on murine model, it was observed that, obese mice have decreased Bacteroides/Firmicutes ratio as opposed to their lean counterparts.

5.3 Maintaining a healthy human microbiome

The microbial entities stay in harmony with host cells and maintain a healthy equilibrium, in which the microbes play a key role to promote well-being and avoid disease, a condition known as homeostasis Age, diet, lifestyle, hormonal variations, inherited genes, and underlying disease are all important factors in preserving homeostasis, as described in the previous sections. However, in addition to the "natural" dynamics that exist in such ecosystems, there are also "artificial" alterations in the microflora structure and cell density that ultimately led to pathological states. One such example is the use of antimicrobials which drastically lowers the quantity and variety of microorganisms. Antipsychotics, nonsteroidal anti-inflammatory pharmaceuticals, antimetabolites, calcium-channel antagonists, antidiabetic (metformin), antiseptics, antivirals, proton pump inhibitors, and conventional antipsychotics have all been demonstrated to affect the makeup of the gut microbiota.[138]

The use of special nutrients (prebiotics) or even live microorganisms (probiotics) to maximize the amount of chosen, potentially beneficial microorganisms confer beneficial health benefits to the host. Prebiotics are nondigestible carbohydrate molecules present in fibrous foods that aid in the growth of beneficial bacteria in the gastrointestinal tract. One method of action is that the gut commensal feeds up on the prebiotics to produce SCFA, such as butyrate, which is beneficial to bacteria. Butyrate maintains the gut barrier and promotes mucin synthesis. Additionally, by promoting the synthesis of defensin protein and IgA in the host, probiotics can also modify gut microbial populations and limit pathogen development. Generally, probiotic-mediated immunomodulation occurs via signaling pathways that influence cytokine production. Thus, probiotics have been advocated as preventative and therapeutic methods to restore the gut microbiome's healthy composition and function.

Carbohydrates, proteins, and fats are the most common components in human meals. Microbial diversity in the gastrointestinal system is reduced by high-fat, low-fiber diets. *Proteobacteria* thrive on high-energy, simple sugar-dominated diets, whereas *Prevotella copri* and *Xylanobacteria* flourish on high-energy, simple sugar-dominated diets.[39] Bacteroides thrive in protein-rich environments.[139] Experiments have shown that dietary changes may cause massive, transitory microbial shifts, indicating that nutrition plays an important role in influencing the microbiome.[140] The impacts of the healthy human microbiota on health are complex, and researchers and health professionals are attempting to educate consumers by incorporating new scientific findings into their practice, particularly for advantages other than digestive health. As a result, pre/probiotics are likely to be coupled with other dietary substances to generate a more powerful health benefit.

6. Conclusion

Human microbiomes are engines of several important processes linked with human development, physiology, immune maturation, resistance to infections, and protection from metabolic diseases. Massive microbial genome sequencing, cultivation, and characterization of thousands of microbial species decoded the compositions, diversity, and dynamics of microbial communities associated with the human body and showed that they are constantly evolving by modulating their functional potency concerning environmental clues. To achieve and maintain healthy states, it is critical to adopt a balanced lifestyle to keep the endogenous microbial ecosystem of our body diverse and unaltered. For a better understanding of the human microbiome and anticipating its role in health and disease, it is extremely crucial to identify factors that positively or negatively modulate the sustainability of microbial taxa and the functional contents of its genome in a complex community. Such understanding will fuel the discovery and development of early diagnostic kits for prevalent and rare diseases and microbiome-based therapy for infectious and communicable diseases.

Acknowledgments

We thank all the members in the Molecular Genetics Laboratory (MGL), THSTI for their valuable feedback. We acknowledge the support of the Executive Director, THSTI. This study received financial support from the Dept. of Biotechnology (DBT), Govt. of India (Grant No. BT/PR38173/MED/97/474/2020) and Translational Health Science and Technology Institute (Intramural, 2021-2022). T.A. received research fellowship from Dept. of Biotechnology (DBT), Govt. of India.

References

1. Witherspoon DJ, Wooding S, Rogers AR, et al. Genetic similarities within and between human populations. *Genetics*. 2007;176(1):351–359.
2. Cho I, Blaser MJJ. The human microbiome: at the interface of health and disease. *Nat Rev Genet*. 2012;13(4):260–270.
3. Lloyd-Price J, Abu-Ali G, Huttenhower C. The healthy human microbiome. *Genome Med*. 2016;8(1):51. https://doi.org/10.1186/s13073-016-0307-y.
4. Whiteside SA, Razvi H, Dave S, Reid G, Burton JPJNRU. The microbiome of the urinary tract—a role beyond infection. *Net Rev Urol*. 2015;12(2):81.
5. Busi SB, de Nies L, Habier J, et al. Persistence of birth mode-dependent effects on gut microbiome composition, immune system stimulation and antimicrobial resistance during the first year of life. *ISME Commun*. 2021;1(1):8. https://doi.org/10.1038/s43705-021-00003-5.

6. Leeming ER, Johnson AJ, Spector TD, Le Roy CI. Effect of diet on the gut microbiota: rethinking intervention duration. *Nutrients*. 2019;11(12). https://doi.org/10.3390/nu11122862.
7. Mahdavinia M, Rasmussen HE, Botha M, et al. Effects of diet on the childhood gut microbiome and its implications for atopic dermatitis. *J Allergy Clin Immunol*. 2019;143(4):1636–1637 e1635. https://doi.org/10.1016/j.jaci.2018.11.034.
8. Morgan XC, Huttenhower CJPCB. Human microbiome analysis. *Microbiol Ecol*. 2012;8(12):e1002808.
9. Pascal M, Perez-Gordo M, Caballero T, et al. Microbiome and allergic diseases. *Front Immunol*. 2018;9:1584.
10. Sarkar S, Das B, Banerjee S. Insights into the human gut microbiome and cardiovascular diseases. *J Pract Cardiovasc Sci*. 2018;4:10. https://doi.org/10.4103/jpcs.jpcs_18_18.
11. Khan Mirzaei M, Deng L. New technologies for developing phage-based tools to manipulate the human microbiome. *Trends Microbiol*. 2022;30(2):131–142. https://doi.org/10.1016/j.tim.2021.04.007.
12. Rooks MG, Garrett WS. Gut microbiota, metabolites and host immunity. *Nat Rev Immunol*. 2016;16(6):341–352. https://doi.org/10.1038/nri.2016.42.
13. Hoeppli RE, Wu D, Cook L, Levings MKJF, i. i. The environment of regulatory T cell biology: cytokines, metabolites, and the microbiome. *Front Immunol*. 2015;6:61.
14. Yilmaz P, Parfrey LW, Yarza P, et al. The SILVA and "all-species living tree project (LTP)" taxonomic frameworks. *Nucleic Acids Res*. 2014;42(D1):D643–D648.
15. Lozupone CA, Stombaugh JI, Gordon JI, Jansson JK, Knight RJN. Diversity, stability and resilience of the human gut microbiota. *Nature*. 2012;489(7415):220–230.
16. O'Hara AM, Shanahan F. The gut flora as a forgotten organ. *EMBO Rep*. 2006;7(7):688–693. https://doi.org/10.1038/sj.embor.7400731.
17. Sekirov I, Russell SL, Antunes LC, Finlay BB. Gut microbiota in health and disease. *Physiol Rev*. 2010;90(3):859–904. https://doi.org/10.1152/physrev.00045.2009.
18. Morton ER, Lynch J, Froment A, et al. Variation in rural African gut microbiota is strongly correlated with colonization by Entamoeba and subsistence. *PLoS Genet*. 2015;11(11):e1005658. https://doi.org/10.1371/journal.pgen.1005658.
19. Pinna NK, Anjana RM, Saxena S, et al. Trans-ethnic gut microbial signatures of prediabetic subjects from India and Denmark. *Genome Med*. 2021;13(1):36. https://doi.org/10.1186/s13073-021-00851-9.
20. Kaplan RC, Wang Z, Usyk M, et al. Gut microbiome composition in the Hispanic Community Health Study/Study of Latinos is shaped by geographic relocation, environmental factors, and obesity. *Genome Biol*. 2019;20(1):219. https://doi.org/10.1186/s13059-019-1831-z.
21. Yang Y, Zheng W, Cai Q, et al. Racial differences in the oral microbiome: data from low-income populations of african ancestry and European ancestry. *mSystems*. 2019;4(6). https://doi.org/10.1128/mSystems.00639-19.
22. Chaudhari DS, Dhotre DP, Agarwal DM, et al. Gut, oral and skin microbiome of Indian patrilineal families reveal perceptible association with age. *Sci Rep*. 2020;10(1):5685. https://doi.org/10.1038/s41598-020-62195-5.
23. Belstrom D, Holmstrup P, Nielsen CH, et al. Bacterial profiles of saliva in relation to diet, lifestyle factors, and socioeconomic status. *J Oral Microbiol*. 2014;6. https://doi.org/10.3402/jom.v6.23609.
24. Hoffman KL, Hutchinson DS, Fowler J, et al. Oral microbiota reveals signs of acculturation in Mexican American women. *PLoS One*. 2018;13(4):e0194100. https://doi.org/10.1371/journal.pone.0194100.
25. Ogai K, Nana BC, Lloyd YM, et al. Skin microbiome profile of healthy Cameroonians and Japanese. *Sci Rep*. 2022;12(1):1364. https://doi.org/10.1038/s41598-022-05244-5.

26. Ring HC, Thorsen J, Saunte DM, et al. The follicular skin microbiome in patients with hidradenitis suppurativa and healthy controls. *JAMA Dermatol.* 2017;153(9):897–905. https://doi.org/10.1001/jamadermatol.2017.0904.
27. Callewaert C, Ravard Helffer K, Lebaron P. Skin microbiome and its Interplay with the environment. *Am J Clin Dermatol.* 2020;21(suppl 1):4–11. https://doi.org/10.1007/s40257-020-00551-x.
28. Fettweis JM, Brooks JP, Serrano MG, et al. Differences in vaginal microbiome in African American women versus women of European ancestry. *Microbiology (Reading).* 2014;160(pt 10):2272–2282. https://doi.org/10.1099/mic.0.081034-0.
29. Mehta O, Ghosh TS, Kothidar A, et al. Vaginal microbiome of pregnant indian women: Insights into the genome of dominant lactobacillus species. *J Infect Dis.* 2020;1-13.
30. Hugerth LW, Pereira M, Zha Y, et al. Optimal protocols for sequence-based characterization of the human vaginal microbiome. *bioRXiv.* 2020.
31. MacIntyre DA, Chandiramani M, Lee YS, et al. The vaginal microbiome during pregnancy and the postpartum period in a European population. *Sci Rep.* 2015;5:8988. https://doi.org/10.1038/srep08988.
32. Bamola VD, Ghosh A, Kapardar RK, et al. Gut microbial diversity in health and disease: experience of healthy Indian subjects, and colon carcinoma and inflammatory bowel disease patients. *Microb Ecol Health Dis.* 2017;28(1):1322447. https://doi.org/10.1080/16512235.2017.1322447.
33. Das B, Ghosh TS, Kedia S, et al. Analysis of the gut microbiome of rural and urban healthy Indians living in sea level and high altitude areas. *Sci Rep.* 2018;8(1):10104. https://doi.org/10.1038/s41598-018-28550-3.
34. Arumugam M, Raes J, Pelletier E, et al. Enterotypes of the human gut microbiome. *Nature.* 2011;473(7346):174–180.
35. Yatsunenko T, Rey FE, Manary MJ, et al. Human gut microbiome viewed across age and geography. *Nature.* 2012;486(7402):222–227.
36. Hoffmann C, Dollive S, Grunberg S, et al. Archaea and fungi of the human gut microbiome: correlations with diet and bacterial residents. *PLoS One.* 2013;8(6):e66019.
37. Pareek S, Kurakawa T, Das B, et al. Comparison of Japanese and Indian intestinal microbiota shows diet-dependent interaction between bacteria and fungi. *NPJ Biofilms Microbiomes.* 2019;5(1):1–13.
38. Schnorr SL, Candela M, Rampelli S, et al. Gut microbiome of the Hadza hunter-gatherers. *Nat Commun.* 2014;5(1):1–12.
39. De Filippo C, Cavalieri D, Di Paola M, et al. Impact of diet in shaping gut microbiota revealed by a comparative study in children from Europe and rural Africa. *Proc Natl Acad Sci USA.* 2010;107(33):14691–14696. https://doi.org/10.1073/pnas.1005963107.
40. Planer JD, Peng Y, Kau AL, et al. Development of the gut microbiota and mucosal IgA responses in twins and gnotobiotic mice. *Nature.* 2016;534(7606):263–266.
41. Das B, Nair GB. Homeostasis and dysbiosis of the gut microbiome in health and disease. *J Biosci.* 2019;44(5).
42. Frank DN, St Amand AL, Feldman RA, Boedeker EC, Harpaz N, Pace NR. Molecular-phylogenetic characterization of microbial community imbalances in human inflammatory bowel diseases. *Proc Natl Acad Sci USA.* 2007;104(34):13780–13785. https://doi.org/10.1073/pnas.0706625104.
43. Caporaso JG, Lauber CL, Costello EK, et al. Moving pictures of the human microbiome. *Genome Biol.* 2011;12(5):1–8.
44. Deo PN, Deshmukh R. Oral microbiome: unveiling the fundamentals. *J Oral Maxillofac Pathol.* 2019;23(1):122–128. https://doi.org/10.4103/jomfp.JOMFP_304_18.
45. Zhao H, Chu M, Huang Z, et al. Variations in oral microbiota associated with oral cancer. *Sci Rep.* 2017;7(1):11773. https://doi.org/10.1038/s41598-017-11779-9.

46. Perera M, Al-Hebshi NN, Speicher DJ, Perera I, Johnson NW. Emerging role of bacteria in oral carcinogenesis: a review with special reference to perio-pathogenic bacteria. *J Oral Microbiol.* 2016;8:32762. https://doi.org/10.3402/jom.v8.32762.
47. Sharma N, Bhatia S, Sodhi AS, Batra NJA. Oral microbiome and health. *J Oral Sci.* 2018;4(1):42.
48. Zaura E, Nicu EA, Krom BP, Keijser BJJF. Acquiring and maintaining a normal oral microbiome: current perspective. *Front Cell Infect Microbiol.* 2014;4:85.
49. Petti S, Polimeni A, Berloco P, Scully CJOD. Orofacial diseases in solid organ and hematopoietic stem cell transplant recipients. *Oral Dis.* 2013;19(1):18–36.
50. Callahan BJ, DiGiulio DB, Goltsman DSA, et al. Replication and refinement of a vaginal microbial signature of preterm birth in two racially distinct cohorts of US women. *Proc Natl Acad Sci USA.* 2017;114(37):9966–9971.
51. Aagaard K, Riehle K, Ma J, et al. A metagenomic approach to characterization of the vaginal microbiome signature in pregnancy. *PLoS One.* 2012;7(6):e36466.
52. Noyes N, Cho K-C, Ravel J, Forney LJ, Abdo ZJP. Associations between sexual habits, menstrual hygiene practices, demographics and the vaginal microbiome as revealed by Bayesian network analysis. 2018;13(1):e0191625.
53. Schwebke JR, Richey CM, Weiss HLJTJ. Correlation of behaviors with microbiological changes in vaginal flora. *J Infect Dis.* 1999;180(5):1632–1636.
54. Culhane JF, Rauh V, McCollum KF, Elo IT, Hogan VJA. Exposure to chronic stress and ethnic differences in rates of bacterial vaginosis among pregnant women. *Am J Obstet Gynecol.* 2002;187(5):1272–1276.
55. Gupta VK, Paul S, Dutta CJF. Geography, ethnicity or subsistence-specific variations in human microbiome composition and diversity. *Front Microbiol.* 2017;1162.
56. Ravel J, Gajer P, Abdo Z, et al. Vaginal microbiome of reproductive-age women. *Biol Sci.* 2011;108(suppl 1):4680–4687.
57. Nelson TM, Borgogna J, Michalek R, et al. Cigarette smoking is associated with an altered vaginal tract metabolomic profile. *Sci Reports.* 2018;8(1):1–13.
58. Costello EK, Lauber CL, Hamady M, Fierer N, Gordon JI, Knight R. Bacterial community variation in human body habitats across space and time. *Science.* 2009;326(5960):1694–1697. https://doi.org/10.1126/science.1177486.
59. Findley K, Oh J, Yang J, et al. Topographic diversity of fungal and bacterial communities in human skin. *Nature.* 2013;498(7454):367–370. https://doi.org/10.1038/nature12171.
60. Grice EA, Kong HH, Conlan S, et al. Topographical and temporal diversity of the human skin microbiome. *Science.* 2009;324(5931):1190–1192. https://doi.org/10.1126/science.1171700.
61. Grice EA, Segre JA. The skin microbiome. *Nat Rev Microbiol.* 2011;9(4):244–253. https://doi.org/10.1038/nrmicro2537.
62. Oh J, Byrd AL, Park M, Program NCS, Kong HH, Segre JA. Temporal stability of the human skin microbiome. *Cell.* 2016;165(4):854–866. https://doi.org/10.1016/j.cell.2016.04.008.
63. Dominguez-Bello MG, Costello EK, Contreras M, et al. Delivery mode shapes the acquisition and structure of the initial microbiota across multiple body habitats in newborns. *Proc Natl Acad Sci USA.* 2010;107(26):11971–11975. https://doi.org/10.1073/pnas.1002601107.
64. Mueller NT, Bakacs E, Combellick J, Grigoryan Z, Dominguez-Bello MG. The infant microbiome development: mom matters. *Trends Mol Med.* 2015;21(2):109–117. https://doi.org/10.1016/j.molmed.2014.12.002.
65. Laborel-Preneron E, Bianchi P, Boralevi F, et al. Effects of the Staphylococcus aureus and Staphylococcus epidermidis secretomes isolated from the skin microbiota of atopic children on CD4+ T cell activation. *PLoS One.* 2015;10(10):e0141067. https://doi.org/10.1371/journal.pone.0141067.

66. Oh J, Byrd AL, Deming C, et al. Biogeography and individuality shape function in the human skin metagenome. *Nature*. 2014;514(7520):59–64. https://doi.org/10.1038/nature13786.
67. Clavaud C, Jourdain R, Bar-Hen A, et al. Dandruff is associated with disequilibrium in the proportion of the major bacterial and fungal populations colonizing the scalp. *PLoS One*. 2013;8(3):e58203. https://doi.org/10.1371/journal.pone.0058203.
68. Park HK, Ha MH, Park SG, Kim MN, Kim BJ, Kim W. Characterization of the fungal microbiota (mycobiome) in healthy and dandruff-afflicted human scalps. *PLoS One*. 2012;7(2):e32847. https://doi.org/10.1371/journal.pone.0032847.
69. Soares RC, Camargo-Penna PH, de Moraes VC, et al. Dysbiotic bacterial and fungal communities not restricted to clinically affected skin sites in dandruff. *Front Cell Infect Microbiol*. 2016;6:157. https://doi.org/10.3389/fcimb.2016.00157.
70. Karlsson FH, Tremaroli V, Nookaew I, et al. Gut metagenome in European women with normal, impaired and diabetic glucose control. *Nature*. 2013;498(7452):99–103. https://doi.org/10.1038/nature12198.
71. Christl SU, Murgatroyd PR, Gibson GR, Cummings JHJG. Production, metabolism, and excretion of hydrogen in the large intestine. *Gasteroenterology*. 1992;102(4): 1269–1277.
72. Duncan SH, Holtrop G, Lobley GE, Calder AG, Stewart CS, Flint HJJBJ. Contribution of acetate to butyrate formation by human faecal bacteria. *Br J Nutr*. 2004;91(6):915–923.
73. Falony G, Vlachou A, Verbrugghe K, De Vuyst LJA. Cross-feeding between Bifidobacterium longum BB536 and acetate-converting, butyrate-producing colon bacteria during growth on oligofructose. *Appl Environ Microbiol*. 2006;72(12): 7835–7841.
74. Rowland I, Gibson G, Heinken A, et al. Gut microbiota functions: metabolism of nutrients and other food components. *Eur J Nutr*. 2018;57(1):1–24. https://doi.org/10.1007/s00394-017-1445-8.
75. Bajaj JS, Heuman DM, Hylemon PB, et al. Altered profile of human gut microbiome is associated with cirrhosis and its complications. *J Hepatol*. 2014;60(5):940–947. https://doi.org/10.1016/j.jhep.2013.12.019.
76. Romano KA, Vivas EI, Amador-Noguez D, Rey FEJM. Intestinal microbiota composition modulates choline bioavailability from diet and accumulation of the proatherogenic metabolite trimethylamine-N-oxide. *mBio*. 2015;6(2).
77. Macfarlane GT, Macfarlane S. Bacteria, colonic fermentation, and gastrointestinal health. *J AOAC Int*. 2012;95(1):50–60. https://doi.org/10.5740/jaoacint.sge_macfarlane.
78. Steliou K, Boosalis MS, Perrine SP, Sangerman J, Faller DV. Butyrate histone deacetylase inhibitors. *Biores Open Access*. 2012;1(4):192–198. https://doi.org/10.1089/biores.2012.0223.
79. De Vadder F, Kovatcheva-Datchary P, Goncalves D, et al. Microbiota-generated metabolites promote metabolic benefits via gut-brain neural circuits. *Cell*. 2014;156(1-2):84–96. https://doi.org/10.1016/j.cell.2013.12.016.
80. Brown AJ, Goldsworthy SM, Barnes AA, et al. The Orphan G protein-coupled receptors GPR41 and GPR43 are activated by propionate and other short chain carboxylic acids. *J Biol Chem*. 2003;278(13):11312–11319.
81. Tazoe H, Otomo Y, Kaji I, Tanaka R, Karaki S, Kuwahara AJJPP. Roles of short-chain fatty acids receptors, GPR41 and GPR43 on colonic functions. *J Physiol Pharmacol*. 2008;59(suppl 2):251–262.
82. Frost G, Sleeth ML, Sahuri-Arisoylu M, et al. The short-chain fatty acid acetate reduces appetite via a central homeostatic mechanism. *Nat Commun*. 2014;5(1):1–11.
83. Vital M, Howe AC, Tiedje JMJM. Revealing the bacterial butyrate synthesis pathways by analyzing (meta)Genomic Data. *mBio*. 2014;5(2).

84. Louis P, Flint HJJE. Formation of propionate and butyrate by the human colonic microbiota. *Environ Microbiol.* 2017;19(1):29–41.
85. Bjerrum JT, Wang Y, Hao F, et al. Metabonomics of human fecal extracts characterize ulcerative colitis, Crohn's disease and healthy individuals. *Metabolomics.* 2015;11 (1):122–133.
86. Thauer RK, Jungermann K, Decker KJB, r. Energy conservation in chemotrophic anaerobic bacteria. *Bacteriol Rev.* 1977;41(1):100.
87. Gibson GR, Macfarlane GT. *Human Colonic Bacteria: Role in Nutrition, Physiology and Pathology.* CRC Press; 1995.
88. Lajoie SF, Bank S, Miller TL, Wolin MJJA. Acetate production from hydrogen and [13C] carbon dioxide by the microflora of human feces. *Appl Environ Microbiol.* 1988;54(11):2723–2727.
89. Suarez F, Furne J, Springfield J, Levitt MJAJ. Insights into human colonic physiology obtained from the study of flatus composition. *Am J Physiol.* 1997;272(5):G1028–G1033.
90. Han M, Wang C, Liu P, Li D, Li Y, Ma X. Dietary fiber gap and host gut microbiota. *Protein Pept Lett.* 2017;24(5):388–396. https://doi.org/10.2174/0929866524666170220113312.
91. Kakiyama G, Pandak WM, Gillevet PM, et al. Modulation of the fecal bile acid profile by gut microbiota in cirrhosis. *J Hepatol.* 2013;58(5):949–955. https://doi.org/10.1016/j.jhep.2013.01.003.
92. Bajaj JS, Ridlon JM, Hylemon PB, et al. Linkage of gut microbiome with cognition in hepatic encephalopathy. *Am J Physiol Gastrointest Liver Physiol.* 2012;302(1):G168–G175. https://doi.org/10.1152/ajpgi.00190.2011.
93. Hofmann AFJH. Chemistry and enterohepatic circulation of bile acids. *Hepatology.* 1984;4(S2):4S–14S.
94. Li T, Chiang JYJP. Bile acid signaling in metabolic disease and drug therapy. *Pharmacol Rev.* 2014;66(4):948–983.
95. Ridlon JM, Kang DJ, Hylemon PB, Bajaj JS. Bile acids and the gut microbiome. *Curr Opin Gastroenterol.* 2014;30(3):332–338. https://doi.org/10.1097/MOG.0000000000000057.
96. Torres-Fuentes C, Schellekens H, Dinan TG, Cryan JFJT. The microbiota–gut–brain axis in obesity. *Lancet Gasteroenterol Hepatol.* 2017;2(10):747–756.
97. Macfarlane G, Cummings J, Allison CJM. Protein degradation by human intestinal bacteria. *J Gen Microbiol.* 1986;132(6):1647–1656.
98. Metchnikoff E. *The Prolongation of Life; Optimistic Studies.* London, UK: William Heinemann; 1909.
99. Organ CL, Otsuka H, Bhushan S, et al. Choline diet and its gut microbe–derived metabolite, trimethylamine N-oxide, exacerbate pressure overload–induced heart failure. *Circ Heart Fail.* 2016;9(1):e002314.
100. Bennett BJ, de Aguiar Vallim TQ, Wang Z, et al. Trimethylamine-N-oxide, a metabolite associated with atherosclerosis, exhibits complex genetic and dietary regulation. *Cell Metab.* 2013;17(1):49–60.
101. Zhang Y, Wang Y, Ke B, Du JJTR. TMAO: how gut microbiota contributes to heart failure. *Transl Res.* 2020.
102. Seldin MM, Meng Y, Qi H, et al. Trimethylamine N-oxide promotes vascular inflammation through signaling of mitogen-activated protein kinase and nuclear factor-κB. *J Am Heart Assoc.* 2016;(2):e002767.
103. Savi M, Bocchi L, Bresciani L, et al. Trimethylamine-N-oxide (TMAO)-induced impairment of cardiomyocyte function and the protective role of urolithin B-glucuronide. *Molecules.* 2018;23(3):549.
104. Shi Y, Hu J, Geng J, et al. Berberine treatment reduces atherosclerosis by mediating gut microbiota in apoE-/-mice. *Biomed Pharmacother.* 2018;107:1556–1563.

105. Chen S, Henderson A, Petriello MC, et al. Trimethylamine N-oxide binds and activates PERK to promote metabolic dysfunction. *Cell Metab.* 2019;30(6). 1141-1151. e1145.
106. Hill MJE. Intestinal flora and endogenous vitamin synthesis. *Eur J Cancer Prev.* 1997;6: S43–S45.
107. LeBlanc JG, Milani C, de Giori GS, Sesma F, van Sinderen D, Ventura M. Bacteria as vitamin suppliers to their host: a gut microbiota perspective. *Curr Opin Biotechnol.* 2013;24(2):160–168. https://doi.org/10.1016/j.copbio.2012.08.005.
108. Magnusdottir S, Ravcheev D, de Crecy-Lagard V, Thiele I. Systematic genome assessment of B-vitamin biosynthesis suggests co-operation among gut microbes. *Front Genet.* 2015;6:148. https://doi.org/10.3389/fgene.2015.00148.
109. Mishra S, Lin Z, Pang S, Zhang W, Bhatt P, Chen S. Recent advanced technologies for the characterization of xenobiotic-degrading microorganisms and microbial communities. *Front Bioeng Biotechnol.* 2021;9:632059. https://doi.org/10.3389/fbioe.2021.632059.
110. Abdelsalam NA, Ramadan AT, ElRakaiby MT, Aziz RK. Toxicomicrobiomics: the human microbiome vs. pharmaceutical, dietary, and environmental xenobiotics. *Front Pharmacol.* 2020;11:390. https://doi.org/10.3389/fphar.2020.00390.
111. Das A, Srinivasan M, Ghosh TS, Mande SS. Xenobiotic metabolism and gut microbiomes. *PLoS One.* 2016;11(10), e0163099. https://doi.org/10.1371/journal.pone.0163099.
112. Bäckhed F, Roswall J, Peng Y, et al. Dynamics and stabilization of the human gut microbiome during the first year of life. *Cell Host Microbe.* 2015;17(5):690–703.
113. Bauer H, Horowitz RE, Levenson SM, Popper H. The response of the lymphatic tissue to the microbial flora. Studies on germfree mice. *Am J Pathol.* 1963;42(4):471.
114. Bevins CL, Salzman NHJNRM. Paneth cells, antimicrobial peptides and maintenance of intestinal homeostasis. *Nat Rev Microbiol.* 2011;9(5):356–368.
115. Ehmann D, Wendler J, Koeninger L, et al. Paneth cell α-defensins HD-5 and HD-6 display differential degradation into active antimicrobial fragments. *Proc Natl Acad Sci USA.* 2019;116(9):3746–3751.
116. Cabrera-Rubio R, Collado MC, Laitinen K, Salminen S, Isolauri E, Mira AJTA. The human milk microbiome changes over lactation and is shaped by maternal weight and mode of delivery. *Am J Clin Nutr.* 2012;96(3):544–551.
117. Grölund M-M, Lehtonen O-P, Eerola E, Kero PJJ. Fecal microflora in healthy infants born by different methods of delivery: permanent changes in intestinal flora after cesarean delivery. *J Pediatr Gastroenterol Nutr.* 1999;28(1):19–25.
118. Brailsford S, Sheehy E, Gilbert S, et al. The microflora of the erupting first permanent molar. *Caries Res.* 2005;39(1):78–84.
119. Cephas KD, Kim J, Mathai RA, et al. Comparative analysis of salivary bacterial microbiome diversity in edentulous infants and their mothers or primary care givers using pyrosequencing. *PLoS One.* 2011;6(8):e23503.
120. Stewart CJ, Ajami NJ, O'Brien JL, et al. Temporal development of the gut microbiome in early childhood from the TEDDY study. *Nature.* 2018;562(7728):583–588.
121. Cundell AMJM. Microbial ecology of the human skin. *Microb Ecol.* 2018;76(1): 113–120.
122. Dayan N. *Skin Microbiome Handbook: From Basic Research to Product Development.* John Wiley & Sons; 2020.
123. Buerger S. *The Skin and Oral Microbiome: An Examination of Overlap and Potential Interactions Between Microbiome Communities*; 2020:45–57.
124. Li Z, Bai X, Peng T, et al. New insights into the skin microbial communities and skin aging. *Front Microbiol.* 2020;2603.

125. Gupta S, Kakkar V, Bhushan IJM. Crosstalk between vaginal microbiome and female health: a review. *Microb Pathog.* 2019;136:103696.
126. Pybus V, Onderdonk ABJM. Microbial interactions in the vaginal ecosystem, with emphasis on the pathogenesis of bacterial vaginosis. *Microbes Infect.* 1999;1(4):285–292.
127. Martin Jr HL, Richardson BA, Nyange PM, et al. Vaginal lactobacilli, microbial flora, and risk of human immunodeficiency virus type 1 and sexually transmitted disease acquisition. *J Infect Dis.* 1999;180(6):1863–1868.
128. Blaser MJ, Webb GFJM. Host demise as a beneficial function of indigenous microbiota in human hosts. *mBio.* 2014;5(6):e02262–02214.
129. Findley K, Grice EAJP. The skin microbiome: a focus on pathogens and their association with skin disease. *PLoS Pathog.* 2014;10(11):e1004436.
130. Baba T, Bae T, Schneewind O, Takeuchi F, Hiramatsu K. Genome sequence of Staphylococcus aureus strain Newman and comparative analysis of staphylococcal genomes: polymorphism and evolution of two major pathogenicity islands. *J Bacteriol.* 2008;190(1):300–310. https://doi.org/10.1128/JB.01000-07.
131. Nakabayashi A, Sei Y, Guillot J. Identification of Malassezia species isolated from patients with seborrhoeic dermatitis, atopic dermatitis, pityriasis versicolor and normal subjects. *Med Mycol.* 2000;38(5):337–341. https://doi.org/10.1080/mmy.38.5.337.341.
132. Plottel CS, Blaser MJ. Microbiome and malignancy. *Cell Host Microbe.* 2011;10 (4):324–335. https://doi.org/10.1016/j.chom.2011.10.003.
133. Lazarova DL, Bordonaro M, Carbone R, Sartorelli AC. Linear relationship between Wnt activity levels and apoptosis in colorectal carcinoma cells exposed to butyrate. *Int J Cancer.* 2004;110(4):523–531. https://doi.org/10.1002/ijc.20152.
134. Wu S, Rhee KJ, Albesiano E, et al. A human colonic commensal promotes colon tumorigenesis via activation of T helper type 17 T cell responses. *Nat Med.* 2009;15 (9):1016–1022. https://doi.org/10.1038/nm.2015.
135. Reikvam DH, Erofeev A, Sandvik A, et al. Depletion of murine intestinal microbiota: effects on gut mucosa and epithelial gene expression. *PLoS One.* 2011;6(3):e17996. https://doi.org/10.1371/journal.pone.0017996.
136. Bajaj JS. Alcohol, liver disease and the gut microbiota. *Nat Rev Gastroenterol Hepatol.* 2019;16(4):235–246. https://doi.org/10.1038/s41575-018-0099-1.
137. Wiest R, Albillos A, Trauner M, Bajaj JS, Jalan R. Targeting the gut-liver axis in liver disease. *J Hepatol.* 2017;67(5):1084–1103. https://doi.org/10.1016/j.jhep.2017.05.007.
138. Forslund K, Hildebrand F, Nielsen T, et al. Disentangling type 2 diabetes and metformin treatment signatures in the human gut microbiota. *Nature.* 2015;528 (7581):262–266. https://doi.org/10.1038/nature15766.
139. Wu GD, Chen J, Hoffmann C, et al. Linking long-term dietary patterns with gut microbial enterotypes. *Science.* 2011;334(6052):105–108. https://doi.org/10.1126/science.1208344.
140. Singh RK, Chang HW, Yan D, et al. Influence of diet on the gut microbiome and implications for human health. *J Transl Med.* 2017;15(1):73. https://doi.org/10.1186/s12967-017-1175-y.

CHAPTER FOUR

An overview of cancer and the human microbiome

Soumendu Mahapatra[a,b,†], **Smrutishree Mohanty**[a,†], **Rasmita Mishra**[a,†], **and Punit Prasad**[a,*]
[a]Chromatin and Epigenetic Unit, Institute of Life Sciences, Bhubaneswar, Odisha, India
[b]Kalinga Institute of Industrial Technology (KIIT), School of Biotechnology, Bhubaneswar, Odisha, India
*Corresponding author: e-mail address: punit@ils.res.in

Contents

† Equal contribution.

Progress in Molecular Biology and Translational Science, Volume 191
ISSN 1877-1173
https://doi.org/10.1016/bs.pmbts.2022.07.007

Abstract

Mutual beneficial associations with the microbial consortia are an essential requisite of human life. Microbial communities have both a symbiotic and a pathogenic standpoint, which portrays a context-dependent scenario of the human microbiome. The symbiotic assemblage works to develop indispensable functions of the human body such as immune system, digestive system, defense against colonization by pathobionts and their toxins, etc. Furthermore, any deviation in the resource utilization by the symbionts due to host factors comprising lifestyle changes, diet, drugs, immunocompromised states, and co-morbidities could perturb beneficial microbes communities and promote the invasion by opportunistic pathogens thus, disrupting the homeostatic state. Microbial infestations have proved to be carcinogenic but this does not spontaneously establish a cancer hallmark, rather they initiate a cascade of events that disturbs the normal cellular activities finally these defective machineries invade distant sites of the body, submitting to a devastative transformed internal milieu. Significant technological and system biology advances have been made in elucidating a lucid but complex basis of such microbe-associated malignancies. This chapter discusses the recent advances, without compromising the concepts of the inception studies, including a brief version of the microbial status in cancer generation, mechanistic approaches adapted, therapeutic interventions, system biology approaches with special mention on the study design gaps, challenges in addressing the drawbacks and finally with a perspective of the future targeted studies, has been a focus of this piece of work.

1. Introduction

The coevolution of humans and the associated microbiome has resulted in commensalism, parasitism, and mutualism between the host and the microbiota, sharing benefits at multiple levels of development and survival. Microbes confer anatomical, physiological, immunological, and metabolic signatures on their host habitat from the inception through disease until death. Eubiosis is consistent with the balance between microbial diversity and the health status of the host. Niche-specific interaction of these organisms in the human body has to be in a steady state and develop immune tolerance to ensure eubiosis, even at the expense of intrusion by some foreign invaders. This relationship is constantly evolving in response to changing host factors and through time, the microbiota is considered an essential part/organ of the human body. Colonizing organisms interact with the host directly through cells (enterocytes, immunocytes) or metabolically active substances.[1,2]

Host-microbe co-existence is sustained by the anatomical separation between the two partners by multi-level barriers. Any factors that breach these barriers may collapse the homeostatic state and thus lead to dysbiosis. Microbial disequilibrium or dysbiosis is the state of imbalance in the native microbial consortia or abundance and their function. Diseases often lead to alterations in the microbial compositions and aberrant regulation of microbial vitality, compromised immune and metabolic states of the body, which pose unfavorable conditions for commensals and promote the colonization of ecological niches by opportunistic pathogens thereby paving the way to dysbiosis. The general course of the dysbiosis-derived infection follows a pathway including inflammatory responses, immune cells infiltration, upregulation of cell survival/proliferative signaling, inhibition of malignancy restricting factors, and derailment in the functioning of the defense system, evasion from the immune regulation, and accumulation of mutations. These microbial tactics over a period of time end up transforming the stability at the level of the cellular system into a phenotype, which would be beyond the grip of the host immune system.

Technological advancements are now at work in unwrapping the unseen communications and developing an apparent anecdote if not distinct, regarding the line of interaction between the host and its associated symbiotic partner both in a healthy as well as in an altered condition. High-throughput sequencing, alternative microbial culturing practices, bioinformatics pipelines, and system biology approaches are making it possible to understand microbial biomarkers for a better prognosis and clinical outcomes of the associated disease. Extensive model systems are developed both in-vitro and in-vivo animal models for having a concrete vision of the factors, processes, and systems involved in generating host-microbe complex relationships, formulating treatment regimens, and more importantly, customized/personalized healthcare services.

Research and validation studies are to be directed towards demystifying the microbial diversity and whether microbes directly intervene in host functions or through mediators is under investigation. Despite the crucial findings which support the microbial theory of dysbiosis, several questions remain unanswered. The two key questions that need further investigation are 1. How microbial dysfunction promotes a diseased state and how do non-microbial factors deviate from cellular homeostasis disrupting the host-microbiome axis? This chapter will present a review of the microbial diversity in several key malignancies, mechanistic insights of the microbial mode of action with special reference to different cancers followed

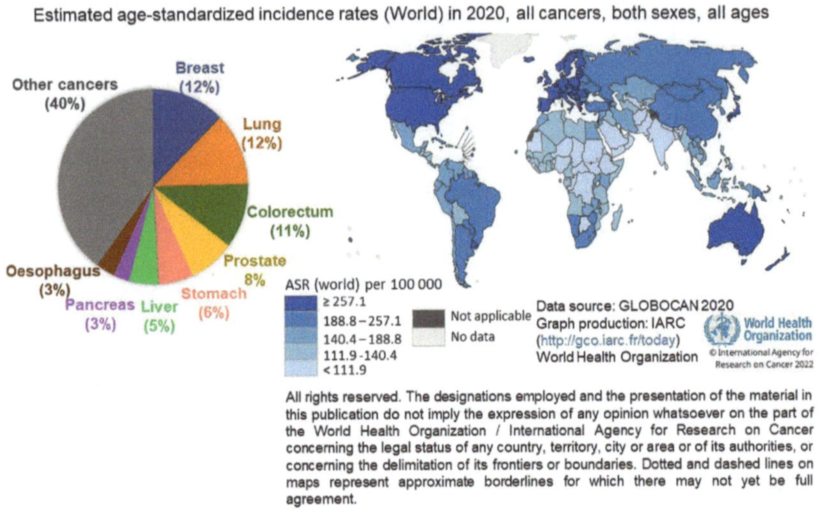

Fig. 1 Pie chart and Global cancer prevalence map. Estimated number of cancer cases in 2020 with reference to statistical considerations worldwide, for both sexes and all age groups. *Adapted from Globocan 2020, world health organisation, and International Agency for Research on Cancer 2022.*

by the therapeutic approaches undertaken to combat these infections including engineered microbial therapy, a system biology strategy for identifying, characterizing and studying the behavior of microbial community with integration of machine learning and artificial intelligence technologies, the study gaps and future directions to answer the fields of concern (Fig. 1).

2. Microbiome associated with cancers

A symbiotic relationship between microbiota and host forms a complex 'super-organism' in which dysbiosis and alteration of the microbial community lead to diseases. A growing body of evidence suggests that the bacterial microbiota plays an active role in the development of cancer.[3] In this section, we will discuss key aspects of the microbiota and their association with different cancers (Fig. 2).

2.1 Colorectal cancer

Dysbiosis in the gut microbial community stimulates sustained inflammation in the colon, a key factor for the initiation and progression of colorectal cancer (CRC).[4] An increase in the permeability of the intestinal barrier due to gut microbial dysbiosis provides an opportunity for bacterial penetration

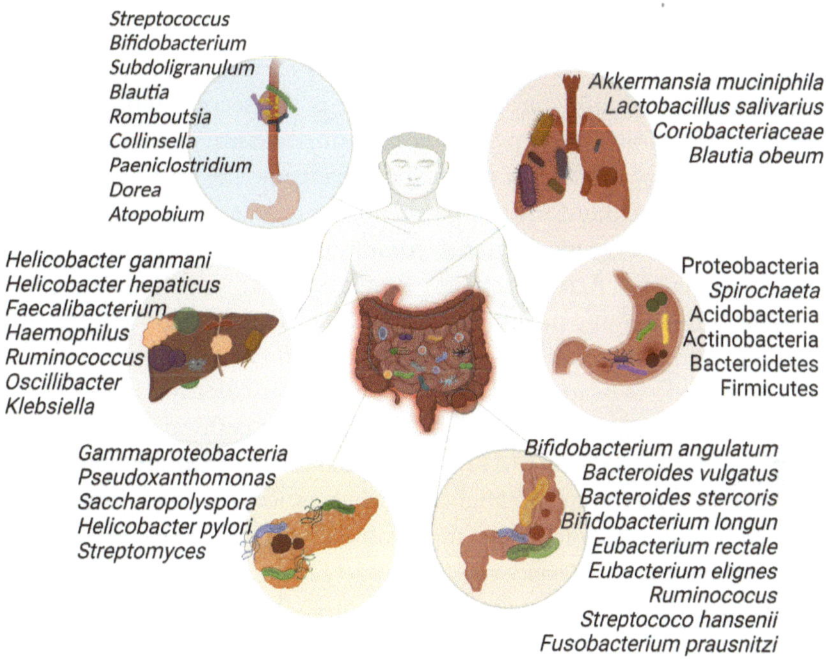

Fig. 2 Cancer-associated bacteriome in the human gut. Different gut bacterial associations with lung cancer, stomach cancer, colorectal cancer, pancreatic cancer, liver cancer, and esophageal cancer.

that triggers cytokine secretion and eventually develops inflammation and promotes tumor progression.[5] A study in the mice model indicated the role of microbiota in CRC development, where microflora-induced inflammation promotes excessive epithelial cell proliferation and a high level of local IL-6 production leading to increased proliferation of tumor progenitors in the pathogenesis of CRCs.[6]

Pathogenic microbes such as *Streptococo hansenii, Fusobacterium prausnitzi, Streptococcus gallolyticus, Ruminococcus obeum,* and *Fusobacterium nucleatum* are involved in promoting CRC.[7] *Fusobacterium nucleatum* can produce oncogenic responses by binding with different immune receptors such as E-cadherin and TIGIT (T cell immune receptor with Ig and ITIM domains) and resist the effects of a chemotherapeutic agent such as oxaliplatin.[8,9] Interestingly, studies show that microbial structure in proximal and distal regions from the tumor site may also be different. In CRC, the abundance of *Pyramidobacter*, *Selenomonas*, *Presvotella*, and *Peptostreptoccus* were more prevalent in the proximal region while the distal region was enriched with *Fusobacterium*, *Leptotrichia*, and *Escherichia, Shigella.*[10]

2.2 Gastric cancer

Gastric cancer (GC) is classified into two main categories, gastric cardia cancer, and gastric non-cardia cancer. GC is the second most prevalent malignant carcinoma in terms of cancer death.[11] Gastric cancer has a multifactorial etiology in addition to dysbiotic changes in the gastric microbiota, which result in the transformation from normal gastric mucosa to malignant one (Fig. 3). The gastric bacterial community is primarily composed of phyla Firmicutes, Actinobacteria, Proteobacteria, and Fusobacteria.[12] *H. pylori* causes chronic inflammation in the gastric mucosa, many of which are associated with stomach cancer. *H. pylori* stimulate inflammatory responses and regulate β-catenin to induce achlorhydria, epithelial atrophy, and dysplasia-like signaling.[13] Oncogenic pathways such as sonic hedgehog, NF-κB, ERK/MAPK, STAT3, and PI3K/Akt are upregulated by Cag+ *H. pylori* strains.[14] Higher stomach cancer risk is associated with *H. pylori's* ability to methylate the promoter regions of E-cadherin and tumor suppressor genes such as GATA4 GATA5, FOXD3, and TFF2.[15,16]

The abundance and association of other bacterial species such as *P. acnes* and *P. copri* in GC patients indicate their involvement in GC progression.[17] Apart from bacterial diversity, fungal diversity shows a high level of *Candida albicans* in GC patients' indicating its role in GC progression and

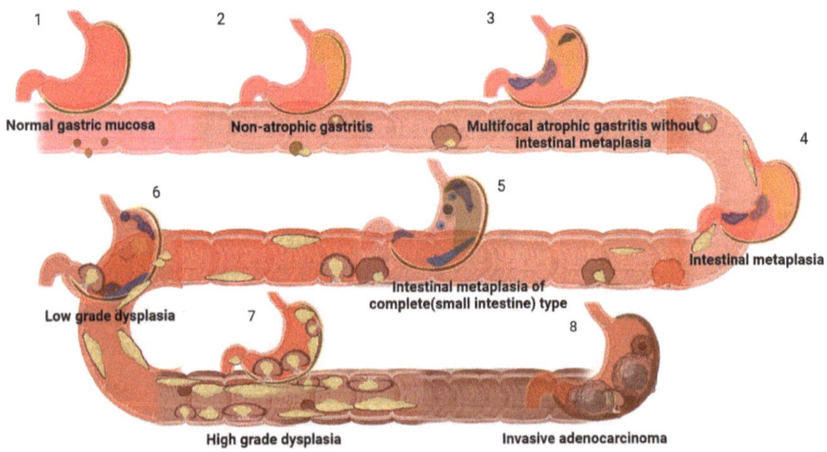

Fig. 3 Stages of gastric cancer development. Different stages of gastric cancer development from initial inflammation to preliminary cancer are normal gastric mucosa, non-atrophic gastritis, multifocal atrophic gastritis without intestinal metaplasia, intestinal metaplasia, intestinal metaplasia of complete (small intestine) type, low-grade dysplasia, high-grade dysplasia, and invasive adenocarcinoma.

may act as a potential microbial biomarker.[18] However, evidence suggests that microbial dysbiosis is a key player in GC, but a deeper functional and mechanistic analysis is needed for comprehending the correlation between microbial dynamics and GC pathogenesis.

2.3 Esophageal cancer

Esophageal cancer (EC) is a cancer of the upper gastrointestinal tract. Esophageal squamous cell carcinoma (ESCC) and esophageal adenocarcinoma (EAC) are the two most common types of EC. The EAC prevalence is increased in developed countries in the last decades, however, ESCC remains the predominant subtype.[19] Microbiome structure is documented and studied on esophagitis and Barrett's esophagus using 16S rRNA sequencing. Actinobacteria, Proteobacteria, and Fusobacteria were found in the esophagus mucosal lining. A distinctive microorganism community such as Clostridiales and Erysipelotrichales was found in the esophagus of individuals with esophageal squamous dysplasia (ESD) and ESCC (stage I-II).[20] Interestingly, the conversion of Gram-positive bacteria to Gram-negative bacteria in oesophagitis and Barrett's esophagus may contribute to oesophageal cancer pathogenesis.[21] Gastroesophageal reflux disease (GERD) and Barrett's esophagus had significantly higher levels of *Campylobacter* compared to controls and esophageal adenocarcinoma and this association displayed higher levels of carcinogenic cytokines (e.g. IL-18).[22] In rats model with Barrett's esophagus and esophageal adenocarcinoma, *E. coli* has greater prevalence, which was accompanied by a stark increase in *TLR 1-3, 6, 7,* and *9* expressions in esophageal adenocarcinoma compared to normal epithelium. Hence, an association between TLRs and *E. coli* in a rat model of esophageal adenocarcinoma provided a hint that early molecular changes were possibly induced by *E. coli.*[23] Evidence support that the oral pathogen *Porphyromonas gingivalis* was associated with high risk of ESCC.[24]

2.4 Hepatocellular carcinoma

The most common histological types of liver cancer are hepatocellular carcinoma (HCC) and cholangiocarcinoma (CCA). The etiology of HCC is linked to several factors including Hepatitis B and C viruses as well as alcohol-induced liver disease (ALD), non-alcoholic fatty liver disease (NAFLD), and aflatoxin B1 (AFB1) drug abuse, and autoimmunity.[25,26] Even though the liver is generally regarded as sterile, its environment is profoundly affected by the microbiota of the GI tract through the portal vein. The liver filters blood, digests, and neutralizes toxins derived from intestinal microbes to protect the host's microbial community. Since the liver regulates the metabolism of microbial metabolites, microbial dysbiosis significantly influences hepatic function and leads to hepatocarcinogenesis.[27] The three major factors linked with gut microbiota inducing hepatocarcinoma are gut microbial metabolites, obesity-related gut microbes, and accumulation of bacteria from the *Helicobacteraceae* family. Further, Gram-positive bacteria produce secondary bile acids, such as deoxycholate and ω-muricholate, which promote HCC by inducing senescence-associated secretory phenotype (SASP) in liver cancer models driven either by hepatocarcinogen diethylnitrosamine (DEN) or overexpression of c-Myc.[28] Lipopolysaccharides (LPS) derived from intestinal microbiota stimulate TLR4, which promotes HCC by increasing cell proliferation and anti-apoptotic signals in non-bone marrow-derived resident liver cells.[29]

Gut microbiome analysis of early HCC showed a higher prevalence of Actinobacteria in early diagnosed HCC patients compared to liver cirrhosis patients. The abundance of butyrate-producing bacterial genera such as *Ruminococcus*, *Coprococcus*, *Clostridium IV*, *Oscillibacter*, and *Faecalibacterium* was reduced while the abundance of LPS-producing *Klebsiella* and *Haemophilus* were enhanced in the early HCC group.[30] Cholesterol modulated-microbiota in the NAFLD mouse model study revealed changes in gut microbiota composition during NAFLD–HCC progression with enrichment of *Helicobacter ganmanii* and depletion of Bacteroides.[31] *Helicobacter hepaticus* further impacts HCC by activating NF-κB signaling and Wnt signaling, oxidative stress, and hepatocyte turnover pathways.[32] MIcrobial genus *Dialister*, *Megasphera*, and *Veillonella* were enriched in the distal duodenum of liver cirrhosis patients thus promoting hepatocarcinogenesis.[33]

2.5 Lung cancer

Histopathologically lung cancer (LC) can be classified as (1) non-small cell lung cancer (NSCLC) and (2) small cell lung cancer (SCLC).[34] The short-chain fatty acids (SCFAs) produced by the gut microbiome, which are the mediators of immune function modulate the bidirectional gut-lung axis.[35,36] By regulating systemic immune response, the distal gut microbiota may impact lung tumor development. The abundance of bacteria from the distal gut, such as *Bacillus*, *Streptococcus*, and *Anaerotruncus* were enriched in SCLC while the abundance of *Lachnospira* and *Ruminococcus* was high in NSCLC.[37] Studies on gut microbiota in LC patients identified several OTUs as potential microbial markers where *Ruminococcus*, *Enterobacteriaceae*, and *Lachnospiraceae* were highly enriched in the cancer group while *Faecalibacterium*, *Streptococcus*, *Bifidobacterium*, and *Veillonella* were significantly enriched in the healthy group. In addition, species belonging to phyla Firmicutes were significantly decreased in the cancer group which is known for promoting the production of colonial luminal SCFA thus modulating inflammation and tumor formation in mice and humans. Several microbes associated with lung cancer such as *Blautia obeum*, *Akkermansia muciniphila*, *Lactobacillus salivarius*, and the family of *Coriobacteriaceae* show enhanced abundance only in metastatic patients.[38] The oral microbiota may be associated with the development of lung cancer according to several lines of observational evidence. Studies show a strong association between lung cancer incidence in people suffering from periodontal disease.[39] This may be caused by oral microbial pathogens and dysbiotic oral microbiota.[40] People with specific oral bacteria, such as *Chlamydia pneumoniae*, are more vulnerable to lung cancer.[41,42] Firmicutes were abundant in the lower lobes, which could increase the pathogenicity of lung tumors.[43] Researchers found that individuals who developed lung cancer have low alpha microbial diversity compared to healthy control and healthy individuals with low alpha diversity have a high risk of developing LC.[44,45]

2.6 Pancreatic cancer

The most common and lethal form of pancreatic cancer (PC) is pancreatic ductal carcinoma (PDAC).[46] Based on *in-silico* prediction, PDAC tumors from long-term survivors exhibit high levels of neoantigens, enhanced cell infiltration, and increased CD8+ T cell activation. Neoantigens exhibit a molecular mimicry with microbial epitopes, pointing to neoantigens as microbe-derived peptides.[47] In light of these data, it appears that microbial host factors may have an impact on tumor behavior and patient outcomes. A study by Geller et al. found Gammaproteobacteria (GP) as a member among other bacteria in human PDAC, which can metabolize gemcitabine (2′,2′-difluorodeoxycytidine) into its inactive counterpart (2′, 2′-difluorodeoxyuridine), indicating that the presence of this bacterium in PDAC is possibly responsible for tumor resistance to gemcitabine.[48] Pushalkar et al.

observed in PDAC murine model that intestinal bacteria may be translocated into the tumor microenvironment during tumor formation.[49] In a recent study, 16S gut microbial profiling of different PDAC patient cohorts suggested that the bacterial population enriched in the PDAC milieu of short-term survivors (STS) and long-term survivors (LTS) harbored significant changes in bacterial diversity. The LTS patients showed enrichment of Proteobacteria (*Pseudoxanthomonas*) and Actinobacteria (*Saccharopolyspora* and *Streptomyces*), while no predominant genus was detected in the STS tumors. At the species level, it was found that *Saccharopolyspora rectivirgula* and *Bacillus clausii* were significantly increased in LTS patients in comparison to the STS patients. Further, this study also demonstrated that the genus *Pseudoxanthomonas, Saccharopolyspora,* and *Streptomyces* in the tumor may contribute to the anti-tumor immune response by promoting the recruitment and activation of CD8+ T cells. In summary, PDAC patients with diverse tumor microbiome have a higher survival rate than patients with restricted diversity in microbial composition. In addition to shaping favorable tumor micro-ecosystem promote recruitment and activation of CD8+ T cells, which may serve in predicting a patient's outcome.[50] *H. pylori* infection further enhance risk factor for PDAC development.[50] Numerous pathogenic constituents derived from *H. pylori*, such as ammonia, LPS, and large amounts of inflammatory cytokine secretion cause damage to the pancreas. *H. pylori* infections activate both NF-κB and AP-1, which causes deregulation of cellular processes. The increase in IL-8 levels accentuates inflammation and eventually leads to pancreatic carcinogenesis. LPS from *H. pylori* is confirmed for hyper-stimulating mutations in *KRAS* and initiating the process of pancreatic carcinogenesis.[51,52] Enrichment of *Enterococcus faecalis* in proximal mucosa is associated with pancreatitis and promotion of carcinogenesis.[53] A consistent link between PC risk and periodontal disease or tooth loss has been found, which is related to alterations in oral microbiota.[54–59] Multiple studies have reported the presence of *Porphyromonas gingivalis* and *Fusobacterium nucleatum* in the oral cavity of patients suffering from PC.[55,60,61] Mechanistically, oral bacteria can communicate to other tissues through the digestive system via blood circulation, lymphokines, or biliary conduction,[62] and they can cause cancers throughout the digestive system.[63] Researchers have previously identified *Aggregatibacter actinomycetemcomitans* as an oral microbe that plays a significant role in the development of PC, along with other bacteria such as *Neisseria elongata, Porphyromonas gingivalis,* and *Streptococcus mitis.*[64–66]

Though association-based studies show the role of oral microbes in PC, their pathogenic mechanisms are yet to be adequately explored.[55,60]

2.7 Oral cancer

Oral squamous cell carcinoma (OSCC), one of the most common cancers of the oral cavity, is promoted by host-related lifestyles such as smoking, alcohol abuse, tobacco usage, and human papilloma virus (HPV) infections.[67–69] The role of bacteria in the metabolism of tobacco-specific products and their role in OSCC is beginning to understand.[70] Cumulative data have strongly suggested the role of oral microbiome in the initiation and progression of oral cancer by releasing carcinogens, such as a cytolethal distending toxin, typhoid toxin, and colibactin in conjunction with proinflammatory responses.[71] Significant alterations of oral microflora such as enrichment for genus *Lactobacillus* and the loss of genus *Rothia, Leptotrichia, Haemophilus, Aggregatibacter,* and *Neisseria* have been noticed in OSCC patients[72] (Fig. 4).

2.8 Head and neck cancer

Head and neck cancer (HNC) is a rare and heterogeneous disease with disparities in incidence and survival rates between Europe and the United States.[73,74] Based on the EUROCARE-5 population-based study, the incidence rate of oral cavity carcinoma (OCC), oropharyngeal carcinoma (OPC), and larynx carcinoma (LC) subsites are higher than that of nasopharyngeal carcinoma (NPC).[73] In addition to alcohol consumption and tobacco use, HPV type-16 and Epstein-Barr virus infections are also linked to OPCs and NPCs respectively.[75,76] In recent studies, head and neck squamous cell carcinoma (HNSCC) and their association with oral microbiome composition revealed that the species linked with periodontal diseases such as *Porphyromonas gingivalis, Tannerella forsythia,* and *Aggregatibacter actinomycetemcomitans* were increased in high-risk HNCC group. In contrast, species such as *Prevotella nanceiensis, Capnocytophaga leadbetteri,* and *Selenomonas sputigena,* were inversely related to HNSCC progression. In addition, *Neisseriaceae,* and *Kingella* were significantly associated with a lower risk of HNSCC. Commensal bacterial genera such as *Corynebacterium* and *Kingella* were associated with a reduced risk of HNSCC. Interestingly, no difference in bacterial α/β-diversity between participants and control groups was observed[77] (Table 1).

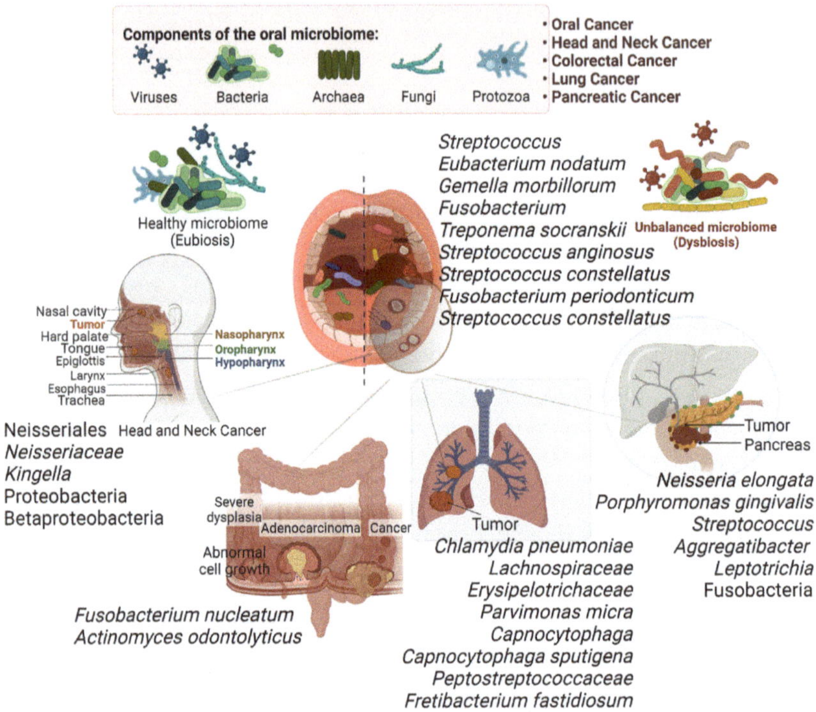

Fig. 4 Oral microbiome composition with different cancers. Oral microbiome associated with oral cancer, head and neck cancer, colorectal cancer, lung cancer, and pancreatic cancer.

Table 1 List of microbiome associated with cancers.

Cancer	Microbiota	Potential association/ function	References
Oral cancer (OC)	*Capnocytophaga gingivalis*, *Prevotella melaninogenica* and *Streptococcus mitis*	Abundant the saliva of OSCC patients; can act as a marker for OSCC onset	78
	Proteobacteria, Fusobacteria and Actinobacteria	The distinct microbial composition between cancer and normal specimens	79
	Streptococcus, *Gemella*, *Rothia*, *Peptostreptococcus*, *Lactobacillus*, and *Porphyromonas*	Salivary microbiota in OSCC differs from normal controls; Fifteen unique phylotypes were present in all three OSCC subjects	80

Table 1 List of microbiome associated with cancers.—cont'd

Cancer	Microbiota	Potential association/ function	References
	Peptostreptococcus stomatis, Streptococcus salivarius, Streptococcus gordonii, Gemella haemolysans, Gemella morbillorum, Johnsonella ignava and Streptococcus parasanguinis	Clonal analysis characterized 80 bacterial species/phylotypes that showed microbial alteration in the tumor samples	81
	Actinomyces, Parvimonas, Fusobacterium nucleatum, Pseudomonas aeruginosa, Dialister	Associated with the development of carcinogenesis	82,83
Head and neck cancer (HNC)	*Streptococcus, Neisseria Leptotrichia, Pseudomonas*	Microbial separation was found not only between NPC patients and healthy controls but also between pre- and post-chemoradiation therapy	84
	Actinomyces, Parvimonas	No significant difference in alpha diversity but abundances deviated from tumor relative to paired normal	85
	Streptococcus, Dialister, and Veillonella	HNSCC patients had a significant loss in bacterial richness and diversity	72
Lung cancer (LC)	*Prevotella intermedia, Prevotella tannerae, Prevotella oris, and Prevotella nigrescens*	Bacterial diversity significantly varied in different anatomical sites of the throat, and between carcinoma and polyps	86
	Ruminococcus, Enterobacteriaceae, Lachnospiraceae *Bacillus*, and *Akkermansia muciniphila*	Bacterial abundance was elevated and established gut microbial signature for lung cancer prediction	38
	Proteobacteria and *Verrucomicrobia*	Highly expressed in small cell lung cancer (SCLC)	87
	Bacteroides, Veillonella, and *Fusobacterium*	Microbial community altered in patients with lung cancer	

Continued

Table 1 List of microbiome associated with cancers.—cont'd

Cancer	Microbiota	Potential association/ function	References
Pancreatic Cancer (PC)	*Neisseria elongata* and *Streptococcus mitis*	Significant variation of salivary microflora between patients with pancreatic cancer and healthy controls	66
	Porphyromonas gingivalis, Aggregatibacter actinomycetemcomitans and *Leptotrichia*	Pancreatic cancer is associated with changes abundant in the microbial community	65
	Leptotrichia, Porphyromonas	The ratio of Leptotrichia to Porphyromonas in the saliva may be used as a biological marker of PC	65
Colorectal Cancer	*Proteobacteria, Dorea spp., Faecalibacterium spp.*	Higher diversity and richness in cases with colorectal adenoma compared with controls	88
	Bilophila, Desulfovibrio, Mogibacterium, Veillonella, Mogibacterium, and Sutterella	Microbial alteration in adenomatous tissues that correlate with alterations in bacterial metabolism in colorectal adenoma patients	89
	Bacteroides, Roseburia, Ruminococcus, Oscillibacter, Porphyromonas, Peptostreptococcus, Parvimonas, Fusobacterium	Mucosal microbiota differs in cases and controls, particularly if the lesion is proximal or distal; fecal and mucosal microbiota differ in CRC	90
Gastric cancer	*Citrobacter, Clostridium, and Rhodococcus*	Gastric carcinoma exhibit a dysbiotic microbial community with genotoxic potential, which is distinct from that of patients with chronic gastritis	91
	Peptostreptococcus, Streptococcus, Parvimonas, Prevotella, Rothia and Granulicatella	Identified the gastric microbes which contribute to the progression of gastric carcinogenesis	92

Table 1 List of microbiome associated with cancers.—cont'd

Cancer	Microbiota	Potential association/ function	References
	Acinetobacter, Bacteroides, Haemophilus parainfluenzae	Serum microbiota differed between gastric cancer lymphatic metastasis and non-lymphatic metastasis	93
Esophageal cancer (EC)	*Actinomyces cardiffensis, Selenomonas, Tannerella forsythia*	Microbes-associated findings have potential implications for the early detection and prevention of EAC and ESCC	24
	Streptococcus, Bifidobacterium, Subdoligranulum, Blautia, Romboutsia, Collinsella, Paeniclostridium, Dorea, and Atopobium	Characterized the gut microbiota composition of EC patients and screened out the optimal potential microbiota biomarkers for EC diagnosis	94
Hepatocellular carcinoma	*Enterobacter ludwigii, Clostridia, Veillonella* Lactobacillales, Bacilli and Gammaproteobacteria	Dysbiosis of the gut microbiota might be involved in PLC development and progression	95
	Streptococcus, Porphyromonas, and Actinomyces	Identified the oral microbiota dysbiosis in liver disease and possible diagnostic biomarkers for liver patients	96

3. Bacterial mechanisms of carcinogenesis

Bacterial infection and their carcinogenic contributions have been associated with multiple organs of the human body, including the lungs, colon, and cervix. Host bacterial dysbiosis is one of the primary causes associated with the carcinogenic effects of the pathogenic bacteria. These bacterial pathobionts trigger different signaling pathways and employ several pathogenic strategies to invade the host body checkpoints and establish malignancies (Fig. 5).

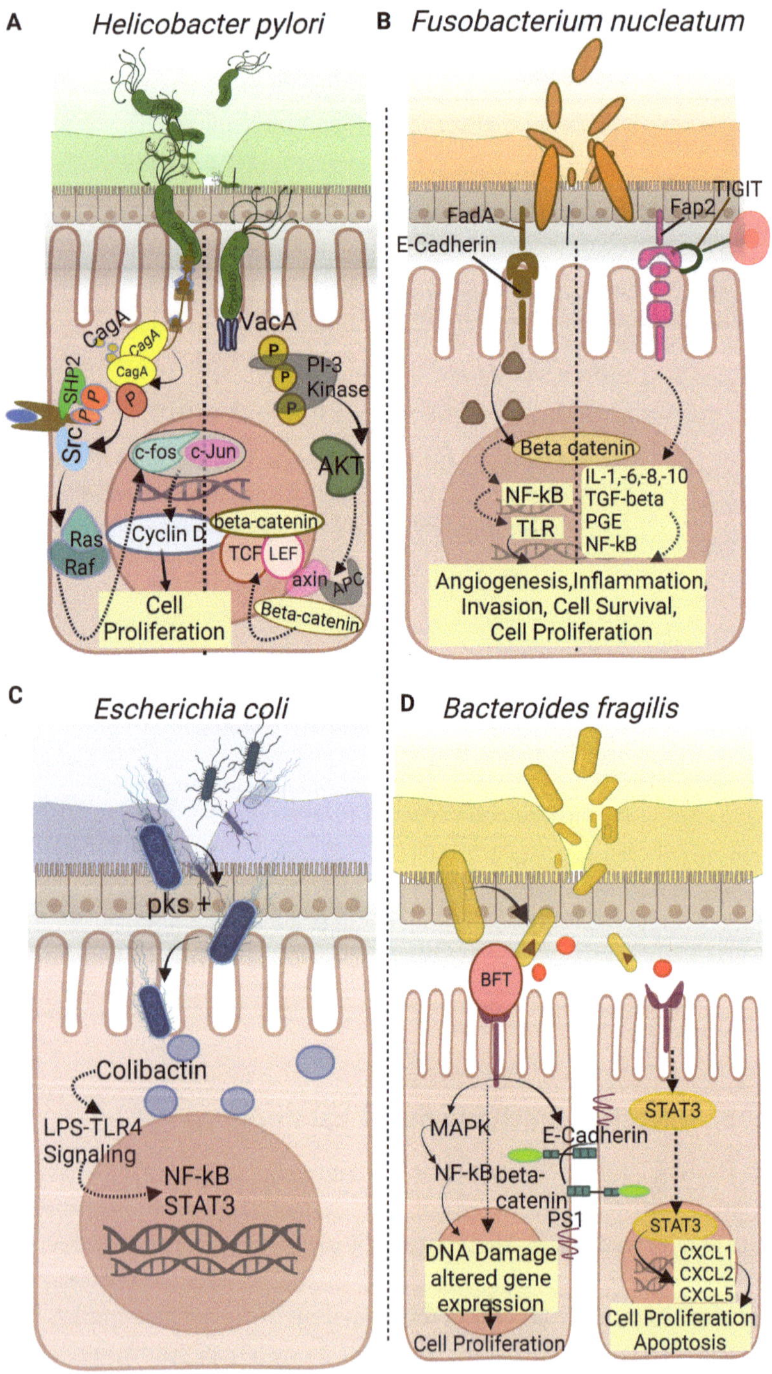

Fig. 5 See figure legend on opposite page.

Fig. 5 Mechanistic insights into bacterial-induced carcinogenesis. *H. pylori* attaches to the gastric epithelial cell surface receptor and injects its virulence factor, CagA directly inside the cell. EPIYA (glutamine-proline-isoleucine-tyrosine-alanine) motif of the CagA component is phosphorylated by host Src/Abl tyrosine kinases, which then interacts with the SH2 domain of SHP2 protein tyrosine phosphatase for activation. Activated SHP2 further activates the Ras-Erk signaling pathway resulting in a mitogenic response. Furthermore, Erk inside the nucleus interacts with the transcription activator Elk which when activated associates with SRF and binds to serum response elements to induce the expression of immediate early genes of activator protein-1 (AP-1) transcription factor family c-Fos and c-Jun, which can promote the transcription of late genes and result into cellular proliferation. The expression of late genes includes the cyclin CDK complexes such as the cyclin D-CDK4/6 which promotes cells to enter the S-phase and cyclin E-CDK2 which initiates replication events and thus all these events lead to abnormal cell proliferation and cellular transformation. Vac A virulence factor attaches to the host cell surface and is internalized via endocytosis. Following internalization, the Vac A channel is formed by the oligomerization on the surface of the plasma membrane, endocytic vesicles, and mitochondria, through which ions and organic molecules are released. Mitochondrial membrane integrity is lost through toxin channels, releases cytochrome c, and pro-apoptotic proteins, and results in apoptosis. VacA virulence factor acts via the activation PI3K-Akt pathway with β-catenin accumulation and subsequent nuclear translocation, followed by Cyclin D gene activation cancer promotion (A). *F. nucleatum* works by its virulence factors, FadA adhesin, interaction of which with E-cadherin, leads to phosphorylation and internalization of the later, which in turn reduces the phosphorylation of the β-catenin and thus its accumulation in the cytoplasm. β-catenin translocate into the nucleus where it enhances the transcription and activation of several of the wnt genes (wnt7a, wnt7b, wnt9a), oncogenes (myc, cyclin D1), transcription factors (lymphoid enhancer factor, LEF-1, NF-κB, T cell factor TCF1, TCF3, TCF4) and proinflammatory cytokines (IL-6, IL-8, IL-18), increasing the transcription of cell proliferation genes and tumor progression. Fap2 adhesin promotes a pro-inflammatory response by the cytokine and chemokine release and it also binds to TIGIT on NK cells and T-cells inhibiting their actions (B). *E. coli* possesses a pathogenic island, pks, which encodes for a genotoxin called colibactin. It induces DNA strand cross-linking and damage with the elevated expression of NF-kB, STAT3 which promotes cancer. *E. coli.* The pks+ adherent invasive *E.coli* (AIEC) strains may promote reactive oxygen species (ROS) production which has the potential to create double-strand DNA breaks (DSBRs) (C). Enterotoxigenic *B. fragilis* upon secreting BFT stimulates the colonic epithelial cells to express COX-2 which then activates the prostaglandin-mediated inflammatory consequences. PGE2 then imposes its carcinogenic effects by binding to its G-protein coupled receptor, EP1-4, through, which it transactivates the epidermal growth factor receptor (EGFR) switching on the downstream PI3K-AKT signaling pathway, which stimulates cell migration resulting in tumor metastasis. *B. fragilis* toxin, destroys the E-cadherin links, acts of STAT3, RAS-MAPK pathway and thus results in altered gene expression (D).

3.1 *Helicobacter pylori*

Gastric adenocarcinoma and mucosal-associated lymphoid tissue (MALT) lymphomas have been reported in some cases as a consequence of *Helicobacter pylori* infection.[97] *H. pylori* is a gram-negative, microaerophilic bacterial inhabitant of the acidic environment in the stomach.[98] Lofgren et. al. showed that gastric atrophy by *H. pylori* infestation may lead to colonization of commensal enteric microbes along with *H. pylori* due to hypochlorhydria (lowering of stomach pH) leading to an increased concentration of mutagens such as nitric oxide from dietary fibers/nitrate metabolism, which induces mutations thus transforming normal cells to malignant ones thereby inducing gastric cancer.[99–101] Evading the mucus layer and host gastric epithelium is the first step in *H. pylori* infection where it adheres to the host gastric epithelium via multiple adhesins and virulence factors. The CagA pathogenicity island (PAI) encodes for the CagA virulence factor and a specific type IV secretion system (Cag-T4SS) through which adherent *H. pylori* translocate CagA into the host epithelial cells[102] or through an interaction between outer membrane protein HopQ of *H. pylori* and the host carcinoembryonic antigen-related cell adhesion molecules (CEACAMs).[103,104]

Downstream signaling cascade, upregulates the RAS-MAPK cell proliferation pathway. This pathway enhances the expression of c-fos, c-jun, and cyclin D1 leading to uncontrolled cell proliferation.[105] CagA also promotes gastric disease by enhancing the secretion of the inflammatory chemokine IL-8 by the gastric epithelial cells.[106,107] IL-8 causes a massive neutrophilic infiltration which generates a strong and prolonged inflammatory response resulting in gastric cancer. Another virulence factor, the vacuolating cytotoxin protein A (Vac A), upon proteolytic cleavage yields 88 KDa functional factor.[108,109] It is secreted by the bacterial TypeVa auto-transport secretion system and is internalized into the host cell via endocytosis.[108] The toxin works by activating the Akt pathway with the accumulation of the β-catenin in the cytoplasm and subsequent nuclear translocation and activation of the cell cycle-specific genes.[109] Other effects imposed by VacA internalization include inhibition of T-lymphocyte activation, proliferation, and viability, induction of autophagy, and upregulated expression of vascular endothelial growth factor.[108–110]

3.2 *Fusobacterium nucleatum*

F. nucleatum is a gram-negative, fusiform, anaerobe inhabiting oral mucosa. Initially established as a symbiotic bacterium, its association with various pathological conditions revised its status to a pathogenic and tumor-associated microbe.[111,112] It aids in the biofilm formation of other bacterial members by enhancing the expression of many of its adhesive structures such as RadD, Aid1, and FomA.[112–114] Virulence factors that are responsible for the pathogenicity of *F. nucleatum* is a complex of FadA, containing an immature pre-FadA and surface mature mFadA and the other is Fap2, which is a galactose-sensitive outer membrane adhesin protein and promotes adherence to cancer cells.[112,115,116] Chen et. al reported that the presence of *F. nucleatum* induces enhanced TLR4 signaling activating the β-catenin pathway and overexpression of cyclin D1 and c-myc with enhanced carcinogenesis burden in CRC.[117] Studies have also shown an association of FadA with E-cadherin, β-catenin, and a membrane protein on the surface of cancerous cells Annexin A1, which stimulate colorectal cancer cell proliferation. Annexin A1 promotes FadA attachment and invasion while FadA and E-cadherin enhance Annexin A1 expression on cancer cells. With Annexin A1 and E-cadherin, FadA aids in β-catenin translocation into the nucleus and subsequent expression of inflammatory genes and oncogenes commencing into cell proliferation and tumorigenic growth.[8] Zang et al. showed *F. nucleatum* infection upregulate the expression of long non-coding RNA, lnc MIR4435-2HG, which interact with mir-296-5p and enhance the expression of SNAI1 promoting EMT transition indicating the novel mechanism for oral cancer progression.[118]

The metabolites secreted upon *F. nucleatum* infection such as short-chain fatty acids (acetate, propionate, and butyrate), short peptides (formylmethionyl-leucyl-phenylalanine) attract and recruits the myeloid lineage infiltrating cells such as myeloid-derived suppressor cells (MDSCs, granulocytic MDSCs and monocytic MDSCs) tumor-associated macrophages (TAMs), tumor-associated neutrophils (TANs), M2 like TAMs, CD11b+, myeloid dendritic cells, CD103+ regulatory dendritic cells.[119–121] These cells form a dampening immune environment by secreting growth factors and chemokines, suppressive T cell activity, inducible nitric oxide synthase, and arginase-1 expression thus enhancing the proliferation rate, metastasis, and angiogenesis in the tumor microenvironment.[122]

F. nucleatum also poses damage to DNA through an increase in the generation of reactive oxygen species.[123] Oxidative stress results in p53 guardian mutation and aberrant hypermethylation[124] leading to carcinogenesis.[125] A microbial surface and galactose binding lectin, fibroblast activation protein 2 (Fap2) interacts with the T cell immunoreceptor, immunoglobin G, and immunoreceptor tyrosine-based inhibitory motif

(TIGIT) domain receptor on the surface of natural killer cells (NK), T regulatory (Treg) cells, CD4+ and CD8+ T cells and suppresses their immune functions, thereby stabilizing an immune-compromised state.[115,126–128] Hence, *F. nucleatum* employs multiple mechanisms to elicit successful and sustained infection resulting in carcinogenesis.

3.3 *Bacteroides fragilis*

Bacteroides is a gram-negative, biofilm-producing, anaerobic intestinal flora. The *Bacteroides fragilis* toxin (BFT) is well known for its damaging effects such as intestinal inflammation, tissue injury, disruption in the tight junction, and increased intestinal permeability.[129,130] Upon ETBF (enterotoxigenic *Bacteroides fragilis*) intestinal colonization and subsequent BFT secretion, carcinogenesis is promoted either through inflammatory signaling pathways by activating STAT3, which enhances the secretion of proinflammatory cytokines (e.g. IL-17) or by the production of reactive oxygen species (ROS).[129–131]

BFT stimulates COX2 expression from colonic epithelial cells which are actively involved in the conversion of arachidonic acid into PGE2.[132] PGE2 is a potential activator of the RAS-MAPK pathway, which stimulates COX-2 expression and in turn enhances the PGE2 levels thereby directly interacting with RAS to switch on the pathway. This positive feedback loop between PGE2 and RAS is responsible for cellular transformation, proliferation, and cancer cell survival by triggering the downstream signaling pathways.[133] Thus, a systematic study is warranted to understand the role of *B. fragilis* enterotoxin in promoting carcinogenesis.

Bacteroides fragilis toxin works through multiple strategies to invade the host intestinal cells and exert its transforming effects. It exerts oncogenic effects via phosphorylation and nuclear translocation of the STAT3 transcription factor in the colonic mucosal immune cells followed by colonic epithelial cells.[129,131] STAT3 overexpression displays its proto-oncogenic behavior which is linked with cancer cell proliferation, survival, inflammation, cellular transformation, metastasis, and angiogenesis. BFT and STAT3 prevent the rejection of cancer by the host immune system via activation and accumulation of Treg cells.[134] Activated Tregs uplift the inhibitions on the Th17 cells. Th17 upregulates IL-17 and TNFα which promotes the inflammatory responses commencing into carcinogenesis. IL-17 upon binding to its receptors on the tumor cells releases IL-6 which is responsible for cancer progression, anti-apoptotic properties, and vascularization via STAT3 activation.[135–137]

E-cadherin is a 120 KDa structural glycoprotein, with tumor suppressor activity, cleavage of which results in loss of membrane association, colonic injury inflammation, and increase in cellular metastatic potential. BFT exerts its metastatic potential by cleaving, degrading, and releasing the 80 KDa extracellular E-cadherin ectodomain with the processing of the intracellular E-cadherin fragment by the host cell presenilin-1/gamma secretase, which renders the detachment of the colonic epithelial cells.[138,139] Following the cleavage of the intracellular domain of the E-cadherin, β-catenin which remains attached and inhibited is thus set free to translocate to the nucleus and aid in the progression of the cancer hallmarks- epithelial-mesenchymal transition, cell proliferation, and enhanced c-myc protooncogene expression.[140]

3.4 *Escherichia coli*

Escherichia coli, a gram-negative, facultative anaerobe is an active participant in maintaining intestinal homeostasis.[141] Some strains of *E. coli* from the B2 phylogroup possess a pathogenic island termed *pks*, encoding for a secondary metabolite, colibactin. This genotoxin has the potential to trigger genomic instability by inducing DNA damage (interstrand DNA crosslinks) and cell cycle arrest.[142–144] Colibactin is a secondary metabolite produced by a 54 kb *pks* gene that encodes a hybrid consisting of non-ribosomal peptide megasynthases (*NRPS*), polyketide megasynthase (*PKS*) assembly line, and accessory proteins clbA-S.[144] *E. coli* bearing a bacterial artificial chromosome encoding a complete *pks* gene island (BACpks) stimulate the ATM and RAD53-related ATR-dependent DNA damage responses (DDR) along with phosphorylation of Checkpoint kinase2 (CH2), replication protein A (RPA), gamma H2AX leading to the accumulation of cyclin-dependent kinase 1 (CDC2), culminating into G2 cell cycle arrest.[144] An elegant study by Manzano et al. showed how genotoxic strain *pks*$^+$ *E. coli* can induce genetic lesions potentially leading to the cancer phenotype. Briefly, they isolated *pks*$^+$ *E. coli* (colibactin producing) from a CRC patient biopsy and injected them into a single cell organoid model, where the single cell was cultured for five months. Whole genome sequencing of single cells revealed a substantially increased number of single-base substitution (SBS) signature (T>N) in ATA, ATT, and TTT trinucleotide sequences.[145] Interestingly, these SBS signatures were much reduced in cells that were injected with *pksΔ* *E. coli* strain (colibactin absent).[146] These studies provide a direct and indirect link between bacterial toxins with carcinogenesis potential and a systematic study is warranted for therapeutic insights.

3.5 Other bacterial infections potentiating carcinogenesis

In this section, we are discussing the role of bacterial species that have scanty evidence for their role in tumor establishment and progression and therefore future studies should focus on these bacterial species. *Salmonella typhi*, a causative agent of typhoid is also reported to be involved in promoting and sustaining gall bladder carcinogenesis.[147,148] Chronic inflammation and typhoid toxin manifest into gall cholelithiasis and further DNA damage result in deregulation of the cell cycle, which leads to cancer progression.[149] Apart from toxin, the extracellular polymeric substances (EPS) of *S. typhi* promotes cholecystitis and epithelial exposure to toxins.[147]

Chlamydophila pneumoniae, an obligate intracellular invader of the lung epithelium, is an important causal agent of pneumonia, sinusitis, pharyngitis, and bronchitis in humans.[149] Oncogenic proteins released upon Chlamydia infection mostly target mitochondrial and other cellular activities leading to perturbations in the cellular growth events and apoptotic pathways.[149] These toxic proteins also stimulate the secretion of cytokines IL-8, IL-10, and TNF in the lung alveolar macrophages which promote tumorigenic growth.[150] Interestingly, *C. pneumoniae* evade the host immune system by veiling themselves in apoptotic polymorphic neutrophils and alveolar macrophages for their propagation.[150]

Microbial populations of the urinary bladder usually comprise *Proteus mirabilis*, *Klebsiella* spp., *Staphylococcus albus hemolytic*, *Staphylococcus aureus*, and *E. coli*. These pathobionts are involved in the production of mutagen, N-nitrosamines from dietary nitrates, which triggers carcinogenic transformation.[149] A comparative analysis of the breast microbiota in normal and malignant tissue revealed an abundance of *Methylobacterium radiotolerance* in breast tumors and *Sphingomonas yanoikuyae* in normal breast tissue.[151,152] Anti-bacterial responses such as IL-12 levels, myeloperoxidase, and receptors of the immune responses TLRs 2,5,9 were reduced with *S. yanoikuyae* downregulation in breast cancer tissue.[152]

Porphyromonas gingivalis a potential threat to the oral cavity results in chronic inflammation through interaction with its receptors B7-H1 and B7DC and enhance production of TNF-α, IL-1, IL-6, and IL-8. It is also responsible for EMT transition with enhanced expression of MMP-1 and MMP-10. *P. gingivalis* promotes buccal epithelial cell proliferation, disruption in p53 function, and downregulated mitochondrial apoptotic pathways.[153] More studies are warranted to understand alterations in oral microbial communities in OSCC to define a clinical link, which may help to establish therapeutic interventions.

4. Virome and cancer

Non-bacterial oncomicrobes equally add to the cancer burden with viruses accounting for up to 15–20% of the total human cancers.[154] Oncogenic viruses account for diversity in their genomes, pathogenesis, infection sites, persistence, and extent of transforming potential. Viral oncoproteins and pathways they regulate target the host cell cycle machinery, apoptotic pathways, repair proteins, and the tumor suppressors such as p53 and pRb. Deciphering these checkpoints and the host-virome interactions would support the fight against viral strategies of carcinogenesis[155,156] (Fig. 6).

4.1 Human papilloma virus

HPVs are double-stranded DNA viruses that colonize epithelial or mucosal surfaces with the potential of causing benign warts and transforming them into carcinomas. HPV 16 and 18 are causative agents of cervical cancer. The infection is commonly transmitted by sexual contact and also by vertical transmission from mother to fetus establishing a transplacental and intranasal route.[154] HPV infects mainly the basal or stem epithelial cells and enters the cervical epithelium at injured sites or through the cells of the squamous cellular junction between the ecto-endocervix.[157]

The viral genome is carried with the differentiating basal epithelial cells and its expression progresses throughout the differentiation events from the basal to the upper epithelial layers.[158] The E2 protein is mainly expressed in the basal epithelial cells in the episomal state where it inhibits the transcription of E6/E7 proteins and thus their carcinogenic activities. While in the upper epithelial layers in the region of the differentiated epithelia, the viral genome gets integrated with the host genome with the loss of the E2 gene and subsequent repression of E2/E7 promoters.[159] The E6 and E7 oncoproteins are co-operatively expressed at lower levels during the early stages of the viral replicative cycle in basal epithelial cells.[160–162] While E6 maintains the episomal form of the viral particle,

E7 restricts terminal differentiation of the epithelial cells by stimulating the G1-S checkpoint and promoting the DNA replication events.[160,163,164] E7 also acts in progressing the cell cycle events by regulating the transcriptional factors (c-fos, c-jun, c-myc, NF-κB, SMAD2/3, STAT1) and inhibiting the activities of the pRb and setting E2F free to carry out the activation of cell proliferation genes such as cyclin A and E, which promotes

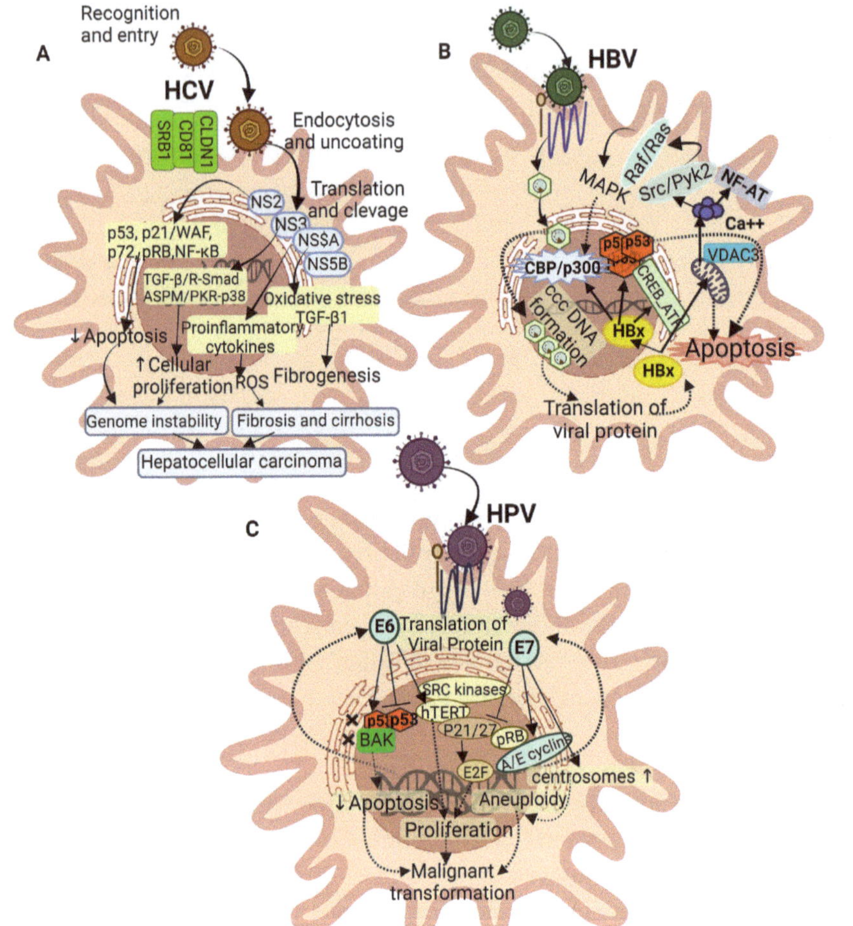

Fig. 6 Virus-mediated cancer induction and progression. HPV after entry into the cells by binding to the cell surface receptors and releasing the genetic material, HPV replicates within a membranous web derived from the endoplasmic reticulum and produces several core and non-structural proteins (NS3, NS4A, NS4B, NS5A, NS5B). These proteins work by inhibiting apoptosis, activation of cell survival signaling, oxidative stress generation, and fibrogenesis. These signatures seemingly lead to genomic instability, fibrosis, and cirrhosis with the onset of neoplastic transformation (A). HBV entry into the cell leads to the formation of covalently closed circular DNA and gives rise to the viral proteins, HBx which plays multiple roles in carcinogenesis by inhibiting p53 gene, apoptosis, CBP associated transcription factor expression, and activating cell proliferation pathways, RAS-MAPK and JAK-STAT (B). There are several viral proteins that HPV employs to produce its damaging effects, mainly E6 and E7 proteins play major roles. E6 protein exerts its effects via activation of SRC kinases, hTERT and telomere elongation, degradation of p53 and pro-apoptotic proteins thus leading to malignant transformation. E7 works similarly by inhibiting p21/p27, pRb releasing the E2F factor for progression through the cell cycle, centrosome amplification, aneuploidy generation, and subsequently cancer formation (C).

entry into S phase for DNA synthesis. It also works by stimulating INK4A (inhibitor of cyclin-dependent kinase) expression[165] and inhibiting p21 and p27 activities, which in turn activates CDK2- Cyclin A/E complexes thus promoting G1-S phase transition and other cell cycle events.[166] E6 and E7 proteins impose their influence by repressing the expression of TLR9 and several other cytokines (TGFβ, TNFα, IFNs), thus downregulating the host's innate immune responses.[167] The E4 and E5 proteins are essential players at the later stage of viral replication and virion particle formation. E4 can arrest cell cycle at G2 stage, and alter cytokeratin filaments thereby making the cells vulnerable enough to release the fully assembled virion particles while E5 can activate the MAPK, EGFR, AKT, and p38 pathways, and regulate apoptosis thus promoting the cell immortalization events.[168]

4.2 Hepatitis B virus

HBV is a blood-borne pathogen and accounts for ~50% of hepatocellular carcinoma (HCC) incidences.[169] HBV is a member of the hepadnavirus with a circular, partially double-stranded DNA genetic material, which changes into a covalently closed circular (cccDNA) form soon after infecting the hepatocytes and replicating via an RNA intermediate.[170] HBV cccDNA after infection in the hepatocytes localizes as the episomal state in the nucleus[170] and remains associated with histone and non-histone proteins forming a minichromosome.[171] HBV genome insertion occurs at specific sites at or around the genes involved in cell proliferation, differentiation, and immune evasion. For instance, the *TERT, SMAD5, MLL4, CYP238, LRP1B, ROCK1, PDGF,* and *CCNE1*[172–175] genes are sensitive to genomic instability, mutations, fusion protein formations, and amplification of genomic DNA.[172,176] HBx is an oncoprotein, which acts via interaction with the promoter element of the genes involved in the regulation of several cellular activities, such that NF-κB, RNA polymerase, IL-8, TNF, EGFR, Jak1/STAT, and protein kinases.[177,178] It generates chromosomal instability by acting on the dynamics of the centrosome, multipolar mitotic spindle formation, and perturbations in centrosome duplication, which may be due to the binding of the Hbx to centrosome regulator, HBx interacting protein (HBXIP).[171,179] HBx imposes its carcinogenic effects by direct association and inhibition of the p53 tumor suppressor protein and thus suppressing apoptosis, activating cell survival pathways and DNA repair genes.[171] This oncoprotein also promotes angiogenesis by upregulation of vascular endothelial growth factor (VEGF), angiopoietin

2 (ANG2) and hypoxia inducible factor HIF-1α.[180] It is clear from these studies that HBV utilizes multiple approaches to transform normal hepatocytes to cancerous by altering epigenetic profile, aberrant signaling and repression of tumor suppressors.

4.3 Hepatitis C virus

Hepatitis C Virus (HCV) is a member of the *Flaviviridae* family with a positive sense of single-stranded RNA genome which does not integrate into the host genome.[181] Studies displayed an indirect HCV association with HCC where alcohol and oncoprotein NS5A were found to act in coordination to enhance tumor formation via induction of TLR4 signaling and Nanog activation.[182,183] Some transgenic mice showed a direct effect of the viral core proteins which developed hepatic steatosis followed by progression to HCC without the involvement of hepatic inflammatory responses.[184] It is a hepatotropic virus and its genome encodes for viral nucleocapsid composed of the structural core and envelope proteins E1 and E2. These proteins are involved in host cell attachment, and entry and also code for a non-structural protein (p7, NS2, NS3, NS4A, NS4B, NS5A, and NS5B), which works in a coordinated fashion for the genome amplification and the maintenance of the viral particles within the host cell.[185]

HCV core protein interacts with and inhibits the activity of tumor suppressor genes (*TP53, TP73,* and *RB1*) and *CDKN1A* (p21) and also aids in generating reactive oxygen species (ROS).[186–189] Structural proteins, NS5A and NS3 inhibit the activities of p53 and NS5B inhibits RB1.[190,191] The RAS/MAPK pathway and E2F1 pathways are activated by the viral oncoproteins core, E2, NS5A/B, which stimulate the cell proliferative events.[191–193] NS5A activates cell proliferation pathways including PI3K/Akt and Wnt-β and inhibits nuclear translocation of SMAD proteins through TGF-β signaling resulting in reduced expression of cell cycle negative regulator, CDKN1A.[194,195] The core protein is also involved in generating oxidative stress leading to mitochondrial and chromosomal disruption, another viral source of HCC induction.[187] The NF-κB pathway is inhibited by the HCV core protein and there is the activation of c-Jun N-terminal kinase pathway by proinflammatory signals, which establishes an inflammatory environment in the hepatocytes thus promoting HCC.[196]

5. Parasites in cancer development

Parasitic etiology of cancer is an emerging field and its characterization could aid in a more promising targeted based therapy in carcinogenesis. Liver, stomach, colorectal and esophageal cancers are mostly associated and progress due to some pathogenic interventions (WHO 2015 https://www.who.int[197] (Fig. 7).

S. haematobium, a blood fluke primarily associated with schistosomiasis and classified as a Group 1 carcinogen by the IARC (IARC 2012, https://www.iarc.who.int[198]) and is primarily associated with urinary bladder cancer. The onset of cancer starts with *S. haematobium* infection with epithelial damage, chronic inflammatory responses, and reactive oxygen species production which enhances the cell proliferative pathways, and induces hyperplasia and metaplasia.[199] Schistosoma releases mutagen, nitrosamines, and an increase in the urinary β-glucuronidase and cyclooxygenase-2, which act as potential carcinogens.[199]

A study from rural China documented the association of *S. japonicum* with HCC, colon cancer, and metastatic lung cancer.[200] It stimulates chronic inflammation, somatic mutations in p53 gene, and an accumulation of arginine missense mutations in schistosomal rectal cancer resulting in genomic instability and tumorigenesis.[201] *S. mansoni* and HCV co-infection promote HCC development. Schistosoma infection inhibits the cell-mediated immune responses which lead to the development of hepatosplenomegaly and targets oncogenes *Bcl-2* and *c-Myc* to promote colorectal cancer.[202] Few studies have also reported the association of *S. mekongi* infection with leiomyosarcoma of the small bowel[203] and *S. intercalatum* in rectosigmoid carcinoma[204] and urinary bladder cancer.[199]

Liver flukes also pose a carcinogenic threat to their hosts namely *Ophisthorchiasis* and *Clonorchiasis* that has widespread infection rates. Their transmission is mainly through the consumption of raw fish already carrying the metacercariae, which are then released from its cyst in the duodenum, pass through the bile ducts consuming biliary epithelia and bile ingredients, and ultimately forming epithelial hyperplasia and fibrosis.[205] Animal model studies have revealed that *Opisthorchiasis* secretions of dimethylnitrosamine are potential inducers of cholangiocarcinoma along with other chemicals such as granulin, which promotes biliary cell proliferation and thioredoxin

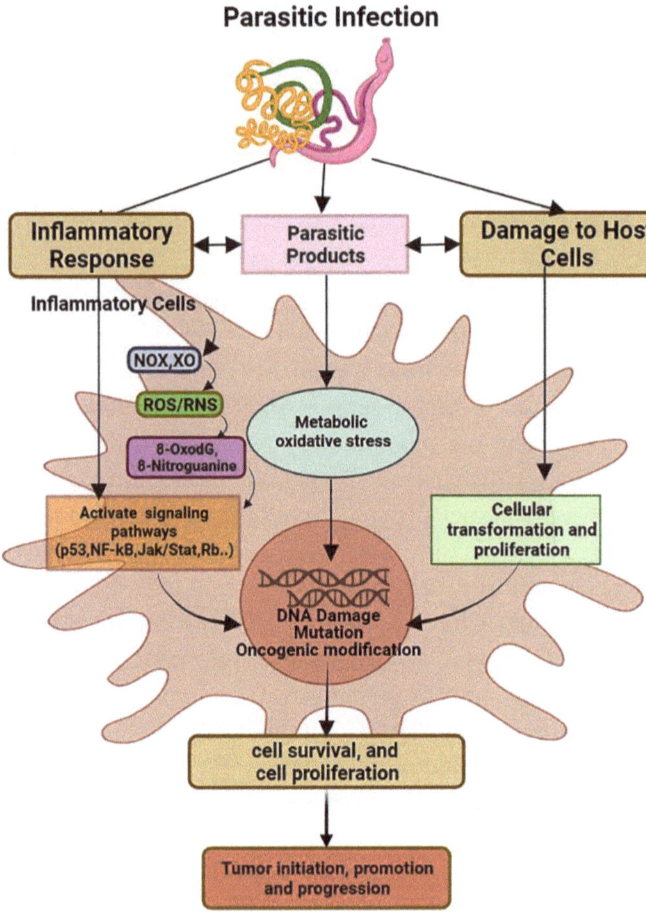

Fig. 7 Role of parasitic infection in tumorigenesis. Parasites such as *Clonorchis*, *Ophisthorchis*, and *Schistosoma* upon infection recruit inflammatory cells and production of reactive oxygen species and oxidizing chemicals causing DNA damage, accumulation of mutations, activation of cell survival pathways, and ultimately oncogenic modifications. Metabolites derived from parasites and any physical damage to the host during parasitic infection followed by wound healing often results in cell proliferation and transformation events.

peroxidase inhibiting apoptosis.[205,206] These derivatives are involved in stimulating inflammatory responses thus activating PI3K/Akt and Wnt/β-catenin pathway, IL-6 production. *O. viverrine* infection inhibits RB1 and p16 expression while promoting cyclin D1 and CDK4 in cholangiocarcinomas.[207] *O. viverrini* infection also results in somatic mutations in p53 and SMAD4 genes disrupting cellular signaling and upregulating carcinogenesis[208] (Table 2).

Table 2 Summary of microbial factors and their mechanisms inducing carcinogenesis.

Microbial signature	Cancer type	Mediators	Mechanism	Reference
Bacteriome				
Helicobacter pylori	Gastric cancer, Mucosal associated lymphoid tissue lymphoma, Colorectal cancer, Pancreatic cancer	Adhesin molecules- BabA, SabA, OipA, AlpA/B; Virulence effectors- CagA, VacA, OMPs	Binds to host cell surface molecules, acts of SHP2 tyrosine phosphatase, and increased expression of cell proliferation genes and pathways (RAS-MAPK); Stimulates IL-8, IL-12 mediated inflammatory reactions; Vac A upon internalization promotes expression of cell cycle factors, loss of mitochondrial membrane integrity and upregulation of autophagy	107–109,117,209–211
Fusobacterium nucleatum	Colorectal cancer Esophageal cancer	Virulence factors- FadA, Fap2	Interacts with cadherin group of adhesion molecules on the host cell surface, internalization and activation of TLR4 signaling, enhanced β catenin translocation and oncogene expression; Promotion of TLR4/MYD88/NF-kB, RAS-MAPK signaling cascades, expression of ncRNAs (miR-21) and recruitment of myeloid-derived cells; ROS production and hypermethylation of repair machinery; Fap2 interacts with inhibitory domains on the surface of immune cells and suppresses immune functions	125,128,212–214

Continued

Table 2 Summary of microbial factors and their mechanisms inducing carcinogenesis.—cont'd

Microbial signature	Cancer type	Mediators	Mechanism	Reference
Bacteroides fragilis	Colorectal cancer	Toxin-BFT	Binds to and activates the STAT3 transcription factor which works either by promoting the inflammatory cascade or by ROS production and DNA damage; PGE2 activates PI3K-AKT and inhibits apoptosis, RAS-MAPK pathway and promotes proliferation, suppresses immune function by inhibiting TH1 and upregulating TH2 activities; Increased production of proinflammatory cytokines such as IL-6, IL-17, TNFα	129–131,133
Escherichia coli	Colorectal cancer, Esophageal adenocarcinoma	Secondary metabolite-colibactin	Binds to the DNA minor groove, alkylates, and generates interstrand crosslinks, DSBRs, ROS production, and cell cycle arrest	215–217
Salmonella typhi	Gall bladder cancer, colorectal adenocarcinoma	Extracellular polymeric substances, Cytolethal distending toxin B, cytotoxic necrotizing factor 1	Toxins bind to and impose DNA damaging effects and also promote inflammation	149,218,219
Chlamydophila pneumoniae	Lung cancer	Oncogenic proteins	Perturbations in the mitochondrial functions, cellular activities, apoptotic events; Cytokine release and chronic inflammation, evasion of immune responses through apoptotic polymorphic neutrophils	149,150,220

Porphyromonas gingivalis	Oral squamous cell carcinoma Pancreatic cancer Esophageal cancer	Nucleoside diphosphate kinase, Gingipains	Invades and resides in the gingival epithelial cells; Overexpression of B7-H1 and B7-DC receptors and enhanced production of TNF-α, IL-1, IL-6, and IL-8; Increased expression of pro-MMP-9, MMP-10, MMP-1 and aids in EMT; Phosphorylation and activation of CDKs thus inducing cellular proliferation, inhibition of TP53 and apoptosis through activation of JAK1/STAT3, PI3K/Akt signaling pathway	153,221,222
Virome				
Human papilloma virus	Cervical cancer	Oncoproteins-E6, E7	Maintained in the episomal state E7 inhibits terminal differentiation of epithelial cells and pRb which sets E2f free thus promoting the expression of cell proliferation genes; E6 sequesters p53, degrades it, inhibits apoptosis via NF-kβ signaling, interacts with myc, NFX1-91 and sustains telomerase in an activated state; Downregulates host innate immune responses by repressing TLR9 signaling	167,168,223–225
	Oral cancer			
	Head and neck cancer			
Hepatitis B virus	Hepatocellular carcinoma	Oncoprotein- Hepatitis BX (HBx)	Interacts with the promoters of cell proliferation genes such as NF-kβ, IL-8, TNF, EGFR, Jak1-STAT; HBXIP binds with centrosome regulator, perturbs centrosome duplication, generation of multipolar mitotic spindle formation; Associates and inhibits p53, suppresses apoptosis, promotes angiogenesis, expression of DNA methyltransferases thus inhibiting the activities of tumor suppressor gene	171,177–179,226

Continued

Table 2 Summary of microbial factors and their mechanisms inducing carcinogenesis.—cont'd

Microbial signature	Cancer type	Mediators	Mechanism	Reference
Hepatitis C virus	Hepatocellular carcinoma	Oncoprotein NS5A	Activates PI3K/Akt, Wnt-β cell proliferation pathways inhibits the cell cycle negative regulator CDKN1A; Proliferation is also enhanced by the interaction with RXR, PPAR α, promotes fatty acid transport and catabolism, maintains telomere lengths	194,195,227
Epstein Barr virus/ Human Herpesvirus Type 4	Hodgkin's Lymphoma	Oncoprotein- LMP1, LMP2A	LMP1 activates STAT, NF-kB transcription factors MAP kinases, PI3K pathways	228,229
	Burkitt's Lymphoma		LMP2A promotes EMT by activation of PLCγ and PI3K pathways	
Human T-Lymphotropic Virus Type 1	Adult T-cell leukemia	Oncoprotein- Tax	Reprogramming of cell cycle events and DNA repair inhibition; Induction of NF-kβ pathway, upregulation of cytokine expression, IL-2, IL-13, IL-15 leading to chronic inflammatory reactions	230,231
Kaposi sarcoma-associated herpesvirus	Kaposi sarcoma	Oncoprotein- Latency-associated nuclear antigen (LANA), K1, K3, K5, v-Cyclin, v-FLIP, v-Bcl2, v-IRF	Cell cycle regulation by LANA 1 and v-Cyclin K1, v-Bcl-2, and v-FLIP inhibits apoptosis; K3, K5, and v-IRF exert immune modulatory effects	232,233
Parasitome				
Schistosoma haematobium	Urinary bladder cancer	Nitrosamines	Initiates chronic inflammatory reactions, generation of reactive oxygen species, DNA damage, inhibition of p53, Rb	199,234,235

Schistosoma japonicum	Hepatocellular carcinoma		Stimulates chronic inflammation, mutations in p53 gene, genomic instability	201
	Colon cancer			
	Metastatic Lung cancer			
Schistosoma mansoni	Hepatocellular carcinoma		Inhibits cell-mediated immune responses	202
Schistosoma mekongi	Leiomyosarcoma of small bowel			203
Schistosoma intercalatum	Rectosigmoid carcinoma			204
	Urinary bladder cancer			
Opisthorchis viverrini	Cholangiocarcinoma	Dimethylnitrosamine, Granulin, Thioredoxin peroxidase	Biliary cell proliferation, inhibition of apoptosis; Activation of PI3K/Akt and Wnt/β-catenin pathways, cytokine production; Inhibition of Rb1 and p16, and promotion of cyclin D1 CDK4	205,206
Clonorchis sinensis	Cholangiocarcinoma	8-oxodG	Promotes squamous metaplasia, mucous gland hyperplasia, inflammatory responses, and DNA damage	199,236

6. System biology tools to understand the complexity of human cancer microbiome interactions

A system biology approach facilitates understanding of the relationship between a host and microbe or cancer-associated microbes interaction (neutralism, commensalism, mutualism, and interactions of isolated molecular and cellular components).[237] Different computational and mathematical methods are used to solve the complexity and heterogeneity of tumorigenesis and understand the microbe's behavior in the process of metastasis and malignant cells.[238] Multiple technologies are applied to integrate the role of the microbiome with genetic/epigenetic lesions, homeostasis of signaling networks, cell-associated microbiota, and microenvironment in the malignant cells. It also helps to develop and implement effective strategies to increase the efficacy of therapeutic interventions (Fig. 8)

Systems microbiology serves as a platform to integrate basic biological information, host-microbes interaction, and microbial community data using omics technologies such as metagenomics, next-generation RNA or DNA sequencing, metatranscriptomics, metaproteomics, single-cell genomics, functional genomics, bioinformatics, mathematical approaches, and computational microbiology. It analyzes complex data sets such as phenomics, genomics, epigenomics, transcriptomics, and metabolomics to

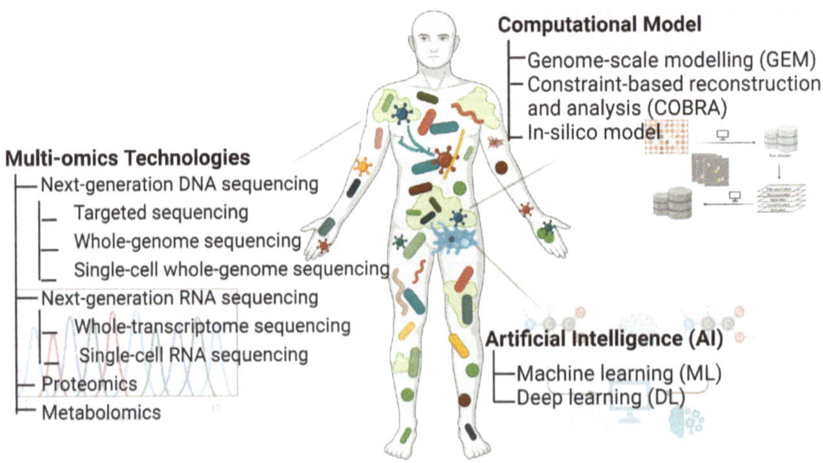

Fig. 8 Overview of system biology approaches in cancer-associated microbiome studies. A comprehensive picture of systems biology approaches, which uses multi-omics technologies, computational model systems, and artificial intelligence (AI) constituting the core components of the system biology-related studies.

develop stochastic models and novel approaches to understanding host-microbe interaction.[239] Understanding 'whole' microbial communities require adequate formulations, which consist of (1) information regarding microbial community size and genome composition to evaluate the phylogenetic and genomic complexity, (2) the role of individual phylotypes within particular environmental niches to identify major microbial signatures, and (3) functional composition and metabolic networks to understand and predict functioning and microbial impact on the environment.[237–240] Hypothesis building and prediction models can be developed to fit the cancer states and cellular microenvironment using systems-based methods. Integrating massive high-throughput multi-omics data from multiple studies and extracting meaningful information is beyond human brain capabilities. Therefore, the assistance of computational tools such as artificial intelligence (AI), which uses deep learning (DL) and machine learning (ML) tools is required. Such approaches are getting integrated efficiently in to understand the complex biological interactions. Adaptive learning techniques such as AI and ML tools using high computational power are need-of-an-hour to investigate the role of the microbiome and cancer. Artificial intelligence is based on strong logical reasoning ability and independent learning ability using large data sets. AI technologies can manage the use of chemotherapy drugs and predict the tolerance of chemotherapy drugs, and radiation treatment, improving the identification of tumor neoantigens and the efficacy of tumor immunotherapy. Machine learning, a type of artificial intelligence (AI), is used to analyze different large-scale data settings and combine them into predictive models for disease risk, diagnosis, prognosis, and appropriate treatments. The workflow of machine learning methods on microbiome data analysis includes (1) collection of sequencing data from bacterial communities associated with various hosts environments, and (2) utilization of these sequences as input into a machine learning model for preprocessing that can help to improve the prediction results of the model, and (3) prediction of machine learning algorithms for the microbial-associated disease after embedding structural information, such as the phylogenetic relationship and the average nucleotide similarity network between sequences.[241] Deep learning can discover multiple representations to achieve better prediction. These approaches may potentiate and predict the role of various factors from host-microbe interactions important for cancer development, progression, and metastasis, which can be validated using experimental models. Such approaches will help in the early detection of progression and therapeutic interventions[242] (Table 3).

Table 3 Summary of computational applications to understand host-microbe interactions.

Computational Tools	Data input	Applications	References
Genome-scale metabolic modeling tools (GEMs)	Experimental validation using in-vitro and in-vivo models	Investigate the roles of individual species in a complex microbial group; Predict microbial metabolic interactions	243,244
	Understanding observational data, microbiota behavior, and interaction in different tumors		
Dynamic Bayesian Network, Boolean network, Ordinary and partial differential equations, Petri network, hybrid and cellular automata, pathway analysis, flux balance analysis (FBA)	Parameters from previous studies, expression, proteomic, metabolic reaction. etc.	Build the signaling pathways leading to cell migration of the metastatic processes; Provide a way to specify highly complex interaction networks between microbes; Petri Nets are used to model the immune system response	238,245
	metabolic rate constants and identity of genes catalyzing each metabolic reaction		
constraint-based reconstruction and analysis (COBRA)	Physical, chemical, and biological constraints for biochemical reaction networks; In the case of metabolism, different flux distributions through the metabolic network	Introduces metabolic behavior and microbes' interaction; Predicts different metabolic phenotypes, including growth rate, microbiota functionality, and gene essentiality	245,246
ODE (Ordinary Differential Equations)	Mathematical differential knowledge; calculus and linear algebra	Modeling large biochemical reaction systems; Concentrations of each molecule type can be described through mass action dynamics; Revealing which extracellular molecule promotes tumor growth and invasion	238,247
CTMC (Continuous Time Markov Chains)	Mathematical statistical knowledge	Capturing the state and the progression of the metastatic process	247
	Automata, topology of biological systems		

In silico model	In-depth understanding of the tumor biology and clinical manifestations of the patients	Enabling a better understanding of protocol parameters, exploring different possible study designs; Reducing the uncertainty of the outcomes	248,249
	Computational, mathematical representations of biological phenomena and mechanisms	Faster and more efficient development of the drugs	
Artificial Intelligence-based tools	Machine learning algorithm,	Generate non-trivial causal, testable hypotheses from raw data; Analyze large-scale data and combine them into predictive models; Predict the tolerance of chemotherapy drugs, and radiation treatment, improving the identification of tumor neoantigens and the efficacy of tumor immunotherapy	241,242
	Supervised Learning (Random Forest, Support Vector Machines, Logistic Regression, K-NN, Artificial-neural-networks)		
	Unsupervised Learning (clustering, Dimensionality Reduction, Non-negative matrix factorization (NMF))		

7. A snapshot of microbial therapeutic approach in cancer

Historically, microbial therapy is used for different diseases. With the advancement of technologies, the host-microbe relationship is beginning to unfold and microbe-based treatment regimens are developed. Microbial therapeutic strategies give new perspectives on cancer treatment. A wide range of treatment options, including cancer vaccines and biological therapies, are being studied in conjunction with conventional regimens for cancer.[250] Microbial therapy can be broadly classified into direct and indirect approaches; direct approaches include engineered microbes, and anticancer immunotherapy and indirect approaches include prebiotics, probiotics, synbiotics, and fecal microbiota transplantation (FMT) (Fig. 9).

Exogenous bacteria and viruses are engineered and genetically modified to harness as potent immunotherapies or neoadjuvant. Talimogene laherparepvec (T-VEC) is an oncolytic virus that selectively replicates in malignant cells and presents antigen for a selective immune response.[251] Several other cancer vaccines are under investigation, such as cellular vaccines where tumor antigens are presented for a higher immune response,[252–255] and peptide vaccines, where autologous immune cells are primed with tumor-specific peptides.[256,257] In microbe-based anticancer immunotherapy, microbes are used to enhance the efficacy of cancer immunotherapeutic modalities.[258] Bacteroides species, such as *Bacteroides thetaiotaomicron* and *B. fragilis* exert anti-cancer and immunostimulatory effects through CTLA-4 blockade.[259] Combination therapy of PD-L1 inhibitors with *Bifidobacterium*, increases its anti-cancer efficacy and retards tumor progression.[260]

Prebiotics can shape the gut microbiome dynamics by providing selective growth advantages to commensal.[259] For example, plant-based fructans preferentially promote the growth of *Flavibacterium* and *Bifidobacterium spp.* gives a positive response to antitumor immunity in melanoma patients.[261] Probiotics containing *Saccharomyces*, *Lactobacillus*, and *Bifidobacterium* spp. have long been known to promote overall well-being.[262] Administration of probiotics (*Bifidobacterium lactis* in combination with *Lactobacillus acidophilus*), can restore the abundance of butyrate-producing bacteria, *Faecalibacterium* and *Clostridiales* spp, with a reduction in CRC-associated genera.[263] Synbiotics are a combination of both prebiotics and probiotics.[263] The administration of synbiotics during neoadjuvant chemotherapy for EC

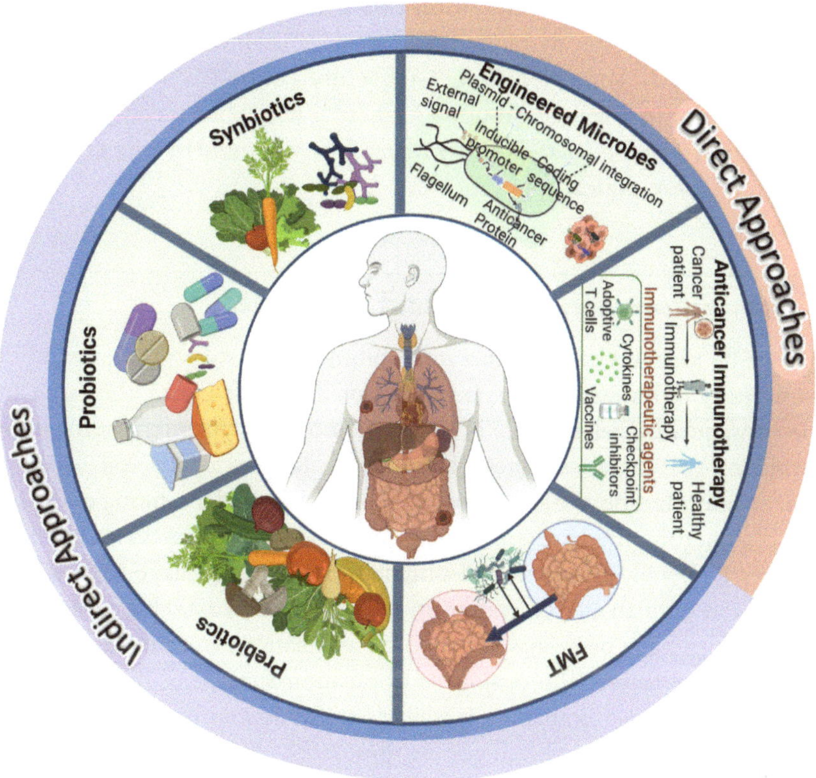

Fig. 9 Possible microbe-based therapeutic interventions. Different strategies/approaches are employed to modulate the microbiome and can be used as interventions to improve cancer treatments, i.e., indirect approaches that include synbiotics, probiotics, prebiotics, and fecal microbiota transplantation (FMT). Direct approaches include engineered microbes and anticancer immunotherapy. Prebiotics, probiotics, and synbiotics approaches include diet and supplement-based treatment. FMT is a procedure in which fecal matter is collected from a healthy donor and placed into the gastrointestinal tract of a patient. It is an effective treatment for *C. difficile* infection. FMT is performed using colonoscopy, Nasogastric (NG) tube, and Oral capsules. Engineered microbe therapy is based on microbial manipulations, which can lead to the selection of new microbial strains and mixture or targeted improvement of existing microbial strains. Anticancer immunotherapy helps the immune system to fight cancer. There are multiple agents used to treat cancer such as checkpoint inhibitors, chimeric antigen receptor (CAR) T-cell therapy, cytokines, immunomodulators, cancer vaccines, etc.

improves gut microbial activity and reduces the side effects of chemotherapy.[264] CRC biomarkers, such as DNA damage and cellular proliferation were improved in synbiotic administered to polypectomized CRC.[265] FMT is a technique to introduce solubilized stool (or the cryopreserved microbial contents within the stool) of a healthy donor into the colon of patients to restore the dysbiotic gut microbiome. FMT has been successfully used for ICI-induced steroid-refractory colitis[266–269] and colorectal tumorigenesis in preclinical murine studies.[270] Microbial therapy for cancer treatment is developing and further research is warranted to fully understand its potential for clinical applications.

8. Future perspectives and challenges

The biggest challenge in a microbiome-based study is to unwind the complexity of host-microbiome interaction and to delineate 'the cause and the consequence' relationship between the host and the microbiome in the formation of diseased states. Despite decades of study on dysbiosis cancer-microbiome, we are far from having a clear understanding of the role of host-microbiome in regulating cancer progression. A range of common diseases (generally based on 16S amplicon sequencing) have suggested an association with an altered composition of gut microbiota. However, these relationships have been mainly based on associations and not substantiated by etiological factors. To establish a population-specific or disease-specific appropriate clinical biomarker, it is necessary to do massive epidemiological screening in terms of different populations, races, ethnicity, clinical manifestations, etc. Microbe-specific disease associations combining both immune and chemotherapy should be validated in a model-based system. Further, well-planned experiments with logical controls are required for both host and associated microbiome profiling, SNPs identification, integration of pathophysiological parameters, along with extensive validation is required. Developing sophisticated tools for predicting immunogenicity would lead to safer, more efficient, and effective cancer treatment. In complex biological systems where thousands of reactions take place simultaneously among a large number of interdependent cells, predictability quickly erodes. Approximately 1,000 bacterial species live within the human gut microbiome, competing and collaborating while communicating with their host. Host-microbes interaction patterns will be more complex in the context of cancer microbiome study because cancers are multifactorial and carcinogenesis is a multi-stage process. Regardless of years of research in this field, we are far from understanding whether microbial dysbiosis is behind the etiology of tumorigenesis or vice-versa. A multi-directional approach is necessary to combine multi-facet study on the microbe-microbe and microbe-host interaction including multi-omics methodology and logical bioinformatics pipeline to develop a comprehensive understanding.

9. Conclusions

Advancements in biomedical research have largely focused on the genetic and epigenetic molecular mechanisms of the diseases. Although the contribution of microbiota to human health is known to mankind for ages, diminutive information is available regarding their role in the overall wellness of human beings. The role of the inseparable association of microbes and hosts for healthy and diseased states is beginning to unfold. Our understanding of dynamic microbial communities impacting human health is rapidly developing, as is the field's scope. The impact of the host microbiome on cancer pathogenesis, development, maintenance, and therapy outcomes has been established through various studies. However, an in-depth understanding of host-microbe interactions and the mechanism behind their 'socialization' warrants further integration of laboratory, etiology, and clinical data along with a robust bioinformatics pipeline. The complexity of study design arises is due to multiple variables such as microbial diversity, abundance, context-dependent interactions, microenvironment contribution, and the list goes on. Therefore, systematic introduction of logical interventions in experimental design, technological development, uniform execution of protocols, validation using appropriate model systems, and practice of multi-omics will enhance our knowledge based on the role of microorganisms in various human diseased states. There are several goals to be achieved, ranging from regulating human metabolism, immunological, and inflammatory responses to preventing carcinogenesis, and boosting the efficacy of personalized cancer treatment. Beneficial and commensal microbiota can work in tandem with chemotherapeutic and immunotherapeutic drugs. Scientists have figured out gut microbiota signatures as potential therapeutic targets, diagnostic, predictive, and prognostic cancer biomarkers. The use of microorganisms that preferentially colonizes within the tumor microenvironment is translated to design targeted therapies against cancer, much like oncolytic viruses. The focus on the emerging field of the microbiome in the context of cancer, to identify best practices for study design, execution, bioinformatics analysis, and interpretation is need of an hour. Standardization of technical, methodological, analytical, and reporting aspects is critical in ensuring the validity and comparability of research results. Differentiating causal correlations from association (part of disease etiology) prerequisites careful planning in validated animal models as well as appropriately controlled clinical trials in patients. Exploring the relationship between cancer and the human microbiome, will uncover new aspects of cancer biology and potentially pave new ways to improve patient care.

Acknowledgments

It gives us immense pleasure to express our gratitude to all the people and organizations who helped us out to carry our research work uninterruptedly. We are greatly indebted to the Department of Biotechnology (DBT), Government of India, for providing the core funding. This work could not have been completed without the support of the ILS flagship project (BT/ILS/Flagship/2019) DBT, Ramalingaswami Re-entry fellowship (BT/RLF/Re-entry/25/2015), and SERB core research grant (CRG/2018/002052). We would also like to thank ILS Flagship project for providing fellowship to S.M. and R.M. and DBT-JRF for providing financial assistance to SSM. A special thanks to Subha Saha and Swati Madhulika for their valuable input.

Appendix A

Glossary

- Alpha diversity—Mean diversity of species in different regions of the same locality.
- Beta diversity—Measures the differences among different populations.
- Bioinformatics—Bioinformatics is the application of computer technology to the understanding and effective use of biological and biomedical data.
- Metagenomics—Study of genomes from mixed microbial communities.
- 16S rRNA—An essential ribosomal RNA required for the synthesis of all prokaryotic proteins, and used for profiling microbial species.
- Amplicon sequencing—A highly targeted approach that enables researchers to analyze genetic variation in specific genomic regions.
- Microbiome—Genome of all the micro-organisms present in all vertebrates.
- Hypervariable regions—hypervariable regions (V1 – V9) that demonstrate considerable sequence diversity among different bacteria in Bacterial 16S ribosomal RNA (rRNA) genes.

- P-value—p-value is the probability of obtaining results at least as extreme as the observed results of a statistical hypothesis test assuming that the null hypothesis is correct.
- System biology—Systems biology is the computational and mathematical analysis and modeling of complex biological systems.
- Metatranscriptomics—Metatranscriptomics is studying the gene expression of microbes within natural environments.
- Metaproteomics—Metaproteomics is the large-scale identification and quantification of all proteins from microbial communities and microbiomes from environmental sources.
- Hyperplasia—An increased number of normal-appearing cells in a tissue, that may or may not lead to cancer.
- Pathobionts—Opportunistic harmful microbes that are otherwise already present in the host.
- Lymphoma—Cancer of the lymphatic system.
- Pathogenicity island—Pathogenicity islands are a class of genomic islands present in pathogenic strains of microbes acquired through horizontal gene transfer and carry genes encoding for various virulence factors.
- Genotoxic effect—Results of a DNA damaging substance and having the ability to cause mutations.
- Inflammation—A localized physical response by the immune system to an injury or infection.
- Apoptosis—A controlled and energy-dependent process of programmed cell death to get rid of unwanted and abnormal cells.
- Neoplastic—An abnormal and uncontrolled growth of cells or tissues that may or may not be cancerous.
- Oncogenic—A property of any substance that has the potential to develop a tumor.
- Episomal—A genetic element in some bacteria possessing the ability to replicate either independently of the host or in association with a chromosome.
- Hepatotrophic—A group of aerobic or anaerobic bacteria that derive energy from organic compounds.
- Fibrosis—The process of healing through the development of fibrous connective tissue in response to any injury or damage.
- Tumorigenesis—A process of formation of a tumor in the body.
- Proliferation—A process of reproduction by cell division that results in an increased number of cells of similar types.

- Differentiation—A process of forming mature cells of different forms and functions from a less specialized cell.
- Neutralism—Neutralism describes the relationship between two species that do interact but do not affect each other.
- Commensalism—Commensalism is a relationship between individuals of two species in which one species obtains food or other benefits from the other without either harming or benefiting the latter.
- Mutualism—Mutualism describes the ecological interaction between two or more species where each species has a net benefit.

Appendix B

Abbreviations

AFB1—aflatoxin B1
AI—artificial intelligence
ALD—alcohol-induced liver disease
ANG2—angiopoietin 2
ANN—artificial neural networks
ASV—amplicon sequence variant
CCA—cholangiocarcinoma
CEACAMs—carcinoembryonic antigen-related cell adhesion molecules
COBRA—constraint-based reconstruction and analysis
COPD—chronic obstructive pulmonary disease
CRC—colorectal cancer
CTMC—continuous timemarkov chains
DEN—diethylnitrosamine
DL—deep learning
EAC—esophageal adenocarcinoma
EC—esophageal cancer
EPS—extracellular polymeric substances
ESCC—esophageal squamous cell carcinoma
ESD—esophageal squamous dysplasia (ESD)
ESMO—European Society of Medical Oncology
FBA—flux balance analysis
FMT—faecal microbiota transplantation
GC—gastric cancer
GEMs—genome-scale metabolic modeling tools
GERD—gastroesophageal reflux disease

GP—gammaproteobacteria
GPU—graphics processing unit
H DEN—diethylnitrosamine
HCC—liver cancer or hepatocellular carcinoma
HNC- head and neck cancer
HNSCC—head and neck squamous cell cancer
HPV—human papilloma virus
HSC—hepatic stellate cells
HSPGs—heparin sulphate proteoglycan
IARC—International Agency for Research on Cancer
ICI—immune-checkpoint inhibitors
KEGG—Kyoto Encyclopedia of Genes and Genomes
k-NN—K-nearest neighbors algorithm
LC—larynx carcinoma
LC—lung cancer
LPS—lipopolysaccharides
LTA—lipoteichoic acid
LTS—long-term survivors
MALT—mucosal-associated lymphoid tissue
MASCC—Multinational Association of Supportive Care in Cancer
MG-RAST—Metagenomic Rapid Annotations using Subsystems Technology (MG-RAST)
mGWAS—microbiome genome-wide association studies
ML—machine learning
MS—mass spectrometry
MSI—microsatellite instability
NAFLD—non-alcoholic fatty liver disease
NCDs—non-communicable diseases
NFM—non-negative matrix factorization
NK—natural killer cells
NMDS—non-metric multidimensional scaling (NMDS)
NMR—nuclear magnetic resonance
NPC—nasopharyngeal carcinoma
NSCLC—non-small cell lung cancer
OCC—oral cavity carcinoma
OPC—oropharyngeal carcinoma
OSCC—oral squamous cell carcinoma
PAI—pathogenicity island
PC—pancreatic cancer

PDAC—pancreatic ductal carcinoma
PGE2—prostaglandin E2
ROS—reactive oxygen species
SASP—senescence-associated secretory phenotype
SASP—senescence-associated secretory phenotype
SBML—systems biology markup language
SCFAs—short-chain fatty acids
SCLC—small cell lung cancer
TAMs—tumor-associated macrophages
TANs—tumor-associated neutrophils
TERT—telomerase reverse transcriptase
TLR4—Toll-like receptor 4
VEGF—vascular endothelial growth factor

References

1. Hooper LV, Littman DR, Macpherson AJ. Interactions between the microbiota and the immune system. *Science*. 2012;336(6086):1268–1273.
2. Arpaia N, et al. Metabolites produced by commensal bacteria promote peripheral regulatory T-cell generation. *Nature*. 2013;504(7480):451–455.
3. Schwabe RF, Jobin C. The microbiome and cancer. *Nat Rev Cancer*. 2013;13 (11):800–812.
4. Dzutsev A, et al. The role of the microbiota in inflammation, carcinogenesis, and cancer therapy. *Eur J Immunol*. 2015;45(1):17–31.
5. Francescone R, Hou V, Grivennikov SI. Microbiome, inflammation, and cancer. *Cancer J*. 2014;20(3):181–189.
6. Hu B, et al. Microbiota-induced activation of epithelial IL-6 signaling links inflammasome-driven inflammation with transmissible cancer. *Proc Natl Acad Sci USA*. 2013;110(24):9862–9867.
7. Moore WE, Moore LH. Intestinal floras of populations that have a high risk of colon cancer. *Appl Environ Microbiol*. 1995;61(9):3202–3207.
8. Rubinstein MR, et al. Fusobacterium nucleatum promotes colorectal cancer by inducing Wnt/beta-catenin modulator Annexin A1. *EMBO Rep*. 2019;20(4).
9. Brennan CA, Garrett WS. Fusobacterium nucleatum—symbiont, opportunist and oncobacterium. *Nat Rev Microbiol*. 2019;17(3):156–166.
10. Gao Z, et al. Microbiota disbiosis is associated with colorectal cancer. *Front Microbiol*. 2015;6:20.
11. Hu Y, Fang JY, Xiao SD. Can the incidence of gastric cancer be reduced in the new century? *J Dig Dis*. 2013;14(1):11–15.
12. Bik EM, et al. Molecular analysis of the bacterial microbiota in the human stomach. *Proc Natl Acad Sci USA*. 2006;103(3):732–737.
13. Doorakkers E, et al. Eradication of Helicobacter pylori and gastric cancer: a systematic review and meta-analysis of cohort studies. *J Natl Cancer Inst*. 2016;108(9).
14. Yong X, et al. Helicobacter pylori virulence factor CagA promotes tumorigenesis of gastric cancer via multiple signaling pathways. *Cell Commun Signal*. 2015;13:30.
15. Sitaraman R. Helicobacter pylori DNA methyltransferases and the epigenetic field effect in cancerization. *Front Microbiol*. 2014;5:115.

16. Sato F, Meltzer SJ. CpG island hypermethylation in progression of esophageal and gastric cancer. *Cancer.* 2006;106(3):483–493.
17. Gunathilake MN, et al. Association between the relative abundance of gastric microbiota and the risk of gastric cancer: a case-control study. *Sci Rep.* 2019;9(1):13589.
18. Zhong M, et al. Candida albicans disorder is associated with gastric carcinogenesis. *Theranostics.* 2021;11(10):4945–4956.
19. Edgren G, et al. A global assessment of the oesophageal adenocarcinoma epidemic. *Gut.* 2013;62(10):1406–1414.
20. Nasrollahzadeh D, et al. Variations of gastric corpus microbiota are associated with early esophageal squamous cell carcinoma and squamous dysplasia. *Sci Rep.* 2015;5:8820.
21. Nardone G, Compare D, Rocco A. A microbiota-centric view of diseases of the upper gastrointestinal tract. *Lancet Gastroenterol Hepatol.* 2017;2(4):298–312.
22. Blackett KL, et al. Oesophageal bacterial biofilm changes in gastro-oesophageal reflux disease, Barrett's and oesophageal carcinoma: association or causality? *Aliment Pharmacol Ther.* 2013;37(11):1084–1092.
23. Zaidi AH, et al. Associations of microbiota and toll-like receptor signaling pathway in esophageal adenocarcinoma. *BMC Cancer.* 2016;16:52.
24. Peters BA, et al. Oral microbiome composition reflects prospective risk for esophageal cancers. *Cancer Res.* 2017;77(23):6777–6787.
25. Shlomai A, de Jong YP, Rice CM. Virus associated malignancies: the role of viral hepatitis in hepatocellular carcinoma. *Semin Cancer Biol.* 2014;26:78–88.
26. Nordenstedt H, White DL, El-Serag HB. The changing pattern of epidemiology in hepatocellular carcinoma. *Dig Liver Dis.* 2010;42(suppl 3):S206–S214.
27. Ohtani N. Microbiome and cancer. *Semin Immunopathol.* 2015;37(1):65–72.
28. Yoshimoto S, et al. Obesity-induced gut microbial metabolite promotes liver cancer through senescence secretome. *Nature.* 2013;499(7456):97–101.
29. Dapito DH, et al. Promotion of hepatocellular carcinoma by the intestinal microbiota and TLR4. *Cancer Cell.* 2012;21(4):504–516.
30. Ren Z, et al. Gut microbiome analysis as a tool towards targeted non-invasive biomarkers for early hepatocellular carcinoma. *Gut.* 2019;68(6):1014–1023.
31. Zhang X, et al. Dietary cholesterol drives fatty liver-associated liver cancer by modulating gut microbiota and metabolites. *Gut.* 2021;70(4):761–774.
32. Fox JG, et al. Gut microbes define liver cancer risk in mice exposed to chemical and viral transgenic hepatocarcinogens. *Gut.* 2010;59(1):88–97.
33. Wang L, Wan YY. The role of gut microbiota in liver disease development and treatment. *Liver Res.* 2019;3(1):3–18.
34. Johnson DH, Schiller JH, Bunn Jr PA. Recent clinical advances in lung cancer management. *J Clin Oncol.* 2014;32(10):973–982.
35. Cait A, et al. Microbiome-driven allergic lung inflammation is ameliorated by short-chain fatty acids. *Mucosal Immunol.* 2018;11(3):785–795.
36. Budden KF, et al. Emerging pathogenic links between microbiota and the gut-lung axis. *Nat Rev Microbiol.* 2017;15(1):55–63.
37. Dong Q, et al. Host-microbiome interaction in lung cancer. *Front Immunol.* 2021;12:679829.
38. Zheng Y, et al. Specific gut microbiome signature predicts the early-stage lung cancer. *Gut Microbes.* 2020;11(4):1030–1042.
39. Chung M, York BR, Michaud DS. Oral health and cancer. *Curr Oral Health Rep.* 2019;6(2):130–137.
40. Sampaio-Maia B, et al. The oral microbiome in health and its implication in oral and systemic diseases. *Adv Appl Microbiol.* 2016;97:171–210.
41. Brenner DR, McLaughlin JR, Hung RJ. Previous lung diseases and lung cancer risk: a systematic review and meta-analysis. *PLoS One.* 2011;6(3):e17479.

42. Zhan P, et al. Chlamydia pneumoniae infection and lung cancer risk: a meta-analysis. *Eur J Cancer.* 2011;47(5):742–747.
43. Bingula R, et al. Characterisation of microbiota in saliva, bronchoalveolar lavage fluid, non-malignant, peritumoural and tumour tissue in non-small cell lung cancer patients: a cross-sectional clinical trial. *Respir Res.* 2020;21(1):129.
44. Hosgood HD, et al. Variation in oral microbiome is associated with future risk of lung cancer among never-smokers. *Thorax.* 2021;76(3):256–263.
45. Yang J, et al. Dysbiosis of the salivary microbiome is associated with non-smoking female lung cancer and correlated with immunocytochemistry markers. *Front Oncol.* 2018;8:520.
46. Cancer Genome Atlas Research Network. Electronic address, a.a.d.h.e. and N. Cancer genome atlas research, integrated genomic characterization of pancreatic ductal adenocarcinoma. *Cancer Cell.* 2017;32(2):185–203 e13.
47. Balachandran VP, et al. Identification of unique neoantigen qualities in long-term survivors of pancreatic cancer. *Nature.* 2017;551(7681):512–516.
48. Geller LT, et al. Potential role of intratumor bacteria in mediating tumor resistance to the chemotherapeutic drug gemcitabine. *Science.* 2017;357(6356):1156–1160.
49. Pushalkar S, et al. The pancreatic cancer microbiome promotes oncogenesis by induction of innate and adaptive immune suppression. *Cancer Discov.* 2018;8 (4):403–416.
50. Riquelme E, et al. Tumor microbiome diversity and composition influence pancreatic cancer outcomes. *Cell.* 2019;178(4):795–806 e12.
51. Huang H, et al. Oncogenic K-Ras requires activation for enhanced activity. *Oncogene.* 2014;33(4):532–535.
52. Daniluk J, et al. An NF-kappaB pathway-mediated positive feedback loop amplifies Ras activity to pathological levels in mice. *J Clin Invest.* 2012;122(4):1519–1528.
53. Li Q, et al. Gut microbiota: its potential roles in pancreatic cancer. *Front Cell Infect Microbiol.* 2020;10, 572492.
54. Maisonneuve P, Amar S, Lowenfels AB. Periodontal disease, edentulism, and pancreatic cancer: a meta-analysis. *Ann Oncol.* 2017;28(5):985–995.
55. Ahn J, Segers S, Hayes RB. Periodontal disease, Porphyromonas gingivalis serum antibody levels and orodigestive cancer mortality. *Carcinogenesis.* 2012;33(5):1055–1058.
56. Arora M, et al. An exploration of shared genetic risk factors between periodontal disease and cancers: a prospective co-twin study. *Am J Epidemiol.* 2010;171(2):253–259.
57. Michaud DS, et al. A prospective study of periodontal disease and pancreatic cancer in US male health professionals. *J Natl Cancer Inst.* 2007;99(2):171–175.
58. Hujoel PP, et al. An exploration of the periodontitis-cancer association. *Ann Epidemiol.* 2003;13(5):312–316.
59. Stolzenberg-Solomon RZ, et al. Tooth loss, pancreatic cancer, and Helicobacter pylori. *Am J Clin Nutr.* 2003;78(1):176–181.
60. Sethi V, et al. The role of the microbiome in immunologic development and its implication for pancreatic cancer immunotherapy. *Gastroenterology.* 2019;156 (7):2097–2115 e2.
61. Castellarin M, et al. Fusobacterium nucleatum infection is prevalent in human colorectal carcinoma. *Genome Res.* 2012;22(2):299–306.
62. Brook I, Frazier EH. Microbiological analysis of pancreatic abscess. *Clin Infect Dis.* 1996;22(2):384–385.
63. Meurman JH. Oral microbiota and cancer. *J Oral Microbiol.* 2010;2.
64. Fan X, et al. Human oral microbiome and prospective risk for pancreatic cancer: a population-based nested case-control study. *Gut.* 2018;67(1):120–127.
65. Torres PJ, et al. Characterization of the salivary microbiome in patients with pancreatic cancer. *PeerJ.* 2015;3:e1373.

66. Farrell JJ, et al. Variations of oral microbiota are associated with pancreatic diseases including pancreatic cancer. *Gut*. 2012;61(4):582–588.
67. Falzone L, et al. Identification of novel MicroRNAs and their diagnostic and prognostic significance in oral cancer. *Cancers (Basel)*. 2019;11(5).
68. Ali J, et al. Genetic etiology of oral cancer. *Oral Oncol*. 2017;70:23–28.
69. Ng JH, et al. Changing epidemiology of oral squamous cell carcinoma of the tongue: a global study. *Head Neck*. 2017;39(2):297–304.
70. Liu M, et al. Effect of smokeless tobacco products on human oral bacteria growth and viability. *Anaerobe*. 2016;42:152–161.
71. Sawant S, et al. Identification & correlation of bacterial diversity in oral cancer and long-term tobacco chewers—a case-control pilot study. *J Med Microbiol*. 2021;70(9).
72. Guerrero-Preston R, et al. 16S rRNA amplicon sequencing identifies microbiota associated with oral cancer, human papilloma virus infection and surgical treatment. *Oncotarget*. 2016;7(32):51320–51334.
73. Gatta G, et al. Prognoses and improvement for head and neck cancers diagnosed in Europe in early 2000s: The EUROCARE-5 population-based study. *Eur J Cancer*. 2015;51(15):2130–2143.
74. Torre LA, et al. Global cancer statistics, 2012. *CA Cancer J Clin*. 2015;65(2):87–108.
75. Chen YP, et al. Nasopharyngeal carcinoma. *Lancet*. 2019;394(10192):64–80.
76. Lydiatt WM, et al. Head and neck cancers-major changes in the American Joint Committee on cancer eighth edition cancer staging manual. *CA Cancer J Clin*. 2017;67(2):122–137.
77. Hayes RB, et al. Association of oral microbiome with risk for incident head and neck squamous cell cancer. *JAMA Oncol*. 2018;4(3):358–365.
78. Mager DL, et al. The salivary microbiota as a diagnostic indicator of oral cancer: a descriptive, non-randomized study of cancer-free and oral squamous cell carcinoma subjects. *J Transl Med*. 2005;3:27.
79. Hooper SJ, et al. A molecular analysis of the bacteria present within oral squamous cell carcinoma. *J Med Microbiol*. 2007;56(pt 12):1651–1659.
80. Pushalkar S, et al. Microbial diversity in saliva of oral squamous cell carcinoma. *FEMS Immunol Med Microbiol*. 2011;61(3):269–277.
81. Pushalkar S, et al. Comparison of oral microbiota in tumor and non-tumor tissues of patients with oral squamous cell carcinoma. *BMC Microbiol*. 2012;12:144.
82. Al-Hebshi NN, et al. Inflammatory bacteriome featuring Fusobacterium nucleatum and Pseudomonas aeruginosa identified in association with oral squamous cell carcinoma. *Sci Rep*. 2017;7(1):1834.
83. Nagy KN, et al. The microflora associated with human oral carcinomas. *Oral Oncol*. 1998;34(4):304–308.
84. Xu Y, et al. Changes of saliva microbiota in nasopharyngeal carcinoma patients under chemoradiation therapy. *Arch Oral Biol*. 2014;59(2):176–186.
85. Wang H, et al. Microbiomic differences in tumor and paired-normal tissue in head and neck squamous cell carcinomas. *Genome Med*. 2017;9(1):14.
86. Gong H, et al. Microbiota in the throat and risk factors for laryngeal carcinoma. *Appl Environ Microbiol*. 2014;80(23):7356–7363.
87. Liu F, et al. Dysbiosis of the gut microbiome is associated with tumor biomarkers in lung cancer. *Int J Biol Sci*. 2019;15(11):2381–2392.
88. Shen XJ, et al. Molecular characterization of mucosal adherent bacteria and associations with colorectal adenomas. *Gut Microbes*. 2010;1(3):138–147.
89. Hale VL, et al. Shifts in the fecal microbiota associated with adenomatous polyps. *Cancer Epidemiol Biomarkers Prev*. 2017;26(1):85–94.
90. Flemer B, et al. Tumour-associated and non-tumour-associated microbiota in colorectal cancer. *Gut*. 2017;66(4):633–643.

91. Ferreira RM, et al. Gastric microbial community profiling reveals a dysbiotic cancer-associated microbiota. *Gut*. 2018;67(2):226–236.
92. Sung JJY, et al. Gastric microbes associated with gastric inflammation, atrophy and intestinal metaplasia 1 year after Helicobacter pylori eradication. *Gut*. 2020;69 (9):1572–1580.
93. Dong Z, et al. Detection of microbial 16S rRNA gene in the serum of patients with gastric cancer. *Front Oncol*. 2019;9:608.
94. Deng Y, et al. Dysbiosis of gut microbiota in patients with esophageal cancer. *Microb Pathog*. 2021;150:104709.
95. Zhang L, et al. Relationship between intestinal microbial dysbiosis and primary liver cancer. *Hepatobiliary Pancreat Dis Int*. 2019;18(2):149–157.
96. Li D, et al. Oral microbial community analysis of the patients in the progression of liver cancer. *Microb Pathog*. 2020;149:104479.
97. de Martel C, et al. Global burden of cancer attributable to infections in 2018: a worldwide incidence analysis. *Lancet Glob Health*. 2020;8(2):e180–e190.
98. Fagoonee S, Pellicano R. Helicobacter pylori: molecular basis for colonization and survival in gastric environment and resistance to antibiotics. A short review. *Infect Dis (Lond)*. 2019;51(6):399–408.
99. Lofgren JL, et al. Lack of commensal flora in Helicobacter pylori-infected INS-GAS mice reduces gastritis and delays intraepithelial neoplasia. *Gastroenterology*. 2011;140 (1):210–220.
100. Sanduleanu S, et al. Non-Helicobacter pylori bacterial flora during acid-suppressive therapy: differential findings in gastric juice and gastric mucosa. *Aliment Pharmacol Ther*. 2001;15(3):379–388.
101. Sanduleanu S, et al. Changes in gastric mucosa and luminal environment during acid-suppressive therapy: a review in depth. *Dig Liver Dis*. 2001;33(8):707–719.
102. Backert S, Tegtmeyer N, Fischer W. Composition, structure and function of the Helicobacter pylori cag pathogenicity island encoded type IV secretion system. *Future Microbiol*. 2015;10(6):955–965.
103. Hamway Y, et al. Cysteine residues in Helicobacter pylori adhesin HopQ are required for CEACAM-HopQ interaction and subsequent CagA translocation. *Microorganisms*. 2020;8(4).
104. Javaheri A, et al. Helicobacter pylori adhesin HopQ engages in a virulence-enhancing interaction with human CEACAMs. *Nat Microbiol*. 2016;2:16189.
105. Selbach M, et al. Host cell interactome of tyrosine-phosphorylated bacterial proteins. *Cell Host Microbe*. 2009;5(4):397–403.
106. Glocker E, et al. Proteins encoded by the cag pathogenicity island of Helicobacter pylori are required for NF-kappaB activation. *Infect Immun*. 1998;66(5):2346–2348.
107. Naumann M, et al. Activation of activator protein 1 and stress response kinases in epithelial cells colonized by Helicobacter pylori encoding the cag pathogenicity island. *J Biol Chem*. 1999;274(44):31655–31662.
108. Chauhan N, et al. Helicobacter pylori VacA, a distinct toxin exerts diverse functionalities in numerous cells: an overview. *Helicobacter*. 2019;24(1):e12544.
109. Kim IJ, Blanke SR. Remodeling the host environment: modulation of the gastric epithelium by the Helicobacter pylori vacuolating toxin (VacA). *Front Cell Infect Microbiol*. 2012;2:37.
110. Gebert B, et al. Helicobacter pylori vacuolating cytotoxin inhibits T lymphocyte activation. *Science*. 2003;301(5636):1099–1102.
111. Ray K. Colorectal cancer: Fusobacterium nucleatum found in colon cancer tissue—could an infection cause colorectal cancer? *Nat Rev Gastroenterol Hepatol*. 2011;8(12):662.

112. Guo P, et al. FadA promotes DNA damage and progression of Fusobacterium nucleatum-induced colorectal cancer through up-regulation of chk2. *J Exp Clin Cancer Res.* 2020;39(1):202.
113. Thurnheer T, et al. Fusobacterium species and subspecies differentially affect the composition and architecture of supra- and subgingival biofilms models. *Front Microbiol.* 2019;10:1716.
114. Liu PF, et al. Vaccination targeting surface FomA of Fusobacterium nucleatum against bacterial co-aggregation: Implication for treatment of periodontal infection and halitosis. *Vaccine.* 2010;28(19):3496–3505.
115. Abed J, et al. Fap2 mediates Fusobacterium nucleatum colorectal adenocarcinoma enrichment by binding to tumor-expressed Gal-GalNAc. *Cell Host Microbe.* 2016;20(2):215–225.
116. Coppenhagen-Glazer S, et al. Fap2 of Fusobacterium nucleatum is a galactose-inhibitable adhesin involved in coaggregation, cell adhesion, and preterm birth. *Infect Immun.* 2015;83(3):1104–1113.
117. Chen Y, et al. Invasive Fusobacterium nucleatum activates beta-catenin signaling in colorectal cancer via a TLR4/P-PAK1 cascade. *Oncotarget.* 2017;8(19):31802–31814.
118. Zhang S, et al. Fusobacterium nucleatum promotes epithelial-mesenchymal transiton through regulation of the lncRNA MIR4435-2HG/miR-296-5p/Akt2/SNAI1 signaling pathway. *FEBS J.* 2020;287(18):4032–4047.
119. Kostic AD, et al. Fusobacterium nucleatum potentiates intestinal tumorigenesis and modulates the tumor-immune microenvironment. *Cell Host Microbe.* 2013;14(2):207–215.
120. Wu J, Li Q, Fu X. Fusobacterium nucleatum contributes to the carcinogenesis of colorectal cancer by inducing inflammation and suppressing host immunity. *Transl Oncol.* 2019;12(6):846–851.
121. Bashir A, et al. Fusobacterium nucleatum, inflammation, and immunity: the fire within human gut. *Tumour Biol.* 2016;37(3):2805–2810.
122. Gabrilovich DI, Ostrand-Rosenberg S, Bronte V. Coordinated regulation of myeloid cells by tumours. *Nat Rev Immunol.* 2012;12(4):253–268.
123. Wang LS, et al. Gene-diet interactions on colorectal cancer risk. *Curr Nutr Rep.* 2012;1:132–141.
124. Jones PA, Baylin SB. The fundamental role of epigenetic events in cancer. *Nat Rev Genet.* 2002;3(6):415–428.
125. Herman JG, et al. Incidence and functional consequences of hMLH1 promoter hypermethylation in colorectal carcinoma. *Proc Natl Acad Sci USA.* 1998;95(12):6870–6875.
126. Hussan H, et al. Fusobacterium's link to colorectal neoplasia sequenced: a systematic review and future insights. *World J Gastroenterol.* 2017;23(48):8626–8650.
127. Parhi L, et al. Breast cancer colonization by Fusobacterium nucleatum accelerates tumor growth and metastatic progression. *Nat Commun.* 2020;11(1):3259.
128. Gur C, et al. Binding of the Fap2 protein of Fusobacterium nucleatum to human inhibitory receptor TIGIT protects tumors from immune cell attack. *Immunity.* 2015;42(2):344–355.
129. Cheng WT, Kantilal HK, Davamani F. The mechanism of bacteroides fragilis toxin contributes to colon cancer formation. *Malays J Med Sci.* 2020;27(4):9–21.
130. Sears CL, Geis AL, Housseau F. Bacteroides fragilis subverts mucosal biology: from symbiont to colon carcinogenesis. *J Clin Invest.* 2014;124(10):4166–4172.
131. Wick EC, et al. Stat3 activation in murine colitis induced by enterotoxigenic Bacteroides fragilis. *Inflamm Bowel Dis.* 2014;20(5):821–834.
132. Buchanan FG, DuBois RN. Connecting COX-2 and Wnt in cancer. *Cancer Cell.* 2006;9(1):6–8.

133. Wang D, et al. Prostaglandin E2 enhances intestinal adenoma growth via activation of the Ras-mitogen-activated protein kinase cascade. *Cancer Res*. 2005;65(5):1822–1829.
134. Rebe C, et al. STAT3 activation: a key factor in tumor immunoescape. *JAKSTAT*. 2013;2(1):e23010.
135. Coleman OI, Nunes T. Role of the microbiota in colorectal cancer: updates on microbial associations and therapeutic implications. *Biores Open Access*. 2016;5(1):279–288.
136. Wang L, et al. IL-17 can promote tumor growth through an IL-6-Stat3 signaling pathway. *J Exp Med*. 2009;206(7):1457–1464.
137. Yang B, et al. The role of interleukin 17 in tumour proliferation, angiogenesis, and metastasis. *Mediators Inflamm*. 2014;2014:623759.
138. Obiso Jr RJ, et al. Proteolytic activity of the Bacteroides fragilis enterotoxin causes fluid secretion and intestinal damage in vivo. *Infect Immun*. 1995;63(10):3820–3826.
139. Wu S, et al. Bacteroides fragilis toxin stimulates intestinal epithelial cell shedding and gamma-secretase-dependent E-cadherin cleavage. *J Cell Sci*. 2007;120(Pt 11):1944–1952.
140. Wu S, et al. Bacteroides fragilis enterotoxin induces c-Myc expression and cellular proliferation. *Gastroenterology*. 2003;124(2):392–400.
141. Tenaillon O, et al. The population genetics of commensal Escherichia coli. *Nat Rev Microbiol*. 2010;8(3):207–217.
142. Bossuet-Greif N, et al. The colibactin genotoxin generates DNA interstrand cross-links in infected cells. *mBio*. 2018;9(2).
143. Xue M, et al. Structure elucidation of colibactin and its DNA cross-links. *Science*. 2019;365(6457).
144. Nougayrede JP, et al. Escherichia coli induces DNA double-strand breaks in eukaryotic cells. *Science*. 2006;313(5788):848–851.
145. Pleguezuelos-Manzano C, et al. Mutational signature in colorectal cancer caused by genotoxic pks(+) E. coli. *Nature*. 2020;580(7802):269–273.
146. Buc E, et al. High prevalence of mucosa-associated E. coli producing cyclomodulin and genotoxin in colon cancer. *PLoS One*. 2013;8(2):e56964.
147. Di Domenico EG, et al. Biofilm producing Salmonella Typhi: chronic colonization and development of gallbladder cancer. *Int J Mol Sci*. 2017;18(9).
148. Shukla R, et al. Roles of Salmonella typhi and Salmonella paratyphi in gallbladder cancer development. *Asian Pac J Cancer Prev*. 2021;22(2):509–516.
149. Sheweita SA, Alsamghan AS. Molecular mechanisms contributing bacterial infections to the incidence of various types of cancer. *Mediators Inflamm*. 2020;2020:4070419.
150. Rupp J, et al. Chlamydia pneumoniae hides inside apoptotic neutrophils to silently infect and propagate in macrophages. *PLoS One*. 2009;4(6):e6020.
151. Xuan C, et al. Microbial dysbiosis is associated with human breast cancer. *PLoS One*. 2014;9(1):e83744.
152. Eslami SZ, et al. Microbiome and breast cancer: new role for an ancient population. *Front Oncol*. 2020;10:120.
153. Chattopadhyay I, Verma M, Panda M. Role of oral microbiome signatures in diagnosis and prognosis of oral cancer. *Technol Cancer Res Treat*. 2019;18:1533033819867354.
154. Vonsky M, et al. Carcinogenesis associated with human papillomavirus infection. *Mech Pot Immunother Biochem (Mosc)*. 2019;84(7):782–799.
155. Mesri EA, Feitelson MA, Munger K. Human viral oncogenesis: a cancer hallmarks analysis. *Cell Host Microbe*. 2014;15(3):266–282.
156. Gaglia MM, Munger K. More than just oncogenes: mechanisms of tumorigenesis by human viruses. *Curr Opin Virol*. 2018;32:48–59.
157. zur Hausen H. Papillomaviruses causing cancer: evasion from host-cell control in early events in carcinogenesis. *J Natl Cancer Inst*. 2000;92(9):690–698.

158. Egawa N, et al. Human papillomaviruses; epithelial tropisms, and the development of neoplasia. *Viruses*. 2015;7(7):3863–3890.
159. Doorbar J. Molecular biology of human papillomavirus infection and cervical cancer. *Clin Sci (Lond)*. 2006;110(5):525–541.
160. Roman A, Munger K. The papillomavirus E7 proteins. *Virology*. 2013;445 (1–2):138–168.
161. Vande Pol SB, Klingelhutz AJ. Papillomavirus E6 oncoproteins. *Virology*. 2013;445 (1-2):115–137.
162. Graham SV. Human papillomavirus: gene expression, regulation and prospects for novel diagnostic methods and antiviral therapies. *Future Microbiol*. 2010;5(10):1493–1506.
163. Wang HK, et al. Robust production and passaging of infectious HPV in squamous epithelium of primary human keratinocytes. *Genes Dev*. 2009;23(2):181–194.
164. Park RB, Androphy EJ. Genetic analysis of high-risk e6 in episomal maintenance of human papillomavirus genomes in primary human keratinocytes. *J Virol*. 2002;76 (22):11359–11364.
165. Songock WK, Kim SM, Bodily JM. The human papillomavirus E7 oncoprotein as a regulator of transcription. *Virus Res*. 2017;231:56–75.
166. Scarth JA, et al. The human papillomavirus oncoproteins: a review of the host pathways targeted on the road to transformation. *J Gen Virol*. 2021;102(3).
167. Westrich JA, Warren CJ, Pyeon D. Evasion of host immune defenses by human papillomavirus. *Virus Res*. 2017;231:21–33.
168. Graham SV. The human papillomavirus replication cycle, and its links to cancer progression: a comprehensive review. *Clin Sci (Lond)*. 2017;131(17):2201–2221.
169. Venook AP, et al. The incidence and epidemiology of hepatocellular carcinoma: a global and regional perspective. *Oncologist*. 2010;15(suppl 4):5–13.
170. Seeger C, Mason WS. Molecular biology of hepatitis B virus infection. *Virology*. 2015;479–480:672–686.
171. Tarocchi M, et al. Molecular mechanism of hepatitis B virus-induced hepatocarcinogenesis. *World J Gastroenterol*. 2014;20(33):11630–11640.
172. Li W, et al. HIVID: an efficient method to detect HBV integration using low coverage sequencing. *Genomics*. 2013;102(4):338–344.
173. Ding D, et al. Recurrent targeted genes of hepatitis B virus in the liver cancer genomes identified by a next-generation sequencing-based approach. *PLoS Genet*. 2012;8(12): e1003065.
174. Sung WK, et al. Genome-wide survey of recurrent HBV integration in hepatocellular carcinoma. *Nat Genet*. 2012;44(7):765–769.
175. Neuveut C, Wei Y, Buendia MA. Mechanisms of HBV-related hepatocarcinogenesis. *J Hepatol*. 2010;52(4):594–604.
176. Bonilla Guerrero R, Roberts LR. The role of hepatitis B virus integrations in the pathogenesis of human hepatocellular carcinoma. *J Hepatol*. 2005;42(5):760–777.
177. Nguyen DH, Ludgate L, Hu J. Hepatitis B virus-cell interactions and pathogenesis. *J Cell Physiol*. 2008;216(2):289–294.
178. Zhang X, Zhang H, Ye L. Effects of hepatitis B virus X protein on the development of liver cancer. *J Lab Clin Med*. 2006;147(2):58–66.
179. Wen Y, et al. Interaction of hepatitis B viral oncoprotein with cellular target HBXIP dysregulates centrosome dynamics and mitotic spindle formation. *J Biol Chem*. 2008;283(5):2793–2803.
180. Sanz-Cameno P, et al. Hepatitis B virus promotes angiopoietin-2 expression in liver tissue: role of HBV x protein. *Am J Pathol*. 2006;169(4):1215–1222.
181. McGivern DR, Lemon SM. Virus-specific mechanisms of carcinogenesis in hepatitis C virus associated liver cancer. *Oncogene*. 2011;30(17):1969–1983.

182. Machida K, et al. Toll-like receptor 4 mediates synergism between alcohol and HCV in hepatic oncogenesis involving stem cell marker Nanog. *Proc Natl Acad Sci USA.* 2009;106(5):1548–1553.
183. Majumder M, et al. Expression of hepatitis C virus non-structural 5A protein in the liver of transgenic mice. *FEBS Lett.* 2003;555(3):528–532.
184. Moriya K, et al. The core protein of hepatitis C virus induces hepatocellular carcinoma in transgenic mice. *Nat Med.* 1998;4(9):1065–1067.
185. Bartenschlager R, Lohmann V, Penin F. The molecular and structural basis of advanced antiviral therapy for hepatitis C virus infection. *Nat Rev Microbiol.* 2013;11(7):482–496.
186. Machida K, et al. Hepatitis C virus causes uncoupling of mitotic checkpoint and chromosomal polyploidy through the Rb pathway. *J Virol.* 2009;83(23):12590–12600.
187. Koike K. Hepatitis C virus contributes to hepatocarcinogenesis by modulating metabolic and intracellular signaling pathways. *J Gastroenterol Hepatol.* 2007;22(Suppl 1): S108–S111.
188. Kao CF, et al. Modulation of p53 transcription regulatory activity and post-translational modification by hepatitis C virus core protein. *Oncogene.* 2004;23(14):2472–2483.
189. Alisi A, et al. Physical and functional interaction between HCV core protein and the different p73 isoforms. *Oncogene.* 2003;22(17):2573–2580.
190. Deng L, et al. NS3 protein of Hepatitis C virus associates with the tumour suppressor p53 and inhibits its function in an NS3 sequence-dependent manner. *J Gen Virol.* 2006;87(Pt 6):1703–1713.
191. Munakata T, et al. Down-regulation of the retinoblastoma tumor suppressor by the hepatitis C virus NS5B RNA-dependent RNA polymerase. *Proc Natl Acad Sci USA.* 2005;102(50):18159–18164.
192. Zhao LJ, et al. Hepatitis C virus E2 protein promotes human hepatoma cell proliferation through the MAPK/ERK signaling pathway via cellular receptors. *Exp Cell Res.* 2005;305(1):23–32.
193. Hayashi J, et al. Hepatitis C virus core protein activates the MAPK/ERK cascade synergistically with tumor promoter TPA, but not with epidermal growth factor or transforming growth factor alpha. *Hepatology.* 2000;32(5):958–961.
194. Choi SH, Hwang SB. Modulation of the transforming growth factor-beta signal transduction pathway by hepatitis C virus nonstructural 5A protein. *J Biol Chem.* 2006;281 (11):7468–7478.
195. Street A, et al. Hepatitis C virus NS5A-mediated activation of phosphoinositide 3-kinase results in stabilization of cellular beta-catenin and stimulation of beta-catenin-responsive transcription. *J Virol.* 2005;79(8):5006–5016.
196. Hui L, et al. Proliferation of human HCC cells and chemically induced mouse liver cancers requires JNK1-dependent p21 downregulation. *J Clin Invest.* 2008;118 (12):3943–3953.
197. Bouvard V, et al. A review of human carcinogens—Part B: biological agents. *Lancet Oncol.* 2009;10(4):321–322.
198. Humans, I.W.G.o.t.E.o.C.R.t. Biological agents. Volume 100 B. A review of human carcinogens. *IARC Monogr Eval Carcinog Risks Hum.* 2012;100(Pt B):1–441.
199. van Tong H, et al. Parasite infection, carcinogenesis and human malignancy. *EBioMed.* 2017;15:12–23.
200. Qiu DC, et al. A matched, case-control study of the association between Schistosoma japonicum and liver and colon cancers, in rural China. *Ann Trop Med Parasitol.* 2005;99 (1):47–52.
201. Zhang R, et al. p53 gene mutations in rectal cancer associated with schistosomiasis japonica in Chinese patients. *Cancer Lett.* 1998;131(2):215–221.
202. Zalata KR, et al. p53, Bcl-2 and C-Myc expressions in colorectal carcinoma associated with schistosomiasis in Egypt. *Cell Oncol.* 2005;27(4):245–253.

203. Cuesta RA, Kaw YT, Duwaji MS. Schistosoma mekongi infection in a leiomyosarcoma of the small bowel: a case report. *Hum Pathol.* 1992;23(4):471–473.
204. Muller MC, van der Werf SD. A young woman from Cameroon with rectal blood loss, intestinal schistosomiasis and rectosigmoid carcinoma. *Ned Tijdschr Geneeskd.* 2008;152(16):951–955.
205. Sripa B, et al. The tumorigenic liver fluke Opisthorchis viverrini—multiple pathways to cancer. *Trends Parasitol.* 2012;28(10):395–407.
206. Smout MJ, et al. A granulin-like growth factor secreted by the carcinogenic liver fluke, Opisthorchis viverrini, promotes proliferation of host cells. *PLoS Pathog.* 2009;5(10): e1000611.
207. Boonmars T, et al. Alterations of gene expression of RB pathway in Opisthorchis viverrini infection-induced cholangiocarcinoma. *Parasitol Res.* 2009;105(5): 1273–1281.
208. Jusakul A, Kongpetch S, Teh BT. Genetics of Opisthorchis viverrini-related cholangiocarcinoma. *Curr Opin Gastroenterol.* 2015;31(3):258–263.
209. Zhang XY, Zhang PY, Aboul-Soud MA. From inflammation to gastric cancer: role of Helicobacter pylori. *Oncol Lett.* 2017;13(2):543–548.
210. Fujimori S, et al. Helicobacter pylori infection increases the risk of colorectal adenoma and adenocarcinoma, especially in women. *J Gastroenterol.* 2005;40(9):887–893.
211. Bulajic M, Panic N, Lohr JM. Helicobacter pylori and pancreatic diseases. *World J Gastrointest Pathophysiol.* 2014;5(4):380–383.
212. Yang Y, et al. Fusobacterium nucleatum increases proliferation of colorectal cancer cells and tumor development in mice by activating toll-like receptor 4 signaling to nuclear factor-kappaB, and up-regulating expression of MicroRNA-21. *Gastroenterology.* 2017;152(4):851–866 e24.
213. Wang S, et al. Fusobacterium nucleatum acts as a pro-carcinogenic bacterium in colorectal cancer: from association to causality. *Front Cell Dev Biol.* 2021;9, 710165.
214. Yamamura K, et al. Human microbiome Fusobacterium nucleatum in esophageal cancer tissue is associated with prognosis. *Clin Cancer Res.* 2016;22(22):5574–5581.
215. Dougherty MW, Jobin C. Shining a light on colibactin biology. *Toxins (Basel).* 2021;13(5).
216. Wernke KM, et al. Structure and bioactivity of colibactin. *Bioorg Med Chem Lett.* 2020;30(15):127280.
217. Maddocks OD, Scanlon KM, Donnenberg MS. An Escherichia coli effector protein promotes host mutation via depletion of DNA mismatch repair proteins. *mBio.* 2013;4(3):e00152-13.
218. Sepe LP, et al. Genotoxic effect of Salmonella paratyphi a infection on human primary gallbladder cells. *mBio.* 2020;11(5).
219. Zhang K, Hornef M, Fulde M. The deadly bite of Salmonella typhi. *EMBO Rep.* 2015;16(8):887–888.
220. Jackson LA, et al. Association of Chlamydia pneumoniae immunoglobulin A seropositivity and risk of lung cancer. *Cancer Epidemiol Biomarkers Prev.* 2000;9 (11):1263–1266.
221. Gao S, et al. Presence of Porphyromonas gingivalis in esophagus and its association with the clinicopathological characteristics and survival in patients with esophageal cancer. *Infect Agent Cancer.* 2016;11:3.
222. Michaud DS, et al. Plasma antibodies to oral bacteria and risk of pancreatic cancer in a large European prospective cohort study. *Gut.* 2013;62(12):1764–1770.
223. Chaturvedi AK, et al. Incidence trends for human papillomavirus-related and -unrelated oral squamous cell carcinomas in the United States. *J Clin Oncol.* 2008;26 (4):612–619.

224. Kreimer AR, et al. Human papillomavirus types in head and neck squamous cell carcinomas worldwide: a systematic review. *Cancer Epidemiol Biomarkers Prev*. 2005;14 (2):467–475.
225. Veldman T, et al. Transcriptional activation of the telomerase hTERT gene by human papillomavirus type 16 E6 oncoprotein. *J Virol*. 2001;75(9):4467–4472.
226. Park IY, et al. Aberrant epigenetic modifications in hepatocarcinogenesis induced by hepatitis B virus X protein. *Gastroenterology*. 2007;132(4):1476–1494.
227. Tanaka N, et al. PPARalpha activation is essential for HCV core protein-induced hepatic steatosis and hepatocellular carcinoma in mice. *J Clin Invest*. 2008;118 (2):683–694.
228. Shair KH, Schnegg CI, Raab-Traub N. EBV latent membrane protein 1 effects on plakoglobin, cell growth, and migration. *Cancer Res*. 2008;68(17):6997–7005.
229. Scholle F, Bendt KM, Raab-Traub N. Epstein-Barr virus LMP2A transforms epithelial cells, inhibits cell differentiation, and activates Akt. *J Virol*. 2000;74(22):10681–10689.
230. Azran I, Schavinsky-Khrapunsky Y, Aboud M. Role of Tax protein in human T-cell leukemia virus type-I leukemogenicity. *Retrovirology*. 2004;1:20.
231. Harhaj EW, et al. Human T cell leukemia virus type I Tax activates CD40 gene expression via the NF-kappa B pathway. *Virology*. 2005;333(1):145–158.
232. Mariggio G, Koch S, Schulz TF. Kaposi sarcoma herpesvirus pathogenesis. *Philos Trans R Soc Lond B Biol Sci*. 2017;372(1732).
233. Cai Q, et al. Molecular biology of Kaposi's sarcoma-associated herpesvirus and related oncogenesis. *Adv Virus Res*. 2010;78:87–142.
234. Botelho MC, et al. Carcinogenic ability of Schistosoma haematobium possibly through oncogenic mutation of KRAS gene. *Adv Cancer Res Treat*. 2013;2013.
235. Santos J, et al. P53 and cancer-associated sialylated glycans are surrogate markers of cancerization of the bladder associated with Schistosoma haematobium infection. *PLoS Negl Trop Dis*. 2014;8(12), e3329.
236. Maeng S, et al. Oxidative stress-mediated mouse liver lesions caused by Clonorchis sinensis infection. *Int J Parasitol*. 2016;46(3):195–204.
237. Thiele I, Heinken A, Fleming RM. A systems biology approach to studying the role of microbes in human health. *Curr Opin Biotechnol*. 2013;24(1):4–12.
238. Suhail Y, et al. Systems biology of cancer metastasis. *Cell Syst*. 2019;9(2):109–127.
239. Yalcin GD, et al. Systems biology and experimental model systems of cancer. *J Pers Med*. 2020;10(4).
240. Levy R, Borenstein E. Metagenomic systems biology and metabolic modeling of the human microbiome: from species composition to community assembly rules. *Gut Microbes*. 2014;5(2):265–270.
241. Hajeebu S, et al. Machine learning as a tool in investigating the possible role of microbiome in development and treatment of cancer. *Cureus*. 2021;13(8):e17415.
242. Cheung H, Yu J. Machine learning on microbiome research in gastrointestinal cancer. *J Gastroenterol Hepatol*. 2021;36(4):817–822.
243. Ji B, Nielsen J. From next-generation sequencing to systematic modeling of the gut microbiome. *Front Genet*. 2015;6:219.
244. Altamirano A, Saa PA, Garrido D. Inferring composition and function of the human gut microbiome in time and space: a review of genome-scale metabolic modelling tools. *Comput Struct Biotechnol J*. 2020;18:3897–3904.
245. Schellenberger J, et al. Quantitative prediction of cellular metabolism with constraint-based models: the COBRA Toolbox v2.0. *Nat Protoc*. 2011;6(9):1290–1307.
246. Heinken A, et al. Systems-level characterization of a host-microbe metabolic symbiosis in the mammalian gut. *Gut Microbes*. 2013;4(1):28–40.
247. Ma J, Chan W, Tilley BC. Continuous time Markov chain approaches for analyzing transtheoretical models of health behavioral change: a case study and comparison of model estimations. *Stat Methods Med Res*. 2018;27(2):593–607.

248. Jean-Quartier C, et al. In silico cancer research towards 3R. *BMC Cancer.* 2018;18 (1):408.
249. Edelman LB, Eddy JA, Price ND. In silico models of cancer. *Wiley Interdiscip Rev Syst Biol Med.* 2010;2(4):438–459.
250. Sarotra P, Medhi B. Use of bacteria in cancer therapy. *Recent Results Cancer Res.* 2016;209:111–121.
251. Conry RM, et al. Talimogene laherparepvec: first in class oncolytic virotherapy. *Hum Vaccin Immunother.* 2018;14(4):839–846.
252. So-Rosillo R, Small EJ. Sipuleucel-T (APC8015) for prostate cancer. *Expert Rev Anticancer Ther.* 2006;6(9):1163–1167.
253. Liau LM, et al. Treatment of a patient by vaccination with autologous dendritic cells pulsed with allogeneic major histocompatibility complex class I-matched tumor peptides. Case report. *Neurosurg Focus.* 2000;9(6):e8.
254. Brossart P. Dendritic cells in vaccination therapies of malignant diseases. *Transfus Apher Sci.* 2002;27(2):183–186.
255. Liau LM, et al. Treatment of intracranial gliomas with bone marrow-derived dendritic cells pulsed with tumor antigens. *J Neurosurg.* 1999;90(6):1115–1124.
256. Sampson JH, et al. An epidermal growth factor receptor variant III-targeted vaccine is safe and immunogenic in patients with glioblastoma multiforme. *Mol Cancer Ther.* 2009;8(10):2773–2779.
257. Speiser DE, et al. In vivo activation of melanoma-specific CD8(+) T cells by endogenous tumor antigen and peptide vaccines. A comparison to virus-specific T cells. *Eur J Immunol.* 2002;32(3):731–741.
258. Leslie M. Microbiome. Microbes aid cancer drugs. *Science.* 2015;350(6261):614–615.
259. Singh RK, et al. Influence of diet on the gut microbiome and implications for human health. *J Transl Med.* 2017;15(1):73.
260. Sivan A, et al. Commensal Bifidobacterium promotes antitumor immunity and facilitates anti-PD-L1 efficacy. *Science.* 2015;350(6264):1084–1089.
261. Gopalakrishnan V, et al. Gut microbiome modulates response to anti-PD-1 immunotherapy in melanoma patients. *Science.* 2018;359(6371):97–103.
262. Sanders ME, et al. Probiotics and prebiotics in intestinal health and disease: from biology to the clinic. *Nat Rev Gastroenterol Hepatol.* 2019;16(10):605–616.
263. Hibberd AA, et al. Intestinal microbiota is altered in patients with colon cancer and modified by probiotic intervention. *BMJ Open Gastroenterol.* 2017;4(1):e000145.
264. Kang KY. Effects of core muscle stability training on the weight distribution and stability of the elderly. *J Phys Ther Sci.* 2015;27(10):3163–3165.
265. Rafter J, et al. Dietary synbiotics reduce cancer risk factors in polypectomized and colon cancer patients. *Am J Clin Nutr.* 2007;85(2):488–496.
266. Wang Y, et al. Fecal microbiota transplantation for refractory immune checkpoint inhibitor-associated colitis. *Nat Med.* 2018;24(12):1804–1808.
267. Juul FE, et al. Fecal microbiota transplantation for primary clostridium difficile infection. *N Engl J Med.* 2018;378(26):2535–2536.
268. Kelly CR, et al. Effect of fecal microbiota transplantation on recurrence in multiply recurrent Clostridium difficile infection: A randomized trial. *Ann Intern Med.* 2016;165(9):609–616.
269. Smits LP, et al. Therapeutic potential of fecal microbiota transplantation. *Gastroenterology.* 2013;145(5):946–953.
270. Bel S, et al. Reprogrammed and transmissible intestinal microbiota confer diminished susceptibility to induced colitis in TMF-/- mice. *Proc Natl Acad Sci USA.* 2014;111 (13):4964–4969.

CHAPTER FIVE

Gut microbiota in gastrointestinal diseases

Gargi Bhattacharjee[a], Khushal Khambhati[a], Nisarg Gohil[a], Priyanka Singh[a], Jigresh Gohil[a], Hiral Gautam[a], Rupesh Maurya[a], Dinh-Toi Chu[b,c], Suresh Ramakrishna[d,e], and Vijai Singh[a,]*

[a]Department of Biosciences, School of Science, Indrashil University, Rajpur, Mehsana, India
[b]Center for Biomedicine and Community Health, International School, Vietnam National University, Hanoi, Vietnam
[c]Faculty of Applied Sciences, International School, Vietnam National University, Hanoi, Vietnam
[d]College of Medicine, Hanyang University, Seoul, South Korea
[e]Graduate School of Biomedical Science and Engineering, Hanyang University, Seoul, South Korea
*Corresponding author: e-mail addresses: vijaisingh15@gmail.com; vijai.singh@indrashiluniversity.edu.in

Contents

Abstract

Gut microbiota is a highly dense population of different kinds of bacteria residing in the gut which co-evolves with the host. It engages in a number of metabolic and immunological activities. Gut microbiota is associated with maintenance of health, and unbalanced microbiota contributes in the development of several diseases. Alteration of beneficial gut microbiota population triggers gastrointestinal diseases including irritable bowel syndrome, inflammatory bowel disease, celiac disease, colorectal cancer, and many others. Gut microbiota can be affected by multiple factors such as diet, stress, genetic variations. In this chapter, we highlight how gut microbiota plays a key role in pathogenesis of gastrointestinal disease.

ISSN 1877-1173
https://doi.org/10.1016/bs.pmbts.2022.06.028

1. Introduction

Gut microbiota constitute the microorganisms residing in the digestive tracts of higher animals. It plays a vital role in ensuring proper functioning of gastrointestinal (GI) tract including conferring resistance to pathogens,[1] colonization, metabolizing dietary and pharmaceutical molecules, maintenance of intestinal epithelium,[2] behavior via gut-brain axis, regulating immune function,[3] and other such functions. The general composition of gut microbiota varies throughout distinct regions of the digestive tract. The highest density of microbiota is observed in the colon wherein resides 300–1000 different species of microbes which reportedly is the largest habitat with highest number of organisms on the earth.[4]

Ninety nine percent of the bacteria residing in the gut belong to 30–40 species only[5] and are predominantly anaerobes. In cecum, the population of aerobic bacteria is comparatively higher than the other parts of the gut.[5] In a study, 2172 species of bacteria were isolated from human samples and classified into 12 different phyla wherein 93.5% belonged to either Proteobacteria, Firmicutes, Actinobacteria, or Bacteroidetes.[6] Imbalance in the composition of microbiota plays a pivotal role in development and progression of gut-associated diseases including irritable bowel syndrome,[7–9] inflammatory bowel diseases,[10,11] celiac disease,[12] colorectal cancer,[13–15] and several others. In this chapter, we highlight how alteration in microbiota leads to disease progression and also provide insights of how homeostasis in microbiota helps to manage GI diseases.

2. Composition and identification of gut microbiota

Human gut microbiome consists of viruses, bacteria, archaea, fungi and protozoans.[16] These are initially colonized in the GI tract, skin surface and airways from the early phase of life. It gradually develops and generates microbial diversity in tandem with the growth of the individual. Every individual has a unique gut microbiota profile which plays a vital role in metabolism, maintenance of structure of gut mucosal barrier, protection from pathogens and immunomodulation.[17] Density of bacterial populations present in the colon is estimated to be around 10^{11} to 10^{12} per mL. Therefore, it is one of the densest habitats of earth.[18] In human gut, only few phyla represents more than 160 species[17,19] and most dominant gut microbial phyla include

Firmicutes, Bacteroidetes, Actinobacteria, Proteobacteria, Fusobacteria, and Verrucomicrobia. Moreover, two of the phyla, namely Firmicutes and Bacteroidetes, covers 90% of gut microbiota. The phylum Firmicutes includes 200 different genera including *Lactobacillus*, *Bacillus*, *Clostridium*, *Enterococcus*, *Ruminicoccus* and many more, of which almost 95% belong to *Clostridium* species. Within the Bacteroidetes, predominant genera are *Bacteroides* and *Prevotella* whereas, Actinobacteria phylum mainly consist of the genus *Bifidobacterium*.[17,20]

There are several underlying challenges in identifying and analyzing gut bacteria considering the unavailability of culture media for growing unculturable bacteria. In the past few years, new technologies have been developed to facilitate the identification of bacteria. Currently, for identification of bacteria, researchers make use of sequencing the ribosomal RNA (16S rRNA) technique, which is a part of the bacterial ribosome and is unique to every species. The 16S rRNA is highly conserved and contains a hypervariable region for species specific signature sequences. It is widely used for the identification and determination of evolutionary relationship as well as diversity analysis study.[21,22] Metagenomics is used for the identification of microbiome diversity and also to uncover a new set of genes or cluster and biosynthetic pathways and open avenues for functional dysbiosis.[13]

Gut bacteria plays a vital role in digestion of food throughout the GI tract whereas commensal bacteria have a role to play in synthesis, extraction, and absorption of nutrients, chemicals and metabolites. It could be lipids, amino acids, bile acids, short chain fatty acids (SCFAs), vitamins or any other such metabolite. Gut microbiota have an important role in immune systems, as in inhibition of colonization of pathogenic bacteria, consuming nutrients and production of bacteriocide such as bacteriosins, commonly known as lantibiotics. Therefore, microbiome protects invasion of pathogenic bacteria by maintaining structural integrity and stability.[17,23] SCFAs have a key role in modulation and alteration of microbiota composition. SCFAs are not only considered as metabolites but also, modulate microbiome population in a dose-dependent manner. It serves to provide better fitness to specific taxa. Prebiotics can help to elevate specific SCFAs for therapeutic applications and help in metabolic flux relating to bacterial fermentative processes. Human gut microbiota is highly dynamic due to physiological conditions, age, lifestyle and diets. Therefore, it has major implications in health and intestinal disorders. Dysbiosis is considered as the alteration of microbiota composition leading to numerous disorders (Fig. 1).[24]

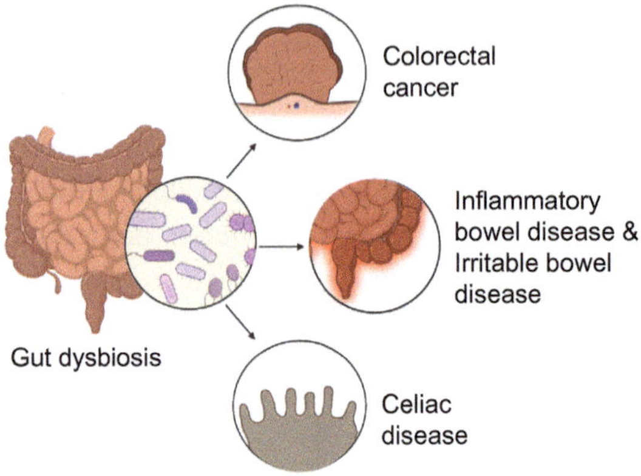

Fig. 1 Different gastrointestinal diseases caused due to gut dysbiosis.

3. Gastrointestinal diseases

3.1 Gut microbiota in Irritable Bowel syndrome

Irritable Bowel Syndrome (IBS) is a common GI disorder that severely upsets the bowel movements either in the form of diarrhoea, constipation or both. Gut microbiota plays a major role in maintenance of gut health and disease conditions. It has been seen that Proteobacteria and Firmicutes are the dominant phylum in IBS patients with a low percentage of Actinobacteria and Bacteroidetes as compared to healthy individuals.[7] Fecal and colonic mucosal biopsy samples taken from IBS patients also display reduction in the number of *Lactobacillus* species in IBS patients against healthy persons.[8] Additionally, the variations in phyla and genera in IBS patients exhibits higher ratio of Firmicutes to Bacteroidetes with lesser Firmicutes (*Lactobacilli*, *Faecalibacterium*) and Actinobacteria (*Bifidobacteria*, *Collinsella*) families. Moreover, there is also an increase in a few specific Firmicutes (*Veillonella*, *Streptococci*, *Ruminococcus* spp.) and Proteobacteria (*Enterobacteriaceae*) families.[25] In addition to this, Lo Presti *et al.*[26] observed that gut microbiota associated with IBS patients exhibit specific changes in distribution and richness of fecal and mucosal associated microbiota. It was found that *Anaerostipes* and *Ruminococcaceae* taxa were abundant in healthy individuals while *Erysipelotrichi* was identified as a potential biomarker for IBS.

Recently, Xie *et al.*[9] investigated the role of probiotics and also low fermentable oligosaccharide, disaccharide, monosaccharide, and polyol

(FODMAP) diet, used for the management of IBS patients. *Lactobacillus* was found suitable and most effective in relieving IBS symptoms while *Bifidobacterium* and *Bacillus* were the potential candidates of the same. However, further investigations are much needed to verify its effectiveness against IBS patients, so as to serve the patients in future with live therapeutics that can help to treat and manage the diseased state to an extent.

3.2 Gut microbiota in inflammatory bowel disease

Inflammatory bowel diseases (IBD) represent a group of intestinal conditions which cause prolonged inflammation in the digestive tract in the form of ulcerative colitis (UC) and Crohn's disease (CRD). IBD arises by the interaction of environment and genetic factors which leads to immunological response and causes inflammation in the intestine.[27] In IBD patients, *Lachnospiraceae* and Bacteroidetes are fewer in number, whereas Proteobacteria are more abundant in IBD patients as compared to healthy individuals.[10] Furthermore, it was also discovered that UC patients have lower number of butyrate producing *Roseburia hominis* and *Faecalibacterium prausnitzii* which reduces microbial diversity.[28]

In CRD patients, *Dialister invisus*, *F. prausnitzii*, and *B. adolescentis* are low in number whereas, *Ruminococcus gnavus* are reportedly high in number.[29] Given the poor composition of *F. prausnitzii* in CRD patients, an oral concoction of live *F. prausnitzii* given to patients have shown to reduce the severity of colitis and help in managing dysbiosis. This helped to understand that probiotics *F. prausnitzii* could be used as an alternative strategy for CRD treatment.[30] Recently, Halaweish *et al.*[31] reported that fecal microbiota transplantation (FMT) using capsule-based fecal microbiota transplantation (cFMT) has proven to be effective for treating *Clostridioides difficile* infection. It helped to restore intestinal microbiota through the oral delivery route and therefore can be applied for numerous other disease treatments.

Dietary supplements plays an important role in preventing IBD. Supplementation of two specific lipids- phosphatidylcholine (PC) and sphingomyelin (SM) to colitis mice was shown to modulate the endogenous metabolism of tryptophan, arginine, proline, purine and regulate bile secretion, digestion and absorption of vitamins. It also could regulate the abundance of certain bacteria in the gut including *Lactobacillus*, *Faecalibacterium*, *Dubosiella*, *Turicibacter*, and *Parasutterella*. It was discovered that PC outperformed SM in terms of preventing colitis in mice. Therefore, dietary supplementation can be an intervention strategy for the prevention of colitis.[32]

Similarly, He *et al.*[33] used dietary supplementation and gut microbiota with the relation to the progression of IBD. It was found that dietary supplements can modulate the occurrence and progression of IBD via several signaling pathways including mitogen-activated protein kinases (MAPKs), mammalian target of rapamycin (mTOR), signal transducer and activator of transcription 3 (STAT3) and nuclear factor kappa B (NF-kappa B). Gut microbiota plays an important role in progression of IBD which in turn can affect the function of enzymes dual oxidase 2 (DUOX2) and apolipoprotein A1 (APOAI), immune function, including the metabolism of SCFAs, bile acids and tryptophan. By dietary supplementation, gut microbiota can modulate SCFAs producing bacteria and *Escherichia coli*. *Faecalibacterium* could be used as a dietary supplement for alleviating IBD. It can be further used as a personalized dietary strategy for prevention of IBD and other related diseases.

3.3 Gut microbiota in celiac disease

Celiac disease (CD), also known as celiac sprue, is a serious autoimmune condition that damages the villi of small intestine and hampers nutrient absorption from foods.[34] Immune reaction elicited by eating gluten (a protein present in wheat, barley and rye), thereby causing celiac disease. It was found that gut microbiota variations play a key role in development of CD that has been linked to inflammation in celiac patients.[12] Dysbiosis can be reduced by using a gluten free diet.[12] Bacterial population of *Bifidobacterium* spp., *B. longum*, *C. histolyticum*, *C. lituseburense*, and *F. prausnitzii* were low in untreated CD patients[35–37] whereas, CD patients showed low concentration of *Lactobacillus* and *Bifidobacterium* and high concentration of pathogenic *Bacteroides* and *E. coli*.[38] Sánchez *et al.*[39] reported *Klebsiella oxytoca*, *Staphylococcus epidermidis*, and *S. pasteuri* to be higher in CD patients, and *Streptococcus anginosus* and *S. mutans* to be lower with respect to healthy individuals.

CD could also be treated with a novel therapy involving gut microbiota. Probiotics have already been used for modulating gut microbiota and thereby treating anti-inflammatory state in CD patients. To further strengthen the proposed study, microbiota formulations could be coupled with clinical trials for treating CD by restoring gut microbiota.[40] Additionally, Olivares *et al.*[41] investigated the effect of breastfeeding in stimulating gut colonization of *Bifidobacterium* spp. *Bifidobacterium* species were less abundant in breast milk of mothers with CD and therefore they can be further supplemented in order to prevent CD in infants by modulating gut microbiota.

Recently, Soheilian-Khorzoghi *et al.*[42] investigated how daily nutritional profiles affect gut microbiota diversity in CD patients against healthy individuals. Stool samples of 31 CD patients (only using gluten-free diet (GFD)) and 20 healthy persons were investigated. They mainly screened gut microbiota such as Bacteroidetes, *Bifidobacterium*, *Clostridium*, *Staphylococcus*, *Enterobacteiaceae*, *Firmicutes*, and *Lactobacillus*, and it was found that proportion of *Bifidobacterium*, *Firmicutes*, and *Lactobacillus* was lower in concentration in CD patients. Also, the consumption of meat and beans was shown to exhibit lesser prevalence of *Lactobacillus* and *Firmicutes*, contrary to which, increased amount of bean consumption revealed low abundance of *Lactobacillus* in CD patients. This study indicates that in some individuals, meat and beans consumption reduces the proportion of beneficial bacteria such as *Firmicutes* and *Lactobacillus* in CD patients. However, dietary supplementation can modulate the gut microbiota for preventing CD.

3.4 Gut microbiota in Colorectal Cancer

Colorectal cancer (CRC) is the third-most common cause of cancer mortality worldwide (500,000 fatalities) which begins in the colon or rectum. Alternatively, it is also known as colon cancer or rectal cancer depending upon its organ specific origin.[43] *B. fragilis*, *Klebsiella*, *Streptococcus*, *Enterococcus*, *Escherichia/Shigella*, and *Peptostreptococcus* are more prevalent against comparatively low population of *Roseburia* and *Lachnospiraceae* (butyrate-producing bacteria). It has been noticed that gut microbiota variations, enriched in *B. vulgatus* and *B. uniformis* while lacking butyrate-producing bacteria, leads to imbalance in CRC patient's microbiota.[13] Similarly, *Proteobacteria*, *Dorea* spp., *Faecalibacterium* spp. are present in larger abundance than Bacteroidetes phyla and *Coprococcus* spp. in CRC patients as compared to healthy individuals.[44] Genomic study hinted the involvement of *Fusobacterium* spp. in promoting colorectal cancer.[45] It was also reported that, in CRC patients, *Fusobacterium* spp. are present in good amount while Bacteroidetes and Firmicutes phyla are present in lesser percentage. Excessive alcohol consumption elevates the risk of CRC, which can severely alter the gut microbiota composition.[14,46] A high percentage of Bacteroidetes and Proteobacteria phyla was observed in people with alcoholism as compared to healthy individuals.[47] It was also observed that, abundance of *Erysipelotrichaceae* (considered to be pro-inflammatory) was higher in chronic alcohol-fed mice as compared to alcohol unfed mice.[48] It was also reported that REG3γ (antimicrobial peptide) expression reduced in alcohol-fed mice, which targeted gram-positive bacteria and allowed *Erysipelotrichaceae* overgrowth.[48]

In mice models, Bai *et al.*[15] investigated how cigarette smoking induces colorectal cancer. When mice exposed to cigarette smoke were compared to mice not exposed to cigarette smoke (control), it was observed that the test mice had an increased tumor occurrence rate and prolific cell proliferation. The gut microbiota of smoke-exposed mice were enriched with *Eggerthella lenta* and depleted of *Parabacteroides distasonis* and *Lactobacillus* spp. It was also reported that metabolites found in colon of smoke-exposed mice had considerably higher amount of bile acids and taurodeoxycholic acid (TDCA), which added to the enhancement of oncogenic MAPK/ERK (mitogen-activated protein kinase/extracellular signal-regulated protein kinase 1/2) signaling (a downstream target of TDCA) and compromised gut barrier function. Indicatively, healthy microbiota is essential for preventing colorectal cancer.

4. Conclusion and future remarks

A symbiotic relationship need to be formed between microbiota and host, imbalance of which could lead to development of several GI diseases. Microbiota can be balanced by changing diet, inclusion of antimicrobials against pathogens, supplementation of probiotics and prebiotics for enrichment of desired species composition for restoring microbiota and thereby controlling disease. More research work is required to investigate gut microbiota to implement them as future therapeutic applications.

Acknowledgment

N.G. and G.B. acknowledge the Indian Council of Medical Research, Government of India for financial assistance as Senior Research Fellowship File No. 5/3/8/63/ITR-F/2020 and File No. 5/3/8/41/ITR-F/2022-ITR, respectively. P.S. gratefully acknowledges the Gujarat State Biotechnology Mission for financial assistance as Junior Research Fellowship (Project ID: 5LY45F). The financial assistance from National Research Foundation of Korea (2021M3A9H3015389) and the Korean Fund for Regenerative Medicine (KFRM) grant funded by the Korea government (the Ministry of Science and ICT, the Ministry of Health & Welfare) (22A0304L1-01) to S.R. is acknowledged.

References

1. Bäumler AJ, Sperandio V. Interactions between the microbiota and pathogenic bacteria in the gut. *Nature*. 2016;535:85–93.
2. Natividad JM, Verdu EF. Modulation of intestinal barrier by intestinal microbiota: Pathological and therapeutic implications. *Pharmacol Res*. 2013;69:42–51.
3. Gensollen T, Iyer SS, Kasper DL, Blumberg RS. How colonization by microbiota in early life shapes the immune system. *Science*. 2016;352:539–544.
4. Guarner F, Malagelada J-R. Gut flora in health and disease. *Lancet*. 2003;361:512–519.
5. Beaugerie L, Petit J-C. Antibiotic-associated diarrhoea. *Baillieres Best Pract Res Clin Gastroenterol*. 2004;18:337–352.
6. Hugon P, Dufour J-C, Colson P, Fournier P-E, Sallah K, Raoult D. A comprehensive repertoire of prokaryotic species identified in human beings. *Lancet Infect Dis*. 2015;15:1211–1219.
7. Krogius-Kurikka L, Lyra A, Malinen E, et al. Microbial community analysis reveals high level phylogenetic alterations in the overall gastrointestinal microbiota of diarrhoea-predominant irritable bowel syndrome sufferers. *BMC Gastroenterol*. 2009;9:1–11.
8. Carroll IM, Chang Y-H, Park J, Sartor RB, Ringel Y. Luminal and mucosal-associated intestinal microbiota in patients with diarrhea-predominant irritable bowel syndrome. *Gut Pathog*. 2010;2:1–9.
9. Xie C-R, Tang B, Shi Y-Z, et al. Low FODMAP diet and probiotics in irritable Bowel syndrome: a systematic review with network meta-analysis. *Front Pharmacol*. 2022;13:853011.
10. Frank DN, Amand ALS, Feldman RA, Boedeker EC, Harpaz N, Pace NR. Molecular-phylogenetic characterization of microbial community imbalances in human inflammatory bowel diseases. *Proc Natl Acad Sci U S A*. 2007;104:13780–13785.
11. Machiels K, Joossens M, Sabino J, et al. A decrease of the butyrate-producing species Roseburia hominis and Faecalibacterium prausnitzii defines dysbiosis in patients with ulcerative colitis. *Gut*. 2014;63:1275–1283.
12. Marasco G, Di Biase AR, Schiumerini R, et al. Gut microbiota and celiac disease. *Dig Dis Sci*. 2016;61:1461–1472.
13. Wang T, Cai G, Qiu Y, et al. Structural segregation of gut microbiota between colorectal cancer patients and healthy volunteers. *ISME J*. 2012;6:320–329.
14. Vassallo G, Mirijello A, Ferrulli A, et al. Alcohol and gut microbiota-the possible role of gut microbiota modulation in the treatment of alcoholic liver disease. *Aliment Pharmacol Ther*. 2015;41:917–927.
15. Bai X, Wei H, Liu W, et al. Cigarette smoke promotes colorectal cancer through modulation of gut microbiota and related metabolites. *Gut*. 2022. https://doi.org/10.1136/gutjnl-2021-325021.
16. Thursby E, Juge N. Introduction to the human gut microbiota. *Biochem J*. 2017;474:1823–1836.
17. Rinninella E, Raoul P, Cintoni M, et al. What is the healthy gut microbiota composition? A changing ecosystem across age, environment, diet, and diseases. *Microorganisms*. 2019;7:14.
18. Ley RE, Turnbaugh PJ, Klein S, Gordon JI. Human gut microbes associated with obesity. *Nature*. 2006;444:1022–1023.
19. Laterza L, Rizzatti G, Gaetani E, Chiusolo P, Gasbarrini A. The gut microbiota and immune system relationship in human graft-versus-host disease. *Mediterr J Hematol Infect Dis*. 2016;8:e2016025.
20. Arumugam M, Raes J, Pelletier E, et al. Enterotypes of the human gut microbiome. *Nature*. 2011;473:174–180.

21. Clarridge III JE. Impact of 16S rRNA gene sequence analysis for identification of bacteria on clinical microbiology and infectious diseases. *Clin Microbiol Rev.* 2004;17:840–862.
22. Gohil N, Panchasara H, Patel S, Singh V. Molecular biology techniques for the identification and genotyping of microorganisms. In: *Microbial Genomics in Sustainable Agroecosystems*. Springer; 2019:203–226.
23. Khosravi A, Mazmanian SK. Disruption of the gut microbiome as a risk factor for microbial infections. *Curr Opin Microbiol.* 2013;16:221–227.
24. Peterson CT, Perez Santiago J, Iablokov SN, Chopra D, Rodionov DA, Peterson SN. Short-chain fatty acids modulate healthy gut microbiota composition and functional potential. *Curr Microbiol.* 2022;79:1–13.
25. Salonen A, de Vos WM, Palva A. Gastrointestinal microbiota in irritable bowel syndrome: present state and perspectives. *Microbiology.* 2010;156:3205–3215.
26. Lo Presti A, Zorzi F, Del Chierico F, et al. Fecal and mucosal microbiota profiling in irritable bowel syndrome and inflammatory bowel disease. *Front Microbiol.* 2019;1655.
27. Baumgart DC, Carding SR. Inflammatory bowel disease: cause and immunobiology. *Lancet.* 2007;369:1627–1640.
28. Hansen CHF, Nielsen DS, Kverka M, et al. Patterns of early gut colonization shape future immune responses of the host. *PLoS One.* 2012;7, e34043.
29. Joossens M, Huys G, Cnockaert M, et al. Dysbiosis of the faecal microbiota in patients with Crohn's disease and their unaffected relatives. *Gut.* 2011;60:631–637.
30. Sokol H, Pigneur B, Watterlot L, et al. Faecalibacterium prausnitzii is an anti-inflammatory commensal bacterium identified by gut microbiota analysis of Crohn disease patients. *PNAS USA.* 2008;105:16731–16736.
31. Halaweish HF, Boatman S, Staley C. Encapsulated fecal microbiota transplantation: development, efficacy, and clinical application. *Front Cell Infect Microbiol.* 2022;271.
32. Li Q, Chen G, Zhu D, et al. Effects of dietary phosphatidylcholine and sphingomyelin on DSS-induced colitis by regulating metabolism and gut microbiota in mice. *J Nutr Biochem.* 2022;105, 109004.
33. He P, Yu L, Tian F, Zhang H, Chen W, Zhai Q. Dietary patterns and gut microbiota: the crucial actors in inflammatory bowel disease. *Adv Nutr.* 2022;, nmac029.
34. Parzanese I, Qehajaj D, Patrinicola F, et al. Celiac disease: from pathophysiology to treatment. *World J Gastroenterol.* 2017;8:27.
35. Collado MC, Donat E, Ribes-Koninckx C, Calabuig M, Sanz Y. Imbalances in faecal and duodenal Bifidobacterium species composition in active and non-active coeliac disease. *BMC Microbiol.* 2008;8:232.
36. Collado MC, Donat E, Ribes-Koninckx C, Calabuig M, Sanz Y. Specific duodenal and faecal bacterial groups associated with paediatric coeliac disease. *J Clin Pathol.* 2009;62:264–269.
37. De Palma G, Nadal I, Medina M, et al. Intestinal dysbiosis and reduced immunoglobulin-coated bacteria associated with coeliac disease in children. *BMC Microbiol.* 2010;10:1–7.
38. Nadal I, Donant E, Ribes-Koninckx C, Calabuig M, Sanz Y. Imbalance in the composition of the duodenal microbiota of children with coeliac disease. *J Med Microbiol.* 2007;56:1669–1674.
39. Sánchez E, Donat E, Ribes-Koninckx C, Fernández-Murga ML, Sanz Y. Duodenal-mucosal bacteria associated with celiac disease in children. *Appl Environ Microbiol.* 2013;79:5472–5479.
40. Chibbar R, Dieleman LA. The gut microbiota in celiac disease and probiotics. *Nutrients.* 2019;11:2375.
41. Olivares M, Albrecht S, De Palma G, et al. Human milk composition differs in healthy mothers and mothers with celiac disease. *Eur J Nutr.* 2015;54:119–128.

42. Soheilian-Khorzoghi M, Rezasoltani S, Moheb-Alian A, et al. Impact of nutritional profile on gut microbiota diversity in patients with celiac disease. *Curr Microbiol.* 2022;79:1–10.
43. Jemal A, Bray F, Center MM, Ferlay J, Ward E, Forman D. Global cancer statistics. *CA Cancer J Clin.* 2011;61:69–90.
44. Shen XJ, Rawls JF, Randall TA, et al. Molecular characterization of mucosal adherent bacteria and associations with colorectal adenomas. *Gut Microbes.* 2010;1:138–147.
45. Kostic AD, Gevers D, Pedamallu CS, et al. Genomic analysis identifies association of Fusobacterium with colorectal carcinoma. *Genome Res.* 2012;22:292–298.
46. Huxley RR, Ansary-Moghaddam A, Clifton P, Czernichow S, Parr CL, Woodward M. The impact of dietary and lifestyle risk factors on risk of colorectal cancer: a quantitative overview of the epidemiological evidence. *Int J Cancer Res.* 2009;125:171–180.
47. Mutlu EA, Gillevet PM, Rangwala H, et al. Colonic microbiome is altered in alcoholism. *Am J Physiol Gastrointest Liver Physiol.* 2012;302:G966–G978.
48. Nagao-Kitamoto H, Kitamoto S, Kuffa P, Kamada N. Pathogenic role of the gut microbiota in gastrointestinal diseases. *Intest Res.* 2016;14:127.

CHAPTER SIX

Gut dysbiosis and metabolic diseases

Purbita Bandopadhyay[a,b] and Dipyaman Ganguly[a,b,*]
[a]IICB-Translational Research Unit of Excellence, CSIR-Indian Institute of Chemical Biology, Kolkata, India
[b]Academy of Scientific and Innovative Research (AcSIR), Ghaziabad, India
*Corresponding author: e-mail address: dipyaman@iicb.res.in

Contents

Abstract

Disruption in the composition of the gut microbial flora, the so-called gut dysbiosis, has been associated with and pathogenically implicated in a number of metabolic disorders such as type 2 diabetes mellitus, non-alcoholic fatty liver disease as well as atherosclerosis. We describe the key finding in this domain in terms of the gut dysbiotic effects of obesogenic diets linked to metabolic dysregulations, the cross-talk between host genetics and gut ecology in preclinical models of metabolic syndrome as well as in humans, effect of circadian rhythm on gut microbiome, the regulation of systemic immunocellular response, participating in the metabolic disorder-associated systemic inflammation, by gut microbiome and metabolites derived from them. Finally, we collate the evidence gathered till date on specific gut dysbiotic features documented in different components of metabolic syndrome in humans. Understanding gut dysbiosis in metabolic syndrome offers new insights into the pathogenesis of the specific clinical contexts as well as provide potential new therapeutic targets that warrant further exploration.

Progress in Molecular Biology and Translational Science, Volume 191
ISSN 1877-1173
https://doi.org/10.1016/bs.pmbts.2022.06.031

Abbreviations

AhR	aryl hydrocarbon receptor
BMI	body mass index
CVD	cardiovascular disease
CYP27A1	cytochrome P450 family 27 subfamily A member 1
CYP7A1	cytochrome P450 family 7 subfamily A member 1
DIO	diet induced obesity
EODF	every other day fasting
FMT	fecal microbial transfer
FXR	farnesoid X receptor
GABA	gamma aminobutyric acid
GLP-1	glucagon like peptide-1
GPCR	G protein coupled receptors
HDL	high density lipids
HFD	high fat diet
LPS	lipopolysaccharides
MetS	metabolic syndrome
NAFLD	nonalcoholic fatty liver disease
NASH	non-alcoholic steatohepatitis
SCFA	short-chain fatty acid
SFB	segmented filamentous bacteria
T2DM	type 2 diabetes mellitus
TMA	trimethylamine
TMAO	trimethylamine-N-oxide
WAT	white adipose tissue
ZO-1	zona occludens-1

1. Introduction

Beneficial gut microbiota has been established to be in a symbiotic relationship with their hosts and several studies have implicated them in the maintenance of metabolic health and a balanced immune system.[1–3] A disruption in the composition of the healthy gut microbiota, the so-called gut dysbiosis, has been associated with and pathogenically implicated in a number of metabolic disorders such as type 2 diabetes mellitus, liver diseases like non-alcoholic fatty liver disease (NAFLD) and non-alcoholic steatohepatitis (NASH) as well as cardiovascular diseases, viz. atherosclerosis.[4–7] Technological advancements in the field of next generation sequencing have led to faster and cost-effective approaches to build a detailed picture of the microbes that colonize the gut and exploring them in different clinical contexts.[8,9]

A number of studies in preclinical models, largely in experimental mice, as well as in humans have implicated high fat diet (HFD) mediated alterations of the gut barrier functions in the induction of endotoxemia due to leakage of bacterial products like lipopolysaccharides (LPS) across gut epithelium into systemic circulation leading to unchecked low-grade inflammation (metaflammation) in different organs that ultimately culminate in the development of metabolic syndrome.[10–12]

Obesity has exponentially increased over the past few years and has reached pandemic status at present times with nearly 650 million people worldwide and over 36.5% of the population in the United States alone suffering from morbid obesity.[13–15] Western diets rich in fat and simple sugars have contributed to increasing body mass index (BMI) which adversely affects health in terms of cardiometabolic disorders.[15] However, just BMI cannot be the only factor to determine the negative impacts of obesity, as ethnicity, regional diversity, dietary intake, age, gender and genetics contribute to increased fat mass or deposition of visceral fat (for the same BMI) that might prove to be equally harmful.[16,17]

Several studies have implicated diet composition in altering the gut microbial communities, for example, germ-free mice were resistant to the development of diet induced obesity (DIO), whereas, fecal microbial transfer (FMT) from obese an individual to germ free mice resulted in the development of obesity and associated glucose intolerance, while that from her non-obese twin did not.[18,19] FMT from lean human individuals to obese individuals, has proven to be beneficial in the amelioration of metabolic syndrome.[20,21]

2. The metabolic syndrome

There have been years of debate on the exact definition of metabolic syndrome (MetS) as it encompasses a vast spectrum of symptoms and associated pathophysiology including expanding waist circumference, elevated levels of triglycerides, lower high-density lipid (HDL) cholesterols or dyslipidemia, hypertension and elevated fasting glucose among others, which ultimately increase the risk of cardiovascular diseases (CVD) and type 2 diabetes (T2DM). Insulin resistance is the pivotal factor that has emerged as the central component from which most of the above comorbidities arise.

A rise in the obesity to epidemic proportions all around the world is of major concern and has highlighted the necessity of paying greater attention to define MetS in order to assess, diagnose and identify therapeutic targets for disease amelioration. Obesity in simple terms may only focuse on weight gain due to increased calorie intake and decreased energy expenditure, however, there are several factors which should be taken into account, such as host genetics, individual's metabolism rates and the composition of their gut microbiota.[22–24]

3. Diet-mediated alteration of gut microbial population regulates obesity and associated metabolic syndrome

Several groups have reported that dietary composition is responsible for temporal changes in the microbial community. A comparative study in humans showed that 5 days of continuous consumption of animal-based diet resulted in an expansion of bile-tolerant bacteria like *Bilophila*, *Bacteroides*, etc. and a reduction of bacteria that digest plant polysaccharides belonging to the Fermicutes phylum like *Roseburia*, *Ruminococcus bromii*, etc. High saturated fats induced activation of type 1 T helper (Th1) cells in a murine preclinical model by altering host bile acid pools due to expansion of sulfur-converting bacteria, *Bilophila wadsworthia*.[25] Other studies have implicated gut dysbiosis caused by increased HFD consumption and decreased fiber intake in the upregulation of metabolic disease associated metabolites like p-cresol, indole, arachidonic acid and LPS and downregulation of beneficial metabolites like SCFAs.[26–28]

Periodic fasting was shown to deplete *Desulfovibrionaceae*, *Hydrogenoanaerobacterium*, etc. as well as regulate the balance of relative abundance of T helper 17 (Th17), Th1 and regulatory T (Treg) cells, among other immunophenotypes, thus reducing MetS associated hypertension, in a clinical trial exploring the immunome and the metabolome in responders as compared to non-responders.[29] In another murine preclinical study involving every other day fasting (EODF) an increased browning of white adipose tissue (WAT) ameliorated weight gain, insulin resistance and steatosis due to modulation of the gut microbe as demonstrated by the nullification of the above phenomena in the absence of the microbial community or replication of abovementioned effects after FMT.[30]

A study by Zou and group showed that inulin, a fermentable fiber, when present abundantly in high fat diet is able to revert gut dysbiosis by upregulating IL-22 production by type 3 innate lymphoid cells (ILC3), enterocyte proliferation and increased antimicrobial peptide Reg3γ expression, which prevents encroachment by harmful gut pathogens and associated low grade inflammation that is responsible for insulin resistance in obesity.[31] In another study, ketogenic diets which are low in carbohydrates but high in fat were also shown to reduce intestinal proinflammatory Th17 cells by depleting *bifidobacteria* through the action of ketone bodies like β-hydroxybutyrate.[32]

In a placebo controlled, randomized, double blind, single center human trial conducted with obese/overweight insulin resistant subjects treated with either placebo or pasteurized *Akkermansia muciniphila* bacteria for 3 months, a significant improvement was observed in terms of increase in insulin sensitivity and reduction in weight gain, plasma cholesterol levels, inflammatory components and liver dysfunction in the *A. muciniphila* recipients as compared to placebo.[33]

4. The triad of host genetics, gut dysbiosis and metabolic derangements

Studies performed in different strains of mice acquired from separate facilities bred under different or similar environmental conditions revealed that correlation of gut microbiota with insulin resistance, weight gain and altered metabolites is largely influenced by genetics.[34–36] 129S6/SvEvTac mice that were prone to obesity and hepatosteatosis became MetS resistant under environment normalization. Antibiotic treatment on the other hand was able to improve MetS symptoms in HFD fed C57Bl/6J mice but not in 129S1(obesity resistant) or 129S6 (obesity prone) mice.[37]

In a genome wide association study by Parks et al. 11 genetic loci were shown to be correlated with obesity traits. They studied more than 100 strains of inbred mice and observed strain specific alterations of gut microbes in response to high fat/high sucrose diet in most of the strains under observation with significant negative correlation of body fat with genera *Akkermansia* (phylum Verrucomicrobia) and positive correlation with the relative abundances of genera *Allobaculum* and *Lactococcus* belonging to *Fermicutes* phylum.[38]

5. Circadian changes in gut microbiota and metabolic syndrome

Other than diet and genetics the environment also plays a prominent role in the development of gut microbial community. One of the major environmental factors is the natural light/dark cycle known as the circadian rhythm.[39,40] The immune system and hormones are regulated by the day/night cycle which can be disrupted by gut microbial disturbances caused due to continuous HFD consumption or antibiotic treatment.[41] Studies by Leone et al. and Thaiss et al. show that circadian clock genes in the brain and liver which ultimately regulate fatty acid metabolism are influenced by microbial composition.[42–44] Altered microbiota also disarrange gut barrier integrity by disrupting rhythmic shifts in gut epithelium thickness and these oscillations increase the risks of T2DM.[43,45] High-throughput screening studies by He et al. identified a novel small molecule compound, nobiletin, found naturally in citrus peels which amplifies the circadian clock by enhancing circadian oscillator ROR nuclear receptor signaling and confers protection against MetS.[46]

6. Changes in the immune landscape in metabolic syndrome

MetS progression may be partially attributed to the inflammatory immune phenotype in the gut of the hosts. Several studies have reported an upregulation of CD4 and CD8 T cell mediated release of IFN-γ, gamma-delta T cell mediated release of IL-17 and downregulation of Treg mediated release of IL-10 in the lamina propria.[47] Diet-induced obesity has been implicated in the increase TNF-α, NF-κB, IL-12p40, IL-1β and TLR-4 in the colon of mice.[48–51] Certain drugs like 5-aminosalicylic acid (5-ASA) have been shown to reduce the inflammatory cell phenotypes in the gut by maintaining gut barrier integrity, increasing PPAR-γ activity which is known to boost Treg cell function, regulating luminal antigen tolerance, reducing VAT metaflammation, all of which contribute to improving insulin sensitivity.[47]

7. Gut microbe derived metabolites in MetS progression

Short chain fatty acids (SCFAs) like acetate, propionate and butyrate are produced by fermentation of complex dietary carbohydrates by anaerobic bacteria in the large intestine.[52,53] They are considered to improve insulin sensitivity, promote gluconeogenesis and maintain gut barrier functions (butyrate mediated).[53–55]

Propionate was shown to induce glucagon like peptide-1 (GLP-1) and peptide YY (PYY) which is a satiety hormone thus reducing food intake and consequent improvement in weight gain, insulin resistance and fat deposition.[56–58] However, under certain experimental conditions increased acetate was considered to be detrimental due to parasympathetic nerve stimulated secretion of ghrelin, increased tissue triglycerides and impairing insulin activity under hyperglycemic states.[59] HFD has been known to cause hypothalamic inflammation to induce leptin resistance and consequent obesity which was shown to be ameliorated on depleting the gut microbiome. This was mediated by increased plasma GLP-1 levels. However, on inhibiting GLP-1 signaling by bocking its receptor GLP-1R the protective effect was lost.[60]

Short chain fatty acids (SCFAs) have controversial reports in terms of their contribution in NAFLD severity. Studies have reported beneficial properties of SCFAs in terms of reduction of pro inflammatory function of T cells.[61] SCFAs signal through G protein coupled receptors (GPCRs) like GPR43, the absence of which results in increased systemic inflammation which are characteristic features of NASH.[62] However, reports suggest greater abundance of SCFAs in the fecal content of NASH patients and concomitant increase in microbiota responsible for SCFA production.[63] These in turn have been associated with weight gain and triglyceride accumulation in the liver as well as an increased Th17 to resting Tregs ratio in the circulation, all of which correlated with an inflammatory phenotype in NASH/NAFLD subjects as compared to healthy controls.[63]

A shift of the community from beneficial to detrimental population in the gut results in the altered levels of metabolites from anti-inflammatory components to endotoxins such as LPS. LPS, identified as one of the major drivers of insulin resistance in various organs, contributes significantly to NASH progression and insulin resistance.[10,12,64,65]

3-(4-hydroxyphenyl) lactate, a gut bacteria derived byproduct of aromatic amino acid metabolism was also found to be increased in NAFLD-associated steatosis.[66] The pathophysiological role of choline metabolism and its downstream products are under investigation as there are anomalous observations in different studies. Choline is an upstream precursor of acetylcholine, which is functionally essential for survival, leaving betaine as one of the byproducts.[67] However, in NASH, there is a switch from betaine to trimethylamine (TMA) and trimethylamine-N-oxide (TMAO) production by altered gut microbiota like *Deferribacteres* and *Tenericutes*, as observed in the gut of mice, as well as increased excretion of choline in the urine, both of which contribute to exacerbation of NASH severity.[67–70]

Bile acid pool maintenance and metabolism is another important factor regulated by gut microbiota that influences NAFLD and T2DM progression.[71,72] Primary bile acids execute various endocrine functions, which are bio-converted to secondary bile acids by gut microbes, which then signal through GPCRs like TGR5 and promote host energy conversion and glucose metabolism.[73–75] Gut dysbiosis characterized by expansion of *Enterobacteriaceae*and decrease of *Lachnospiraceae, Blautia and Ruminococcaceae* reduce the availability of secondary bile acids, that ultimately promote liver cirrhosis.[76,77] Lowering of bile acids is also associated with increased TMAO levels as it inhibits CYP27A1 and CYP7A1, two key enzymes that drive bile acid metabolism.[70,78,79] Secondary bile acids and succinate (a fructose-1,6-bisphosphatase ligand) produced by *Parabacteroides distasonis* was shown to improve obesity, reduced NAFLD symptoms and promoted intestinal gluconeogenesis by stimulating the FXR pathway.[80]

Tryptophan metabolites like tryptamine and indole-3-acetate that serve as aryl hydrocarbon receptor (AhR) ligands are important for maintaining gut homeostasis and microbiota immune system crosstalk. Reduced AhR signaling increases gut permeability, macrophage TNF-α production, hepatocyte lipogenesis, mucosal inflammation, while reducing GLP-1 and IL-22 levels. High fat diet depletes beneficial taxa that are involved in the production of these metabolites. Supplementation with *Lactobacillus reuteri*, a strain known to produce metabolites which are AhR agonists, improved insulin sensitivity and decreased inflammation.[81,82] Thus, gut dysbiosis and associated changes in metabolite levels in circulation and various organs are major contributors of steatosis and liver cirrhosis in NASH and obesity associated insulin resistance (Fig. 1).

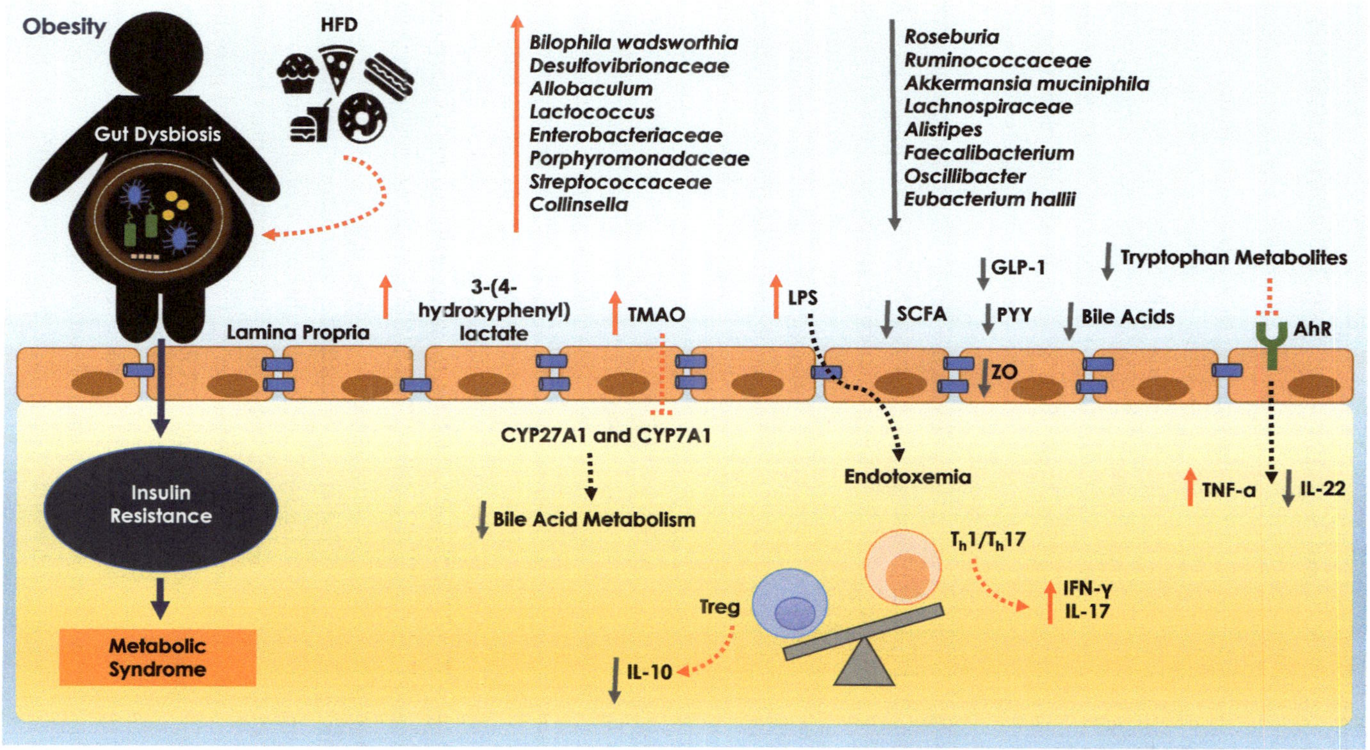

Fig. 1 High fat diet induced gut dysbiosis is characterized by an upregulation of inflammation promoting bacteria and downregulation of beneficial gut microorganisms. This results in altered levels of metabolites which easily pass through the disarranged gut barrier, leading to pro-inflammatory immune cascades that ultimately culminate in the development of insulin resistance and associated metabolic syndrome.

8. Gut microbial signature in type 2 diabetes mellitus

Increasing global burden of T2DM has been a cause of rising concern due to associated systemic inflictions leading to substantial mortality.[83,84] It has been associated with obesity in majority of reported cases, however, non-obese individuals may also be at the risk of developing T2DM.[85] According to the global burden of disease (GBD) data 7079 out of 100,000 individuals will be afflicted with this disease by 2030. Sedentary lifestyle and unhealthy diets contribute greatly in the development of T2DM, characterized by elevated fasting glucose and some degree of insulin resistance among other symptoms.[86,87]

An important factor that plays a role in the development and progression of T2DM is the composition of gut microbiota, however, studies such as this have certain caveats such as the influence of antibiotics and drugs like metformin that are known to significantly alter microbial communities in the gut.[88] In order to understand differences in subjects without the effect of medication, Wu et al. performed cohort studies based on glycemic status in a discovery and a validation cohort comprising Swedish subjects naïve for any T2DM treatment. Gut microbial community was significantly altered in glucose intolerant prediabetic or T2DM groups. There was a decrease in butyrate producing bacteria like *Intestinimonas butyriciproducens*, *Clostridiales bacterium*, *Oscillibacter* spp., *Flavonifractor plautii*, *Alistipes obesi*, *Faecalibacterium* spp., and *Pseudoflavonifractor* spp. as well as genes encoding proteins (atoA/D and 4hbt) and carbohydrates (but and buk) of the butyrate biosynthesis pathway.[89]

Zhou and group collected data of 106 subjects over a span of 4 years who were either healthy or pre-diabetic, and took a multiomics approach to understand the role of gut microbiota in disease onset. They discovered significant differences between host microbe interactions between the two groups and identified several molecules through global coassociation analysis which determine host microbe interaction dissimilarities between insulin sensitive and insulin resistant subjects. They also identified personal molecular markers like IL-1RA and CRP in an individual that preceded clinical T2DM onset.[90]

Studies indicate that the gut microbe reprogramming at a very early stage in development impact the onset of late stage MetS like T2DM. Cox et al. administered low dose penicillin (LDP) to mice at weaning age and observed significant alteration of four microbial groups, namely, *Lactobacillus*, *Allobaculum*, *Candidatus Arthromitus* (SFB) and *Rikenellaceae*which are protective microorganisms that reportedly reduce adiposity. The altered microbiome is involved in immunological or genetic changes at the intestinal and hepatic levels like changes in genes involved in fatty acid metabolism (Cd36, Fabp2 or Pparγ) or altered intestinal Th17 differentiation, all of which contribute to increased adiposity and T2DM at a later stage in life.[91] Other than bacterial community, dsDNA virome was assessed in healthy, obese, or obese school going children with MetS, by a group that performed large scaled sequencing of fecal virus like particles (VLPs), with significant increase in phage diversity and richness in obese groups with or without MetS as compared to the normal weight group.[92]

Garidou et al. have shown how high fat diet disrupts crosstalk between antigen presenting cells (APCs) and T lymphocytes in the small intestine lamina propria by downregulating MHCII, CD86 and ICAM signaling by APCS which reduce IL17 producing Th17/RORγ CD4 T cells in the ileum, promoting insulin resistance and MetS.[93]Switching out casein from high fat high sugar western diet (WD) to whole cell lysates from *Methylococcus capsulatus* Bath improves $FoxP3^+RORγt^+IL\text{-}17^+$ Tregs in the lamina propria of large and small intestine in C57BL/6JRj mice and reverts WD induced gut dysbiosis by expanding *Parabacteroides* and *Parasutterella*, and downregulating *Desulfovibrio,* thus improving weight gain, glucose intolerance and insulin resistance.[94]

Caloric restriction diet has proven to favorably alter the gut microbiota to improve glycemic control and metabolic health. This has been attributed to reduced synthesis of the enzymes that are necessary for Lipid A production, a critical component of LPS in bacteria. Reduction of LPS TLR4 interactions promotes polarization of macrophages to the anti-inflammatory M2 phenotype and development of beige adipose tissue, all of which improve insulin sensitivity and prevent weight gain.[95] Gut microbiota remodeling due to exercise also influences successful weight loss in responders. An expansion of SCFA producing bacteria like *Eubacterium hallii, A. muciniphila* and *Coprococcus* was observed in the responder group.[96]

Certain drugs like metformin have been shown to positively remodel the gut microbiota by upregulating bile acid and SCFA production through expansion of bacteria belonging to Bacteroidetes and Verrucomicrobia phyla and genera *Akkermansia, Escherichia*and *Bacteroides*. An *Eucommia* leaf extract, Asperuloside (ASP), was also shown to alter gut bacteria in mice to improve insulin resistance by upregulating secondary metabolites that reduce the risk of T2DM by increasing the abundance of several beneficial microorganisms like *Akkermansia, Bacteroides, Parabacteroides, Anaerostipes, Roseburia, Coprobacillus* abundance.[97]

9. Gut microbial signature in NAFLD and NASH

Clinical and pathological association between NASH or NAFLD with obesity leads to the exploration of gut microbes and its effect on disease onset and progression.[98–100] Non-alcoholic fatty liver disease has affected more than a quarter of the global population and is considered the leading hepatic manifestation of metabolic syndrome (MetS) due to its close association with dyslipidemia, insulin resistance, type 2 diabetes and cardiovascular disorders.[101–104] NAFLD is defined by deposition of lipid droplets in greater than 5% of the hepatocytes in subjects with no history of alcohol consumption. NAFL is characterized by only steatosis in histological examinations which progresses to NASH when steatosis is accompanied by hepatocyte ballooning, lobular inflammation with or without fibrosis.[105,106]

Association of NAFLD with gut microbiome has been implicated in studies involving FMT from mice with hepatic steatosis to germ free mice, resulting in NAFLD symptom manifestations like increased triglycerides, lipogenesis and fatty acid catabolism in the latter.[107] Henao-Mejia and group showed a critical requirement of NLRP3 and NLRP6 inflammasome and their effector protein IL-18 signaling to prevent progression of NAFL to NASH. $Asc^{-/-}$ mice when subjected to HFD, developed steatosis due to altered gut-liver axis especially characterized by the expansion of members of *Porphyromonadaceae* family which have been previously associated with insulin resistance in humans.[108]

Composition of the gut microbiome determines how the host will respond to HFD as demonstrated in a C57-BL/6 mouse model by an upregulation of Lachnospiraceae bacterium 609 and a *Barnesiella intestinihominis* relative in the responder group that developed hepatic macrovesicular steatosis as compared to non-responders.[109]

On comparing the gut microbial signatures of healthy controls with the NAFLD/NASH subjects with or without fibrosis progression, significant differences were observed at the phylum, family, genera and species levels. Expansion of phylum Verrucomicrobia, Proteobacteria in NAFLD and Fusobacteria in both NAFLD and fibrosis, family Enterobacteriaceae, Streptococcaceae, Pasteurellaceae, Veillonellaceae in both, and genera*Escherichia, Dorea* and *Peptoniphilus*in NAFLD alone, while

Shigella, Bacteroides and *Ruminococcus* in both NAFLD and fibrosis were observed.[109–114] On the other hand, a decrease in Ruminococcaceae and Rikenellaceae family and *Anaerosporobacter, Coprococcus, Eubacterium, Faecalibacterium, Prevotella*, etc. were documented at the genus level.[110,111,113–116] Overall, there was an increase in the proportion of gram-negative bacteria and decrease in gram-positive bacteria in NAFLD and fibrosis progression.[112,115] *Bacteroidesvulgatus,* a species known to be significantly correlated with insulin resistance, along with *Escherichia coli* that has previously been associated with T2DM, has also been found to be overrepresented in advanced stages of fibrosis in NAFLD.[112]

Myrecitin, a naturally occurring flavonol component, was shown to reduce hepatic lipid synthesis and inflammation by expansion of butyric acid producing bacteria, reduction of LPS and upregulation of zonula occludens-1 (ZO-1) which maintains gut barrier integrity, thus preventing LPS mediated endotoxemia.[117]

10. Gut microbial signature in atherosclerosis

Gut microbe has also been implicated in the development of cardiovascular diseases such as atherosclerosis which is characterized by increased systemic cholesterol, macrophage recruitment to inflamed artery walls and the build up of atherosclerotic plaques.[118–120] A gut microbial dysbiosis contributes to inflammatory processes like oxidative stress, altered uric acid and cholesterol metabolism, immune cell activation, etc.[121]

An abundance of *Collinsella* and downregulation of *Roseburia* and *Eubacterium* was observed in atherosclerosis patients compared to healthy controls.[122] Emoto et al. showed that the proportions of matured *Lactobacilli* were higher in patients with coronary heart disease as compared to healthy controls whereas *Prevotella* and *Bifidobacterium* was lowered.[123,124] Elevated TMAO levels have been shown to be positively correlated with the risk of stroke and thrombosis in CVD.[125–127]

Gut microbe immune interplay have been implicated in CVD and associated risks. Increased abundance of inflammatory microbiota in the gut induce release of a number of proinflammatory cytokines like IL-1, IL-17, IL-18 and IL-22 by immune cells as well as induction of anti-microbial peptides and Ig-A which may travel to distal locations in the body and cause immunopathogenesis.[126]

A number of gut microbes such as *Lactobacillus, Enterococcus, Clostridium, Listeria, Eubacterium, Desulfovibrio*, etc. regulate cholesterol metabolism, where microbial production of metabolites like bile acids, TMAO and SCFA are involved in maintaining or disturbing cardiovascular health.[127–129] Probiotics like *Lactobacillus rhamnosus* GR-1and prebiotics like acetic acid and high fiber diets have shown promising results in terms of disease amelioration in mouse models.[130–132] Gut barrier integrity was found to be improved with *Akkermansia muciniphila* supplementation.[133,134] Thus, gut microbes have a prominent role in development and progression of CVDs like atherosclerosis.

11. Promise of fecal microbiota transfer in metabolic diseases

To explore the role of microbiota in the regulation of host phenotypes microbiota maybe directly transferred to donors using different methods like capsules, nasopharyngeal tubes, duodenal/colonic infusions, etc. collectively known as fecal microbiota transfer (FMT).[135–138] Bäckhed et al. were among the first to report successful colonization of germ free C57BL/6 mice with gut microbiome from conventionally raised mice which resulted in 60% increase in body fat and insulin resistance due to increased monosaccharide uptake through gut epithelium and de novo hepatic lipogenesis in just 14 days.[139]

Studies reported allogenic transfer of microbiota from lean donors to MetS recipients altered their gut microbiome, specifically the increase of butyrate producing bacteria like *Roseburia intestinalis* and *E. hallii,* which improved insulin sensitivity and glycemic control within 6 weeks.[21] Another similar study by Kootte, however, showed that this effect is transient, mediated by plasma metabolites like γ-aminobutyric acid (GABA) and reduced microbial diversity of recipients before FMT.[20]

A randomized, double blind, placebo controlled, phase 2 trial in severely obese individuals with MetS showed that daily supplementation of oral FMT with low fermentable (LF) fibers significantly improved HOMA2-IR status at 6 weeks as compared to baseline. This may be attributed to cellulose acting as a binding agent which regulated metabolite concentration in the intestinal lumen and improved intestinal barrier functions.[140] Another randomized, placebo controlled, trial involving FMT from metformin (a well-known drug prescribed to T2DM patients) treated type 2 diabetics to germ free mice showed an upregulation of *Akkermansia muciniphila* associated with improved glucose tolerance in the recipients.[141] FMT, thus, serves as a promising tool in mitigating obesity associated syndromes.[142]

12. Conclusion

Insulin resistance serves as the central focus of almost all of the metabolic disorders and requires extensive attention to reduce the increasing medical burden of obesity and related health risks across the globe. The dysregulated gut microbiota plays a crucial role in mediating obesity associated metaflammation and has vastly unexplored areas that are being discovered as key therapeutic targets by several research groups.[142–146] Supplementation with beneficial microbiota or "probiotics" has emerged as promising treatment options in ameliorating metabolic syndrome in a number of cases and needs further exploration.[147–149] Thus, the extensive research efforts and rapidly accumulating data on alteration of commensal gut microbes in obesity sheds light on the role gut dysbiosis and its contribution to the development of metabolic disorders.

References

1. Lynch SV, Pedersen O. The human intestinal microbiome in health and disease. *N Engl J Med*. 2016;375(24):2369–2379.
2. Kim MH, Yun KE, Kim J, et al. Gut microbiota and metabolic health among overweight and obese individuals. *Sci Rep*. 2020;10:19417.
3. Nogal A, Valdes AM, Menni C. The role of short-chain fatty acids in the interplay between gut microbiota and diet in cardio-metabolic health. *Gut Microbes*. 2021;13 (1):1–24.
4. Fan Y, Pedersen O. Gut microbiota in human metabolic health and disease. *Nat Rev Microbiol*. 2021;19:55–71.
5. Qin J, Li Y, Cai Z, et al. A metagenome-wide association study of gut microbiota in type 2 diabetes. *Nature*. 2012;490:55–60.
6. Leung C, Rivera L, Furness J, et al. The role of the gut microbiota in NAFLD. *Nat Rev Gastroenterol Hepatol*. 2016;13:412–425.
7. Yang T, Santisteban MM, Rodriguez V, et al. Gut dysbiosis is linked to hypertension. *Hypertension*. 2015;65(6):1331–1340.
8. Qin J, Li R, Raes J, et al. A human gut microbial gene catalogue established by metagenomic sequencing. *Nature*. 2010;464(7285):59–65.
9. Li J, Jia H, Cai X, et al. An integrated catalog of reference genes in the human gut microbiome. *Nat Biotechnol*. 2014;32(8):834–841.
10. Cani PD, Amar J, Iglesias MA, et al. Metabolic endotoxemia initiates obesity and insulin resistance. *Diabetes*. 2007;56(7):1761–1772.
11. Kolb H, Mandrup-Poulsen T. The global diabetes epidemic as a consequence of lifestyle-induced low-grade inflammation. *Diabetologia*. 2010;53:10–20.
12. Cani PD, Bibiloni R, Knauf C, et al. Changes in gut microbiota control metabolic endotoxemia-induced inflammation in high-fat diet-induced obesity and diabetes in mice. *Diabetes*. 2008;57(6):1470–1481.
13. The Lancet Gastroenterology Hepatology. Obesity: Another ongoing pandemic. *Lancet Gastroenterol Hepatol*. 2021;6(6):411.
14. Inoue Y, Qin B, Poti J, Sokol R, Gordon-Larsen P. Epidemiology of obesity in adults: Latest trends. *CurrObes Rep*. 2018;7(4):276–288.

15. Martinez KB, Leone V, Chang EB. Western diets, gut dysbiosis, and metabolic diseases: Are they linked? *Gut Microbes*. 2017;8(2):130–142.
16. Després JP. Body fat distribution and risk of cardiovascular disease: An update. *Circulation*. 2012;126(10):1301–1313.
17. Goodpaster BH, Krishnaswami S, Harris TB, et al. Obesity, regional body fat distribution, and the metabolic syndrome in older men and women. *Arch Intern Med*. 2005;165 (7):777–783.
18. Rabot S, Membrez M, Bruneau A, et al. Germ-free C57BL/6J mice are resistant to high-fat-diet-induced insulin resistance and have altered cholesterol metabolism. *FASEB J*. 2010;24(12):4948–4959.
19. Ridaura VK, Faith JJ, Rey FE, et al. Gut microbiota from twins discordant for obesity modulate metabolism in mice. *Science*. 2013;341(6150):1241214.
20. Kootte RS, Levin E, Salojärvi J, et al. Improvement of insulin sensitivity after lean donor feces in metabolic syndrome is driven by baseline intestinal microbiota composition. *Cell Metab*. 2017;26(4):611–619.e6.
21. Vrieze A, Van Nood E, Holleman F, et al. Transfer of intestinal microbiota from lean donors increases insulin sensitivity in individuals with metabolic syndrome. *Gastroenterology*. 2012;143(4):913–916.
22. Alberti KG, Zimmet P, Shaw J. IDF epidemiology task force consensus group. The metabolic syndrome–a new worldwide definition. *Lancet*. 2005;366(9491): 1059–1062.
23. Cornier MA, Dabelea D, Hernandez TL, et al. The metabolic syndrome. *Endocr Rev*. 2008;29(7):777–822.
24. Eckel RH, Alberti KG, Grundy SM, Zimmet PZ. The metabolic syndrome. *Lancet*. 2010;375(9710):181–183.
25. David LA, Maurice CF, Carmody RN, et al. Diet rapidly and reproducibly alters the human gut microbiome. *Nature*. 2014;505(7484):559–563.
26. Wan Y, Wang F, Yuan J, et al. Effects of dietary fat on gut microbiota and faecal metabolites, and their relationship with cardiometabolic risk factors: A 6-month randomised controlled-feeding trial. *Gut*. 2019;68(8):1417–1429.
27. Sonnenburg ED, Smits SA, Tikhonov M, Higginbottom SK, Wingreen NS, Sonnenburg JL. Diet-induced extinctions in the gut microbiota compound over generations. *Nature*. 2016;529(7585):212–215. https://doi.org/10.1038/nature16504.
28. Desai MS, Seekatz AM, Koropatkin NM, et al. A dietary fiber-deprived gut microbiota degrades the colonic mucus barrier and enhances pathogen susceptibility. *Cell*. 2016;167(5):1339–1353.
29. Maifeld A, Bartolomaeus H, Löber U, et al. Fasting alters the gut microbiome reducing blood pressure and body weight in metabolic syndrome patients. *Nat Commun*. 2021;12 (1):1970.
30. Li G, Xie C, Lu S, et al. Intermittent fasting promotes white adipose Browning and Decreases obesity by shaping the gut microbiota. *Cell Metab*. 2017;26(4):672–685.
31. Zou J, Chassaing B, Singh V, et al. Fiber-mediated nourishment of gut microbiota protects against diet-induced obesity by restoring IL-22-mediated colonic health. *Cell Host Microbe*. 2018;23(1):41–53.e4.
32. Ang QY, Alexander M, Newman JC, et al. Ketogenic diets Alter the gut microbiome resulting in decreased intestinal Th17 cells. *Cell*. 2020;181(6):1263–1275.
33. Depommier C, Everard A, Druart C, et al. Supplementation with Akkermansia muciniphila in overweight and obese human volunteers: A proof-of-concept exploratory study. *Nat Med*. 2019;25(7):1096–1103.
34. Sharma M, Li Y, Stoll ML, Tollefsbol TO. The epigenetic connection between the gut microbiome in obesity and diabetes. *Front Genet*. 2020;15(10):1329. https://doi.org/10.3389/fgene.2019.01329 [PMID: 32010189; PMCID: PMC6974692].

35. Kreznar JH, Keller MP, Traeger LL, et al. Host genotype and gut microbiome modulate insulin secretion and diet-induced metabolic phenotypes. *Cell Rep.* 2017;18(7):1739–1750.
36. Fujisaka S, Avila-Pacheco J, Soto M, et al. Diet, genetics, and the gut microbiome drive dynamic changes in plasma metabolites. *Cell Rep.* 2018;22(11):3072–3086.
37. Ussar S, Griffin NW, Bezy O, et al. Interactions between gut microbiota, host genetics and diet modulate the predisposition to obesity and metabolic syndrome. *Cell Metab.* 2015;22(3):516–530.
38. Parks BW, Nam E, Org E, et al. Genetic control of obesity and gut microbiota composition in response to high-fat, high-sucrose diet in mice. *Cell Metab.* 2013;17(1):141–152.
39. Wright Jr KP, McHill AW, Birks BR, Griffin BR, Rusterholz T, Chinoy ED. Entrainment of the human circadian clock to the natural light-dark cycle. *Curr Biol.* 2013;23(16):1554–1558.
40. Mukherji A, Kobiita A, Ye T, Chambon P. Homeostasis in intestinal epithelium is orchestrated by the circadian clock and microbiota cues transduced by TLRs. *Cell.* 2013;153(4):812–827.
41. Zarrinpar A, Chaix A, Yooseph S, Panda S. Diet and feeding pattern affect the diurnal dynamics of the gut microbiome. *Cell Metab.* 2014;20(6):1006–1017.
42. Leone V, Gibbons SM, Martinez K, et al. Effects of diurnal variation of gut microbes and high-fat feeding on host circadian clock function and metabolism. *Cell Host Microbe.* 2015;17(5):681–689.
43. Thaiss CA, Levy M, Korem T, et al. Microbiota diurnal rhythmicity programs host transcriptome oscillations. *Cell.* 2016;167(6):1495–1510.
44. Thaiss CA, Zeevi D, Levy M, et al. Transkingdom control of microbiota diurnal oscillations promotes metabolic homeostasis. *Cell.* 2014;159(3):514–529.
45. Reitmeier S, Kiessling S, Clavel T, et al. Arrhythmic gut microbiome signatures predict risk of type 2 diabetes. *Cell Host Microbe.* 2020;28(2):258–272.
46. He B, Nohara K, Park N, Park YS, Guillory B, Zhao Z. The small molecule nobiletin targets the molecular oscillator to enhance circadian rhythms and protect against metabolic syndrome. *Cell Metab.* 2016;23(4):610–621.
47. Luck H, Tsai S, Chung J, et al. Regulation of obesity-related insulin resistance with gut anti-inflammatory agents. *Cell Metab.* 2015;21(4):527–542.
48. Kim KA, Gu W, Lee IA, Joh EH, Kim DH. High fat diet-induced gut microbiota exacerbates inflammation and obesity in mice via the TLR4 signaling pathway. *PLoS One.* 2012;7(10):e47713.
49. Ding S, Chi MM, Scull BP, et al. High-fat diet: Bacteria interactions promote intestinal inflammation which precedes and correlates with obesity and insulin resistance in mouse. *PLoS One.* 2010;5(8):e12191.
50. Wang N, Wang H, Yao H, et al. Expression and activity of the TLR4/NF-κB signaling pathway in mouse intestine following administration of a short-term high-fat diet. *Exp Ther Med.* 2013;6(3):635–640.
51. Li H, Lelliott C, Håkansson P, et al. Intestinal, adipose, and liver inflammation in diet-induced obese mice. *Metabolism.* 2008;57(12):1704–1710.
52. Sharpton SR, Ajmera V, Loomba R. Emerging role of the gut microbiome in nonalcoholic fatty liver disease: From composition to function. *Clin Gastroenterol Hepatol.* 2019;17(2):296–306.
53. Kolodziejczyk AA, Zheng D, Shibolet O, Elinav E. The role of the microbiome in NAFLD and NASH. *EMBO Mol Med.* 2019;11(2):e9302.
54. den Besten G, Lange K, Havinga R, et al. Gut-derived short-chain fatty acids are vividly assimilated into host carbohydrates and lipids. *Am J PhysiolGastrointest Liver Physiol.* 2013;305(12):G900–G910.

55. Aron-Wisnewsky J, Vigliotti C, Witjes J, et al. Gut microbiota and human NAFLD: Disentangling microbial signatures from metabolic disorders. *Nat Rev Gastroenterol Hepatol.* 2020;17(5):279–297.
56. Tolhurst G, Heffron H, Lam YS, et al. Short-chain fatty acids stimulate glucagon-like peptide-1 secretion via the G-protein-coupled receptor FFAR2. *Diabetes.* 2012;61 (2):364–371.
57. Psichas A, Sleeth ML, Murphy KG, et al. The short chain fatty acid propionate stimulates GLP-1 and PYY secretion via free fatty acid receptor 2 in rodents. *Int J Obes (Lond).* 2015;39(3):424–429.
58. De Vadder F, Kovatcheva-Datchary P, Goncalves D, et al. Microbiota-generated metabolites promote metabolic benefits via gut-brain neural circuits. *Cell.* 2014;156 (1–2):84–96.
59. Perry RJ, Peng L, Barry NA, et al. Acetate mediates a microbiome-brain-β-cell axis to promote metabolic syndrome. *Nature.* 2016;534(7606):213–217.
60. Heiss CN, Mannerås-Holm L, Lee YS, et al. The gut microbiota regulates hypothalamic inflammation and leptin sensitivity in Western diet-fed mice via a GLP-1R-dependent mechanism. *Cell Rep.* 2021;35(8):109163.
61. Sun M, Wu W, Liu Z, Cong Y. Microbiota metabolite short chain fatty acids, GPCR, and inflammatory bowel diseases. *J Gastroenterol.* 2017;52(1):1–8.
62. Maslowski KM, Vieira AT, Ng A, et al. Regulation of inflammatory responses by gut microbiota and chemoattractant receptor GPR43. *Nature.* 2009;461 (7268):1282–1286.
63. Rau M, Schilling AK, Meertens J, et al. Progression from nonalcoholic fatty liver to nonalcoholic steatohepatitis is marked by a higher frequency of Th17 cells in the liver and an increased Th17/resting regulatory T cell ratio in peripheral blood and in the liver. *J Immunol.* 2016;196(1):97–105.
64. Trøseid M, Nestvold TK, Rudi K, Thoresen H, Nielsen EW, Lappegård KT. Plasma lipopolysaccharide is closely associated with glycemic control and abdominal obesity: Evidence from bariatric surgery. *Diabetes Care.* 2013;36(11):3627–3632.
65. Creely SJ, McTernan PG, Kusminski CM, et al. Lipopolysaccharide activates an innate immune system response in human adipose tissue in obesity and type 2 diabetes. *Am J Physiol Endocrinol Metab.* 2007;292(3):E740–E747.
66. Caussy C, Hsu C, Lo MT, et al. Genetics of NAFLD in twins consortium. Link between gut-microbiome derived metabolite and shared gene-effects with hepatic steatosis and fibrosis in NAFLD. *Hepatology.* 2018;68(3):918–932.
67. Dumas ME, Barton RH, Toye A, et al. Metabolic profiling reveals a contribution of gut microbiota to fatty liver phenotype in insulin-resistant mice. *Proc Natl Acad Sci U S A.* 2006;103(33):12511–12516.
68. Wang Z, Klipfell E, Bennett BJ, et al. Gut flora metabolism of phosphatidylcholine promotes cardiovascular disease. *Nature.* 2011;472(7341):57–63.
69. Tang WH, Wang Z, Levison BS, et al. Intestinal microbial metabolism of phosphatidylcholine and cardiovascular risk. *N Engl J Med.* 2013;368(17):1575–1584.
70. Koeth RA, Wang Z, Levison BS, et al. Intestinal microbiota metabolism of L-carnitine, a nutrient in red meat, promotes atherosclerosis. *Nat Med.* 2013;19(5):576–585.
71. Wahlström A, Sayin SI, Marschall HU, Bäckhed F. Intestinal crosstalk between bile acids and microbiota and its impact on host metabolism. *Cell Metab.* 2016;24(1):41–50.
72. Prawitt J, Caron S, Staels B. Bile acid metabolism and the pathogenesis of type 2 diabetes. *Curr Diab Rep.* 2011;11(3):160–166.
73. Ridlon JM, Kang DJ, Hylemon PB, Bajaj JS. Bile acids and the gut microbiome. *CurrOpin Gastroenterol.* 2014;30(3):332–338.
74. Chiang JYL. Bile acid metabolism and signaling in liver disease and therapy. *Liver Res.* 2017;1(1):3–9.

75. Arab JP, Karpen SJ, Dawson PA, Arrese M, Trauner M. Bile acids and nonalcoholic fatty liver disease: Molecular insights and therapeutic perspectives. *Hepatology*. 2017;65(1):350–362.
76. Kakiyama G, Pandak WM, Gillevet PM, Hylemon PB, Heuman DM, Daita K. Modulation of the fecal bile acid profile by gut microbiota in cirrhosis. *J Hepatol*. 2013;58(5):949–955.
77. Chávez-Talavera O, Tailleux A, Lefebvre P, Staels B. Bile acid control of metabolism and inflammation in obesity, type 2 diabetes, dyslipidemia, and nonalcoholic fatty liver disease. *Gastroenterology*. 2017;152(7):1679–1694.
78. Liu H, Hu C, Zhang X, Jia W. Role of gut microbiota, bile acids and their cross-talk in the effects of bariatric surgery on obesity and type 2 diabetes. *J Diabetes Investig*. 2018;9 (1):13–20.
79. Chen YM, Liu Y, Zhou RF, et al. Associations of gut-flora-dependent metabolite trimethylamine-N-oxide, betaine and choline with non-alcoholic fatty liver disease in adults. *Sci Rep*. 2016;8(6):19076.
80. Wang K, Liao M, Zhou N, et al. Parabacteroides distasonis alleviates obesity and metabolic dysfunctions via production of succinate and secondary bile acids. *Cell Rep*. 2019;26(1):222–235.e5.
81. Natividad JM, Agus A, Planchais J, et al. Impaired aryl hydrocarbon receptor ligand production by the gut microbiota is a key factor in metabolic syndrome. *Cell Metab*. 2018;28(5):737–749.
82. Krishnan S, Ding Y, Saedi N, Choi M, Sridharan GV, Sherr DH. Gut microbiota-derived tryptophan metabolites modulate inflammatory response in hepatocytes and macrophages. *Cell Rep*. 2018;23(4):1099–1111.
83. de Souza RJ, Mente A, Maroleanu A, et al. Intake of saturated and trans unsaturated fatty acids and risk of all cause mortality, cardiovascular disease, and type 2 diabetes: Systematic review and meta-analysis of observational studies. *BMJ*. 2015;11(351): h3978.
84. Ford ES. Risks for all-cause mortality, cardiovascular disease, and diabetes associated with the metabolic syndrome: A summary of the evidence. *Diabetes Care*. 2005;28 (7):1769–1778.
85. George AM, Jacob AG, Fogelfeld L. Lean diabetes mellitus: An emerging entity in the era of obesity. *World J Diabetes*. 2015;6(4):613–620.
86. Sardinha LB, Magalhães JP, Santos DA, Júdice PB. Sedentary patterns, physical activity, and cardiorespiratory fitness in association to glycemic control in type 2 diabetes patients. *Front Physiol*. 2017;28(8):262.
87. Schwingshackl L, Hoffmann G, Lampousi AM, et al. Food groups and risk of type 2 diabetes mellitus: A systematic review and meta-analysis of prospective studies. *Eur J Epidemiol*. 2017;32(5):363–375.
88. Thingholm LB, Rühlemann MC, Koch M, Fuqua B, Laucke G. Boehm RObese individuals with and without type 2 diabetes show different gut microbial functional capacity and composition. *Cell Host Microbe*. 2019;26(2):252–264.
89. Wu H, Tremaroli V, Schmidt C, et al. The gut microbiota in prediabetes and diabetes: A population-based cross-sectional study. *Cell Metab*. 2020;32(3):379–390.
90. Zhou W, Sailani MR, Contrepois K, et al. Longitudinal multi-omics of host-microbe dynamics in prediabetes. *Nature*. 2019;569(7758):663–671.
91. Cox LM, Yamanishi S, Sohn J, et al. Altering the intestinal microbiota during a critical developmental window has lasting metabolic consequences. *Cell*. 2014;158 (4):705–721.
92. Bikel S, López-Leal G, Cornejo-Granados F, et al. Gut dsDNA virome shows diversity and richness alterations associated with childhood obesity and metabolic syndrome. *iScience*. 2021;24(8):[102900].

93. Garidou L, Pomié C, Klopp P, et al. The gut microbiota regulates intestinal CD4 T cells expressing RORγt and controls metabolic disease. *Cell Metab.* 2015;22(1):100–112.
94. Jensen BAH, Holm JB, Larsen IS, et al. Lysates of Methylococcuscapsulatus Bath induce a lean-like microbiota, intestinal FoxP3+RORγt+IL-17+ Tregs and improve metabolism. *Nat Commun.* 2021;12(1):1093.
95. Fabbiano S, Suárez-Zamorano N, Chevalier C, et al. Functional gut microbiota remodeling contributes to the caloric restriction-induced metabolic improvements. *Cell Metab.* 2018;28(6):907–921.
96. Liu Y, Wang Y, Ni Y, et al. Gut microbiome fermentation determines the efficacy of exercise for diabetes prevention. *Cell Metab.* 2020;31(1):77–91.
97. Nakamura A, Yokoyama Y, Tanaka K, et al. Asperuloside improves obesity and type 2 diabetes through modulation of gut microbiota and metabolic signaling. *iScience.* 2020;23(9):[101522].
98. Kolodziejczyk AA, Zheng D, Shibolet O, Elinav E. The role of the microbiome in NAFLD and NASH. *EMBO Mol Med.* 2019;11(2):e9302.
99. Canfora EE, Meex RCR, Venema K, Blaak EE. Gut microbial metabolites in obesity, NAFLD and T2DM. *Nat Rev Endocrinol.* 2019;15(5):261–273.
100. Boursier J, Mueller O, Barret M, et al. The severity of nonalcoholic fatty liver disease is associated with gut dysbiosis and shift in the metabolic function of the gut microbiota. *Hepatology.* 2016;63(3):764–775.
101. Chatrath H, Vuppalanchi R, Chalasani N. Dyslipidemia in patients with nonalcoholic fatty liver disease. *Semin Liver Dis.* 2012;32(1):22–29.
102. Gaggini M, Morelli M, Buzzigoli E, DeFronzo RA, Bugianesi E, Gastaldelli A. Non-alcoholic fatty liver disease (NAFLD) and its connection with insulin resistance, dyslipidemia, atherosclerosis and coronary heart disease. *Nutrients.* 2013;5(5):1544–1560.
103. Younossi ZM, Golabi P, de Avila L, et al. The global epidemiology of NAFLD and NASH in patients with type 2 diabetes: A systematic review and meta-analysis. *J Hepatol.* 2019;71(4):793–801.
104. Anstee QM, Targher G, Day CP. Progression of NAFLD to diabetes mellitus, cardiovascular disease or cirrhosis. *Nat Rev Gastroenterol Hepatol.* 2013;10(6):330–344.
105. Younossi ZM, Stepanova M, Rafiq N, et al. Pathologic criteria for nonalcoholic steatohepatitis: Interprotocol agreement and ability to predict liver-related mortality. *Hepatology.* 2011;53(6):1874–1882.
106. Takahashi Y, Fukusato T. Histopathology of nonalcoholic fatty liver disease/nonalcoholic steatohepatitis. *World J Gastroenterol.* 2014;20(42):15539–15548.
107. Le Roy T, Llopis M, Lepage P, et al. Intestinal microbiota determines development of non-alcoholic fatty liver disease in mice. *Gut.* 2013;62(12):1787–1794.
108. Henao-Mejia J, Elinav E, Jin C, et al. Inflammasome-mediated dysbiosis regulates progression of NAFLD and obesity. *Nature.* 2012;482(7384):179–185.
109. Hoyles L, Fernández-Real JM, Federici M, et al. Molecular phenomics and metagenomics of hepatic steatosis in non-diabetic obese women. *Nat Med.* 2018;24(7):1070–1080.
110. Shen F, Zheng RD, Sun XQ, Ding WJ, Wang XY, Fan JG. Gut microbiota dysbiosis in patients with non-alcoholic fatty liver disease. *Hepatobiliary Pancreat Dis Int.* 2017;16(4):375–381.
111. Raman M, Ahmed I, Gillevet PM, et al. Fecal microbiome and volatile organic compound metabolome in obese humans with nonalcoholic fatty liver disease. *Clin Gastroenterol Hepatol.* 2013;11(7):868–875 [e1-3].
112. Loomba R, Seguritan V, Li W, et al. Gut microbiome-based metagenomic signature for non-invasive detection of advanced fibrosis in human nonalcoholic fatty liver disease. *Cell Metab.* 2017;25(5):1054–1062.

113. Zhu L, Baker SS, Gill C, et al. Characterization of gut microbiomes in nonalcoholic steatohepatitis (NASH) patients: A connection between endogenous alcohol and NASH. *Hepatology*. 2013;57(2):601–609.
114. Boursier J, Mueller O, Barret M, et al. The severity of nonalcoholic fatty liver disease is associated with gut dysbiosis and shift in the metabolic function of the gut microbiota. *Hepatology*. 2016;63(3):764–775.
115. Wang B, Jiang X, Cao M, et al. Altered fecal microbiota correlates with liver biochemistry in nonobese patients with non-alcoholic fatty liver disease. *Sci Rep*. 2016;23(6):32002. https://doi.org/10.1038/srep32002 [PMID: 27550547; PMCID: PMC4994089].
116. Del Chierico F, Nobili V, Vernocchi P, et al. Gut microbiota profiling of pediatricnonalcoholic fatty liver disease and obese patients unveiled by an integrated meta-omics-based approach. *Hepatology*. 2017;65(2):451–464.
117. Sun WL, Li XY, Dou HY, et al. Myricetin supplementation decreases hepatic lipid synthesis and inflammation by modulating gut microbiota. *Cell Rep*. 2021;36 (9):109641.
118. Gui T, Shimokado A, Sun Y, Akasaka T, Muragaki Y. Diverse roles of macrophages in atherosclerosis: From inflammatory biology to biomarker discovery. *Mediators Inflamm*. 2012;2012:693083.
119. Jie Z, Xia H, Zhong SL, et al. The gut microbiome in atherosclerotic cardiovascular disease. *Nat Commun*. 2017;8(1):845.
120. Ma J, Li H. The role of gut microbiota in atherosclerosis and hypertension. *Front Pharmacol*. 2018;25(9):1082.
121. Liu H, Zhuang J, Tang P, Li J, Xiong X, Deng H. The role of the gut microbiota in coronary heart disease. *CurrAtheroscler Rep*. 2020;22(12):77.
122. Karlsson FH, Fåk F, Nookaew I, et al. Symptomatic atherosclerosis is associated with an altered gut metagenome. *Nat Commun*. 2012;3:1245.
123. Emoto T, Yamashita T, Kobayashi T, et al. Characterization of gut microbiota profiles in coronary artery disease patients using data mining analysis of terminal restriction fragment length polymorphism: Gut microbiota could be a diagnostic marker of coronary artery disease. *Heart Vessels*. 2017;32(1):39–46.
124. Emoto T, Yamashita T, Sasaki N, et al. Analysis of gut microbiota in coronary artery disease patients: A possible link between gut microbiota and coronary artery disease. *J AtherosclerThromb*. 2016;23(8):908–921.
125. Astudillo AA, Mayrovitz HN. The gut microbiome and cardiovascular disease. *Cureus*. 2021;13(4):e14519.
126. Yeh CF, Chen YH, Liu SF, et al. Mutual interplay of host immune system and gut microbiota in the immunopathology of atherosclerosis. *Int J Mol Sci*. 2020;21 (22):8729.
127. Vourakis M, Mayer G, Rousseau G. The role of gut microbiota on cholesterol metabolism in atherosclerosis. *Int J Mol Sci*. 2021;22(15):8074.
128. Xu J, Yang Y. Gut microbiome and its meta-omics perspectives: Profound implications for cardiovascular diseases. *Gut Microbes*. 2021;13(1):[1936379].
129. Pieczynska MD, Yang Y, Petrykowski S, Horbanczuk OK, Atanasov AG, Horbanczuk JO. Gut microbiota and its metabolites in atherosclerosis development. *Molecules*. 2020;25(3):594.
130. Gan XT, Ettinger G, Huang CX, et al. Probiotic administration attenuates myocardial hypertrophy and heart failure after myocardial infarction in the rat. *Circ Heart Fail*. 2014;7(3):491–499.
131. Marques FZ, Nelson E, Chu PY, et al. High-fiber diet and acetate supplementation change the gut microbiota and prevent the development of hypertension and heart failure in hypertensive mice. *Circulation*. 2017;135(10):964–977.

132. Witkowski M, Weeks TL, Hazen SL. Gut microbiota and cardiovascular disease. *Circ Res*. 2020;127(4):553–570.
133. Li J, Lin S, Vanhoutte PM, Woo CW, Xu A. Akkermansia Muciniphila protects against atherosclerosis by preventing metabolic endotoxemia-induced inflammation in Apoe−/− mice. *Circulation*. 2016;133(24):2434–2446.
134. Katiraei S, de Vries MR, Costain AH, et al. Akkermansia muciniphila exerts lipid-lowering and immunomodulatory effects without affecting neointima formation in hyperlipidemic APOE*3-Leiden.CETP mice. *Mol Nutr Food Res*. 2020;64(15): e1900732.
135. Youngster I, Sauk J, Pindar C, et al. Fecal microbiota transplant for relapsing Clostridium difficile infection using a frozen inoculum from unrelated donors: A randomized, open-label, controlled pilot study. *Clin Infect Dis*. 2014;58 (11):1515–1522.
136. Postigo R, Kim JH. Colonoscopic versus nasogastric fecal transplantation for the treatment of Clostridium difficile infection: A review and pooled analysis. *Infection*. 2012;40 (6):643–648.
137. Allegretti JR, Korzenik JR, Hamilton MJ. Fecal microbiota transplantation via colonoscopy for recurrent C. difficile infection. *J Vis Exp*. 2014;94:52154.
138. Brandt LJ, Aroniadis OC. An overview of fecal microbiota transplantation: Techniques, indications, and outcomes. *GastrointestEndosc*. 2013;78(2):240–249.
139. Bäckhed F, Ding H, Wang T, et al. The gut microbiota as an environmental factor that regulates fat storage. *Proc Natl Acad Sci U S A*. 2004;101(44):15718–15723.
140. Mocanu V, Zhang Z, Deehan EC, et al. Fecal microbial transplantation and fiber supplementation in patients with severe obesity and metabolic syndrome: A randomized double-blind, placebo-controlled phase 2 trial. *Nat Med*. 2021;27(7):1272–1279.
141. Wu H, Esteve E, Tremaroli V, et al. Metformin alters the gut microbiome of individuals with treatment-naive type 2 diabetes, contributing to the therapeutic effects of the drug. *Nat Med*. 2017;23(7):850–858.
142. Dabke K, Hendrick G, Devkota S. The gut microbiome and metabolic syndrome. *J Clin Invest*. 2019;129(10):4050–4057.
143. Le Chatelier E, Nielsen T, Qin J, et al. Richness of human gut microbiome correlates with metabolic markers. *Nature*. 2013;500(7464):541–546.
144. Godoy-Matos AF, Silva Júnior WS, Valerio CM. NAFLD as a continuum: From obesity to metabolic syndrome and diabetes. *DiabetolMetabSyndr*. 2020;12:60.
145. Karlsson FH, Tremaroli V, Nookaew I, et al. Gut metagenome in European women with normal, impaired and diabetic glucose control. *Nature*. 2013;498(7452):99–103.
146. Tilg H, Zmora N, Adolph TE, Elinav E. The intestinal microbiota fuelling metabolic inflammation. *Nat Rev Immunol*. 2020;20(1):40–54.
147. Scheithauer TPM, Rampanelli E, Nieuwdorp M, et al. Gut microbiota as a trigger for metabolic inflammation in obesity and type 2 diabetes. *Front Immunol*. 2020;16 (11):571731.
148. Plovier H, Everard A, Druart C, et al. A purified membrane protein from Akkermansia muciniphila or the pasteurized bacterium improves metabolism in obese and diabetic mice. *Nat Med*. 2017;23(1):107–113.
149. Dewulf EM, Cani PD, Claus SP, et al. Insight into the prebiotic concept: Lessons from an exploratory, double blind intervention study with inulin-type fructans in obese women. *Gut*. 2013;62(8):1112–1121.

CHAPTER SEVEN

Gut microbiome and type 2 diabetes

Sweta Patel, Sriram Seshadri, and Sarat Dalai*
Institute of Science, Nirma University, Ahmedabad, Gujarat, India
*Corresponding author: e-mail address: sarat.dalai@nirmauni.ac.in

Contents

Abstract

Dietary patterns with excess caloric have shaped a complex metabolic disorders like type 2 diabetes (T2D). T2D involves complications in the metabolism of glucose, lipid, cholesterol and their storage. Along with the metabolic dysregulation, systemic inflammation is also the reason for Insulin Resistance and T2D. The importance of gut microbiota has recently been highlighted. It establishes a link between dietary patterns and the types of bacteria that overgrow and modify fermentation bi-products such as SCFA, secondary bile acids, and mucosal immune cells. These changes have a direct impact on the liver's metabolism and immune system. As a result, using Pre-Pro-biotics to manage microbiota can assist overcome or lessening disease symptoms. Antibiotics are currently employed to produce a germ-free environment or to eradicate specific types of bacteria in order to better understand the role of microflora. This chapter covers the basics of good bacteria, as well as the mechanisms that they work on.

Progress in Molecular Biology and Translational Science, Volume 191
ISSN 1877-1173
https://doi.org/10.1016/bs.pmbts.2022.06.029

Abbreviations

BSH bile salt hydrolase
CA cholic acid
CDCA chenodeoxycholic acid
DA deoxycholic acid
FMT fecal microbiota transplantation
HSD high sugar diet
IR insulin resistance
NDCs non-digestible carbohydrates
NLR NOD-like receptors
SCFA short chain fatty acid
T2D type 2 diabetes
TLR Toll-like receptors

1. Introduction

1.1 Diabetes

Diabetes is now ranked 9th among the leading causes of death by WHO as per the 2019 report. Overweight and obesity are also driving forces of T2D which is not always correlated.[1] The main cause of diabetic induction is attributed to a change in eating habits toward junk food, low quality fat and carbohydrates, etc. Thus leading to overall high caloric intake and a sedentary lifestyle. An individual with daily consumption of sugar-sweetened beverages increases 26% risk of getting diabetic. Additionally, sedentary behavior increases risk by 14%.[1] There is a direct correlation between the western diet and T2D. Western diet and excess calorie intake ameliorate adipogenicity and hepatic steatosis. Systemic inflammation is also studied as major pathogenesis of T2D.[2] Systemic inflammation is caused by gut microflora under influence of high sugar and fat diet.[3,4] T2D is characterized by insulin receptor insensitivity toward glucose, i.e., Insulin Resistance (IR) in the liver, muscles and adipose tissues. The cells could not recognize insulin present in the blood and therefore impaired glucose utilization by cells.[5]

1.2 Pathophysiology

Until the role of microflora was discovered in diabetes, the cause of T2D was only excessive energy storage and obesity, for other than hereditary reason. Glucose and fatty acids are major fuels for the cells. Utilization/metabolism of both orchestrated by cells to balance the need of cell and plasma glucose concentration. During the fasting state, cells utilize energy from stored fat.

The free fatty acid increase in serum inhibits glucose utilization by cells, which is explained by the Randle cycle.[6] Basal serum glucose level is maintained at 80–120 mg/dL. Increment in plasma glucose will stimulate insulin secretion which will suppress endogenous production and stimulate its uptake by muscles and liver.[7] This metabolic activity orchestrates continually to maintain fasting glucose or the basal glucose level. A diabetic state is determined when the fasting glucose level (2 h after meal) is higher (more than 150 mg/dL). Muscle, as well as liver cells, do not respond to insulin to reduce serum glucose levels, resulting in IR and glucose tolerance. Additionally, indigenous hepatic glucose production system ameliorate the conditions.[8] Gut microflora is also able to stimulate lipid metabolism in adipose tissues. A high-calorie diet raises fasting glucose levels and encourages the proliferation of Gram-negative bacteria in the gut.[9] As a result, the involvement of gut microbiota in diabetes has been highlighted. The role of microflora-induced systemic inflammation in T2D pathogenesis has evolved.

1.3 Systemic inflammation

Inflammation and activation of the immune system are due to changes in gut microflora-induced systemic inflammation. Infiltration of bacterial products like LPS, and peptidoglycan from the gut to the liver leads to a progression of the disease. These bi-products enter the liver through the hepatic portal vein. The activation of Toll-like receptors (TLR) and NOD-like receptors (NLR) by the translocated bacterial products increase liver inflammation. The binding of LPS to TLR 4 triggers a pro-inflammatory response and disrupts the insulin receptor activation pathway, resulting in systemic inflammation.[10] These events upsurge the secretion of pro-inflammatory cytokines such as IL6 and TNFα. This condition is known as chronic low-grade inflammatory state[11,12] which leads to phosphorylation of serine residue instead of tyrosine of insulin receptor. The insulin receptor is now not capable of binding with insulin and can't activate further metabolic signaling, thus creating IR.[13] Thus systemic inflammation is directly linked with gut dysbiosis and IR.

Microbiota plays a role in IR by influencing energy balance, glucose metabolism, and causing systemic inflammation.[14,15] As a result, it's critical to investigate the role of microflora in T2D and how reducing Gram-negative bacteria can help improve metabolic and immunological balance in the liver and gut.

1.4 Dysbiosis of microflora in T2D

Commensal microflora resides in the intestine, interacts with intestinal epithelium and is involved in metabolic degradation of some undigested materials via fermentation, vitamin uptake and prevents pathogenic bacterial colonization, it also promotes the development of secondary lymphoid organs structure, etc.[16] Commensal microflora evolve in host with age, so there is no blueprint or baseline which can be considered as ideal gut microflora. There are major four phyla present in the gut, i.e., Bacteroidetes, Firmicutes, Actinobacteria and Proteobacteria. Bacteroidetes and Firmicutes cover 90% while Actinobacteria and Proteobacteria account <1–5% of a total bacterial population. The ratio of Firmicutes to Bacteroidetes determines whether the host is thin or obese. Small intestine bacterial overgrowth, or bacterial dysbiosis, is increased by a high-fat, high-carbohydrate Western diet.

The conventional method such as culturing of bacteria would not allow identify many strains also PCR based amplification of the 16S rRNA phylogenetic marker approach has certain drawbacks like biases in sample preparation, and amplicon sequences limited to genus level identification and reproducibility. Newly emerging technique known as shot-gut Metagenomics offers species and strain level identification of bacteria from the environmental sample (here, it will be a fecal sample of a patient or animal understudy). This method also allows to correlate functional relationship between host and bacteria.[17]

Overall, the gut microbiota and its modulation as cause as well as the effect have been reported in Diabetes, which can be caused by extraneous involvement such as diet leads to dysbiosis and imbalance of intestinal microbial bi-products, induces IR by various mechanism.

2. Involvement of bacterial bi-products

2.1 Role of microflora in T2D prevention

Microflora present in symbiosis helps to regulate metabolism and reduce the risk of T2D. Therefore there is upcoming interest is to identifying bacteria that positively influence good health. Bifidobacteria produces bacteriocin that blocks adherence to the mucosa and preserves gut barrier function[18]. The principal gut microorganisms that create short chain fatty acid (SCFAs) are *Eubacterium, Faecalibacterium, Anaerostipes, Marvinbryantia, Blautia, Coprococcus, Megasphaera, Roseburia, and Ruminococcus*. By increasing intestinal synthesis of glucagon-like peptide-1 and peptide YY, the

SCFAs acetate and butyrate help to maintain glucose homeostasis.[19] It is not just about the beneficial effect of bacterial bi-product, they also regulate immune cells to fight against pathogens. Treg cells are critical for maintaining a balance between pro and anti-inflammatory immune responses, as it is tolerogenic in nature. *Bacteroides thetaiotaomicron, Roseburia intestinalis, Clostridium clusters IV,* and *XIVa* all induce Treg cells.[20] Furthermore, butyrate promotes colonic Treg development, along with other SCFAs produced by the gut microflora.[21] Beneficial microorganisms avoid low-grade inflammation by suppression of pro-inflammatory cytokines and chemokines.

2.1.1 SCFA

SCFAs are bi-product of bacterial fermentation in the large intestine from undigested food. Acetate, propionate and butyrate are major SCFA found in colon and serum. A Major source of acetate production is carbohydrates. Amino acids also contribute to SCFA production, mainly acetate and propionate. Acetogenic bacteria are responsible for the production of 70% acetate present in the colon. Bifidobacterium genus, Eubacteriaceae family members (*Eubacterium rectale, Eubacterium hallii*), Ruminococcaceae family (*Faecalibacterium prausnitzii* (Firmicutes)), and a various clusters of clostridium are known for butyrate production. Fermentation of starch by *Ruminococcus bromii* produce butyrate. *Akkermansia muciniphila* is a propionate-producing bacteria[22].

Diet pattern largely affects the gut microflora composition, and so does the SCFA ratio. SCFA ratio can be considered a health status marker. It attributes to various health effects such as a reduction in luminal pH which inhibits pathogenic bacterial growth. Butyrate is fuel for colonocytes and improves tight-junction integrity. SCFA ratio also has a key role in gut barrier function. Absorbed SCFA enters in liver where it gets incorporated into the glucose and lipid synthesis pathway. So, their function in metabolic disorders is fascinating.[23,24] SCFA can also stimulate hepatic TG levels through suppression of Lipoprotein Lipase.[25] SCFA has anti-inflammatory potential. Animals fed with Acetate and propionate combination in the diet showed a reduction in CCL3 marker (a marker for microphage infiltration into adipose) compared to the high-fat diet-fed animals. Systemic inflammation in T2D is in favor of high Th1 and Th17; low Treg cells. SCFAs supplementation reduces anti-inflammatory cytokines and reduces the level of Th1 and Th17 in the spleen.[24]

2.1.2 Bile acids

Bile acid pool act as a metabolic controller. Change in Bile acid pool alters glycemic control in T2D patients.[26] Bile acids are synthesized from the cholesterol in liver. Primary bile acids (cholic acid and chenodeoxycholic acid) are conjugated mainly with glycine and to a lesser extent with taurine. Conjugated bile acids are stored in gallbladder, After meal is ingested, it gets released into duodenum in there they facilitate the emulsification of lipids and absorption of fat soluble vitamins.[27] Conjugated bile salts are deconjugated by gut microbes as they produce an enzyme known as bile salt hydrolase (BSH). The deconjugation process allows the release of free amino acids (glycine and taurine). Further, it gets converted to secondary bile acids, thereby regulating bile and cholesterol biosynthesis.[28] Secondary bile acids are produced by a variety of microorganisms, mostly from the distal ileum. Genus *Clostridium* can carry out bile acid 7a dehydroxylation. Bacterial activity converts cholic acid (CA) and chenodeoxycholic acid (CDCA) to deoxycholic acid (DA) and lithocholic acid.[29] Biotransformation of bile acids by bacteria involves deconjugation, oxidation, epimerization, esterification and desulfation. The BSH activity is found in various species of Lactobacillus and Bifidobacterium. Bacteria can generate oxo-(keto) bile acids by oxidation of hydroxyl group present at 3, 7, and 12th positions of the ring structure by the enzyme hydroxysteroid dehydrogenase. This enzyme is found in Actinobacteria, Proteobacteria, Firmicutes, and Bacteroidetes. Above mentioned oxidation reaction is reversible, resulting in epimerization called as isobile acids can be performed by *Eubacteriumlentum and Clostridium perfringens*[27]. *Bacteroides, Eubacterium and Lactobacillus* can convert bile acid to C24 ethyl ester. Clostridium, *Peptococcus, Fusobacterium and Pseudomonas* are having sulfatase activity.[30] Therefore bacterial composition decides the fate of secondary bile acid products. During diabetes, bile acid pool alters due to changes in lipid metabolism and specific species in the gut. This alteration involves increased concentrations of DA and decreased concentrations of CDCA in T2DM patients (Fig. 1).[31]

The bacterial metabolites such as SCFAs and bile acids could be beneficial to alleviate IR in T2D via modulation of gut microbiota. SCFAs, independently or in the consortium and targeting bile acid, its metabolites or receptors, could be new therapeutic targets for diabetic treatment.

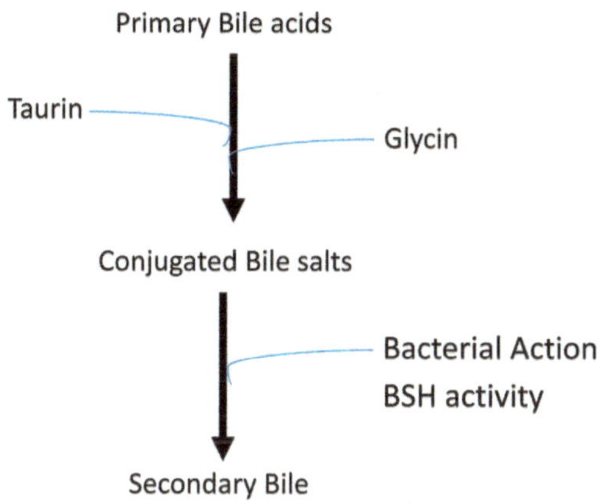

Fig. 1 Bile acids formation. Primary bile acids are synthesized in the liver from cholesterol, which are conjugated with glycine and taurine. Bile acids stored in gall bladder. After the ingestion of food, pool of bile acid is released into small intestine for lipid emulsification. 95% of the conjugated bile acids are reabsorbed. Completing the entero-hepatic circulation, and remaining 5% are altered by gut microbes into secondary bile acids.

3. Prevention/treatment strategy involving microflora alterations by pre-biotic and pro-biotic

Prebiotics are fermentable, non-digestive foods that allow special bacterial flora to grow. It stimulates preferential bacterial growth in the colon such as lactobacillus and Bifidobacterium.[32] Various non-digestible carbohydrates (NDCs) are recognized as pre-biotics such as inulin-type fructans, and galacto-oligosaccharides. They can selectively impact bifidobacteria and are able to improve glucose homeostasis and inflammation. The mechanism could be the regulation of GPRs and PPAR-γ via SCFA produced from the fermentation of NDCs[33].

Probiotics include ingestion of live bacteria in the colon, to exert a beneficial role. It acts as a dietary supplement by suppressing pathogenic bacterial growth thus, manipulating gut microflora, mucosal immunity, and intestinal barrier function.[34] Various species of Lactobacillus were widely studied. *Lactobacillus rhamnose* GG is having anti-inflammatory properties, prevent gut-barrier function, also improves hepatic inflammation and lipid metabolism.[35]

3.1 Impact of antibiotics on the gut microflora and type 2 diabetes

Antibiotics are designed to kill the pathogenic microorganisms, but they affect some members of host-microbiota also, leaving an impression on the gut microflora long after the removal of antibiotics.[36] Antibiotics can preclude the advancement of T2D to some extent. Oxytetracycline treatment significantly restored insulin sensitivity by increasing insulin binding affinity to the hepatocytes cell membrane.[37] Membrez et al. found that glucose sensitivity can be enhanced by modulating gut microflora using combination of broad-spectrum antibiotics like norfloxacin and ampicillin, in mice.[38] Antibiotics formulations such as microspheres can be used to target colon and small intestinal bacterial overgrowth to overcome T2D and reduce antibiotic side effects on the liver compared to pure form of it. Colon-targeted broad-spectrum and Gram-negative antibiotics showed a preventive effect in High sugar diet (HSD) induced T2D rat.[39,40] Formulation of Gram-negative antibiotics to target colonic bacteria in the HSD mice model also reduced T2D pathophysiological conditions with improved Treg/Th17 ratio and liver inflammations.[41]

3.2 Fecal transplantation

Fecal microbiota transplantation (FMT) might be advantageous for insulin resistance, especially when it is done from lean donors in patients with metabolic syndrome resulting in a noteworthy improvement in insulin sensitivity along with improved intestinal microflora diversity, including a diverse increase in butyrate-producing bacterial strains. A fecal transplant from obese animal to germ-free resulted in donor somatotype indicating that altered microbiota plays important role in disease onset.[42] Various studies are being done using FMT to treat metabolic disorders like T2D, type 1 diabetes and intestinal bowel disease, etc., and found to be safe. FMT from healthy mice to T2D mice showed improved insulin sensitivity, and reserved chronic inflammation in pancreatic islet.[43] Very few clinical trials for T2D have been performed. FMT has shown promising results for IBD. Groot et al. have done a human randomized trial with Type 1 Diabetic patient and found that autologous FMT efficiently extended beta-cell function.[44]

4. Conclusion and future remarks

Manipulation of gut microbiota using antibiotics/Prebiotic/Probiotic may help to decrease inflammation and insulin resistance associated with

T2D and to elucidate the physiological and immunological role of gut microbiota for prevention of diseases. Metagenomic approach has revealed many bacterial involvements in diabetes pathogenesis. Many times different types of microflora involve in same type of metabolic disorder. However, most researcher have focused on gut microbiota characterization instead of functional confirmation of specific microbial community that increase T2D risk. Identifying specific origin relevant microbes or their metabolites involved in T2D pathogenesis could open up new possibilities for T2D diagnosis, treatment, and prevention. The role of microbiota-driven disease has been diminished thanks to fecal transplantation. Many bacterial involvements in diabetes pathogenesis have been discovered using a metagenomic approach. Different forms of microflora are frequently involved in the same type of metabolic disease. Mainly *Bacteroides, Bifidobacterium, Akkermansia, Faecalibacterium* and *Roseburia* are healthy microbial genera. The Lactobacillus genus, in particular, is widely reported to have probiotic properties that reduce inflammation caused by Gram-negative bacteria in the colon. The overall approach involves gut microflora to ameliorate the metabolic disordered conditions. In this process, also direct to have a healthy diet and healthy lifestyle of an individual.

References

1. Hu FB. Globalization of diabetes: the role of diet, lifestyle, and genes. *Diabetes Care.* 2011;34(6):1249–1257. https://doi.org/10.2337/dc11-0442.
2. Robertson MD, Pedersen C, Hinton PJ, Mendis ASJR, Cani PD, Griffin BA. Elevated high density lipoprotein cholesterol and low grade systemic inflammation is associated with increased gut permeability in normoglycemic men. *Nutr Metab Cardiovasc Dis.* 2018;28(12):1296–1303. https://doi.org/10.1016/j.numecd.2018.07.006.
3. Cani PD, Bibiloni R, Knauf C, et al. Changes in gut microbiota control metabolic endotoxemia-induced inflammation in high-fat diet-induced obesity and diabetes in mice. *Diabetes.* 2008;57(6):1470–1481. https://doi.org/10.2337/db07-1403.
4. Jena PK, Prajapati B, Mishra PK, Seshadri S. Influence of gut microbiota on inflammation and pathogenesis of sugar rich diet induced diabetes. *Immunome Res.* 2016;12 (1):1–10. https://doi.org/10.4172/1745-7580.10000109.
5. Galicia-Garcia U, Benito-Vicente A, Jebari S, et al. Pathophysiology of type 2 Diabetes Mellitus. *Int J Mol Sci.* 2020;21(17):6275. https://doi.org/10.3390/ijms21176275.
6. Hue L, Taegtmeyer H. The randle cycle revisited: a new head for an old hat. *Am J Physiol Endocrinol Metab.* 2009;297(3):E578–E591. https://doi.org/10.1152/ajpendo.00093.2009.
7. Cersosimo E, Triplitt C, Solis-Herrera C, Mandarino LJ, DeFronzo RA. Pathogenesis of type 2 Diabetes Mellitus. In: Feingold KR, Anawalt B, Boyce A, et al., eds. *Endotext.* MDText.com, Inc.; 2000. http://www.ncbi.nlm.nih.gov/books/NBK279115/. Accessed 28 September 2021.
8. Abdul-Wahed A, Gautier-Stein A, Casteras S, et al. A link between hepatic glucose production and peripheral energy metabolism via hepatokines. *Mol Metab.* 2014;3 (5):531–543. https://doi.org/10.1016/j.molmet.2014.05.005.

9. Escobedo G, López-Ortiz E, Torres-Castro I. Gut microbiota as a key player in triggering obesity, systemic inflammation and insulin resistance. *Rev Invest Clin.* 2014;66 (5):450–459.
10. Rakoff-Nahoum S, Paglino J, Eslami-Varzaneh F, Edberg S, Medzhitov R. Recognition of commensal microflora by toll-like receptors is required for intestinal homeostasis. *Cell.* 2004;118(2):229–241.
11. Seki E, Schnabl B. Role of innate immunity and the microbiota in liver fibrosis: crosstalk between the liver and gut. *J Physiol.* 2012;590(3):447–458. https://doi.org/10.1113/jphysiol.2011.219691.
12. Prajapati B, Jena PK, Rajput P, Purandhar K, Seshadri S. Understanding and modulating the toll like receptors (TLRs) and NOD like receptors (NLRs) cross talk in type 2 diabetes. *Curr Diabetes Rev.* 2014;10(3):190–200.
13. Draznin B. Molecular mechanisms of insulin resistance: serine phosphorylation of insulin receptor substrate-1 and increased expression of p85α: The two sides of a coin. *Diabetes.* 2006;55(8):2392–2397. https://doi.org/10.2337/db06-0391.
14. Dumas ME, Barton RH, Toye A, et al. Metabolic profiling reveals a contribution of gut microbiota to fatty liver phenotype in insulin-resistant mice. *Proc Natl Acad Sci U S A.* 2006;103(33):12511–12516. https://doi.org/10.1073/pnas.0601056103.
15. Wang Z, Klipfell E, Bennett BJ, et al. Gut flora metabolism of phosphatidylcholine promotes cardiovascular disease. *Nature.* 2011;472(7341):57–63. https://doi.org/10.1038/nature09922.
16. Fujimura KE, Slusher NA, Cabana MD, Lynch SV. Role of the gut microbiota in defining human health. *Expert Rev Anti Infect Ther.* 2010;8(4):435–454. https://doi.org/10.1586/eri.10.14.
17. Rausch P, Rühlemann M, Hermes BM, et al. Comparative analysis of amplicon and metagenomic sequencing methods reveals key features in the evolution of animal metaorganisms. *Microbiome.* 2019;7(1):133. https://doi.org/10.1186/s40168-019-0743-1.
18. Rivière A, Selak M, Lantin D, Leroy F, De Vuyst L. Bifidobacteria and butyrate-producing colon bacteria: importance and strategies for their stimulation in the human gut. *Front Microbiol.* 2016;7(979):1–21. https://doi.org/10.3389/fmicb.2016.00979.
19. Huda MN, Kim M, Bennett BJ. Modulating the microbiota as a therapeutic intervention for type 2 diabetes. *Front Endocrinol.* 2021;12:632335. https://doi.org/10.3389/fendo.2021.632335.
20. Hoffmann TW, Pham H-P, Bridonneau C, et al. Microorganisms linked to inflammatory bowel disease associated dysbiosis differentially impact host physiology in gnotobiotic mice. *ISME J.* 2016;10(2):460–477. https://doi.org/10.1038/ismej.2015.127.
21. Furusawa Y, Obata Y, Fukuda S, et al. Commensal microbe-derived butyrate induces the differentiation of colonic regulatory T cells. *Nature.* 2013;504(7480):446–450. https://doi.org/10.1038/nature12721.
22. Louis P, Flint HJ. Formation of propionate and butyrate by the human colonic microbiota. *Environ Microbiol.* 2017;19(1):29–41. https://doi.org/10.1111/1462-2920.13589.
23. Ríos-Covián D, Ruas-Madiedo P, Margolles A, Gueimonde M, de los Reyes-Gavilán CG, Salazar N. Intestinal short chain fatty acids and their link with diet and human health. *Front Microbiol.* 2016;7:1–9. https://doi.org/10.3389/fmicb.2016.00185.
24. Mandaliya DK, Patel S, Seshadri S. The combinatorial effect of acetate and propionate on high-fat diet induced diabetic inflammation or metaflammation and T cell polarization. *Inflammation.* 2021;44(1):68–79. https://doi.org/10.1007/s10753-020-01309-7.
25. Bäckhed F, Ding H, Wang T, et al. The gut microbiota as an environmental factor that regulates fat storage. *Proc Natl Acad Sci U S A.* 2004;101(44):15718–15723.
26. Prawitt J, Caron S, Staels B. Bile acid metabolism and the pathogenesis of type 2 diabetes. *Curr Diab Rep.* 2011;11(3):160–166. https://doi.org/10.1007/s11892-011-0187-x.

27. Wahlström A, Sayin SI, Marschall HU, Bäckhed F. Intestinal crosstalk between bile acids and microbiota and its impact on host metabolism. *Cell Metab.* 2016;24(1):41–50. https://doi.org/10.1016/j.cmet.2016.05.005.
28. Eyssen H. Role of the gut microflora in metabolism of lipids and sterols. *Proc Nutr Soc.* 1973;32(2):59–63. https://doi.org/10.1079/PNS19730016.
29. Ridlon JM, Alves JM, Hylemon PB, Bajaj JS. Cirrhosis, bile acids and gut microbiota. *Gut Microbes.* 2013;4(5):382–387. https://doi.org/10.4161/gmic.25723.
30. Gérard P. Metabolism of cholesterol and bile acids by the gut microbiota. *Pathogens.* 2014;3(1):14–24. https://doi.org/10.3390/pathogens3010014.
31. Wu Y, Zhou A, Tang L, Lei Y, Tang B, Zhang L. Bile acids: key regulators and novel treatment targets for type 2 diabetes. *J Diabetes Res.* 2020;2020:1–11. https://doi.org/10.1155/2020/6138438.
32. Quigley EMM, Quera R. Small intestinal bacterial overgrowth: roles of antibiotics, prebiotics, and probiotics. *Gastroenterology.* 2006;130(2):S78–S90. https://doi.org/10.1053/j.gastro.2005.11.046.
33. Geurts L, Neyrinck AM, Delzenne NM, Knauf C, Cani PD. Gut microbiota controls adipose tissue expansion, gut barrier and glucose metabolism: novel insights into molecular targets and interventions using prebiotics. *Benef Microbes.* 2014;5(1):3–17. https://doi.org/10.3920/bm2012.0065.
34. Seshadri S. Understanding the role of probiotics as a possible treatment strategy for alcoholism induced liver dysfunction. *Alcoholism: Causes, Symptoms, Effects and Treatments.* Manakin Press Private Limited; 2017:85–95.
35. Ritze Y, Bárdos G, Claus A, et al. Lactobacillus rhamnosus GG protects against non-alcoholic fatty liver disease in mice. *PLOS ONE.* 2014;9(1), e80169. https://doi.org/10.1371/journal.pone.0080169.
36. Antunes LCM, Han J, Ferreira RBR, Lolić P, Borchers CH, Finlay BB. Effect of antibiotic treatment on the intestinal metabolome. *Antimicrob Agents Chemother.* 2011;55(4):1494–1503. https://doi.org/10.1128/AAC.01664-10.
37. Bégin-Heick N, Bourassa M, Heick HMC. The effect of oxytetracycline on insulin resistance in obese mice. *Biochem J.* 1974;142(3):465–475. https://www.ncbi.nlm.nih.gov/pmc/articles/PMC1168310/. Accessed 9 October 2018.
38. Membrez M, Blancher F, Jaquet M, et al. Gut microbiota modulation with norfloxacin and ampicillin enhances glucose tolerance in mice. *FASEB J.* 2008;22(7):2416–2426. https://doi.org/10.1096/fj.07-102723.
39. Jena PK, Singh S, Prajapati B, Nareshkumar G, Mehta T, Seshadri S. Impact of targeted specific antibiotic delivery for gut microbiota modulation on high-fructose-fed rats. *Appl Biochem Biotechnol.* 2014;172(8):3810–3826. https://doi.org/10.1007/s12010-014-0772-y.
40. Prajapati B, Jena PK, Patel S, Seshadri S. Divergent outcomes of gut microbiota alteration upon use of spectrum antibiotics in high sugar diet-induced diabetes in rats. *RSC Adv.* 2018;8(46):26201–26211. https://doi.org/10.1039/C8RA03774D.
41. Patel S, Mandaliya D, Prajapati B, Kumar S, Seshadri S. Cefdinir microsphere modulated microflora and liver immunological response to diet induced diabetes in mice. *Endocr Metab Immune Disord Drug Targets.* 2019;19(3):349–357. https://doi.org/10.2174/1871530319666181224122115.
42. Grigorescu I, Dumitrascu DL. Implication of gut microbiota in diabetes mellitus and obesity. *Acta Endocrinol (Buchar).* 2016;12(2):206–214. https://doi.org/10.4183/aeb.2016.206.
43. Wang H, Lu Y, Yan Y, et al. Promising treatment for type 2 diabetes: fecal microbiota transplantation reverses insulin resistance and impaired islets. *Front Cell Infect Microbiol.* 2020;9:455. https://doi.org/10.3389/fcimb.2019.00455.
44. de Groot P, Nikolic T, Pellegrini S, et al. Faecal microbiota transplantation halts progression of human new-onset type 1 diabetes in a randomised controlled trial. *Gut.* 2021;70(1):92–105. https://doi.org/10.1136/gutjnl-2020-322630.

CHAPTER EIGHT

Gut microbiome and non-alcoholic fatty liver disease

Ayushi Purohit[a], Md Jahangir Alam[b], Bharti Kandiyal[a], Shalimar[c], Bhabatosh Das[a], and Sanjay K. Banerjee[b,*]

[a]Molecular Genetics Laboratory, Centre for Human Microbial Ecology, Translational Health Science and Technology Institute, NCR Biotech Science Cluster, Faridabad, Haryana, India
[b]Department of Biotechnology, National Institute of Pharmaceutical Education and Research (NIPER), Guwahati, India
[c]Gastroenterology and Human Nutrition, All India Institute of Medical Sciences, New Delhi, India
*Corresponding author: e-mail address: sanjay@niperguwahati.in

Contents

Abstract

The human gastrointestinal tract (GIT) contains a dynamic and diverse collection of bacteria, archaea, and fungi termed the "gut microbiome." The gut microbiome has a major impact on the host during homeostasis and disease. The connection between both the host and the microbiome is complex, although its manipulation may assist prevent or treating a multitude of morbidities. These microorganisms play a critical role in the

ISSN 1877-1173
https://doi.org/10.1016/bs.pmbts.2022.07.004

host's energy metabolism and homeostasis. According to new research, the microbes in the gastrointestinal tract play a substantial role in host health, and alterations in its composition and function might lead to the emergence of metabolic disorders like non-alcoholic fatty liver disease (NAFLD). The resilience of the GIT microbial ecology and its tolerance to perturbation are robust but not ideal. Several factors may disrupt the GIT microbiome's homeostasis leading to dysbiosis, characterized by an imbalanced equilibrium and perturbations in gut homeostasis. Irritable bowel disease (IBD), malnutrition, and metabolic disorders, such as NAFLD, have been associated with the dysbiotic gut microbiome. Recent evidence suggests that utilizing medications, prebiotics, probiotics, and fecal microbiota transplantation (FMT) to manipulate the microbiome could be a viable method for treating NAFLD.

Abbreviations

ALT	alanine transaminase
AST	aspartate transaminase
BAs	bile acids
FMT	fecal microbiota transplantation
FXR	farnesoid X receptor
GIT	gastrointestinal tract
IR	insulin resistance
ITS	internal transcribed spacer
LBP	LPS binding protein
LPS	lipopolysaccharide
MAMP	microbe-associated molecular pattern
MCT-1	monocarboxylate transporter 1
NAFLD	non-alcoholic fatty liver disease
NASH	non-alcoholic steatohepatitis
SAM	*S*-adenosyl methionine
SCFAs	short-chain fatty acids
SMCT-1	sodium-coupled monocarboxylate transporter 1
TCA	tricarboxylic acid
TLR4	toll-like receptor 4
TMA	trimethylamine
VLDL	very-low-density lipoprotein

1. Introduction

The human body surfaces are inhabited by microbial populations that are at least one order of magnitude higher than human own somatic and germ cells.[1] The interaction between microbes and humans has indeed been precarious for centuries. The human body contains trillions of microorganisms that originate immediately after birth.[2] The microbiome contributes significantly to a variety of biochemical events that are critical to host pathophysiology and the Earth's ecosystem. In recent decades, extensive study has shown the microbiome's role in human health and diseases.[3] The resident microbes are associated with the host in a symbiotic manner and play a central role in metabolic functions, immune system maturation, metabolizing bile salts metabolites, repression of virulence factors in enteric pathogens, conversion of prodrugs into active drugs, and chemical transformation of xenobiotic compounds such as antibiotics.[4] Furthermore, the gut microbiota is often used to synthesize vitamins, neurotransmitters, and signaling molecules, which are critical components of human physiology and may retrain the host's immune system, restrict enteric pathogen colonization, and release cytokine and immunoglobulin.[5] The microbiota's association with the host is critical for longevity and disease vulnerability. The commensal and symbiotic microbial communities are well-known as the first-line defense against disease.[6] Changes in microbiome composition or intestinal permeability, on either hand, may result in bacteria becoming translocated directly into the portal vein. The gut and liver are anatomically and physiologically linked, and their interaction with the gut microbiota constitutes the "gut-liver axis," which has the potential to alter the liver function and increase the prognosis of the disease. The gut microbiota is made up of microorganisms that live naturally in the human intestine and help maintain the host in a state of homeostasis. However, metabolism is affected when microbial components in dysbiotic conditions can translocate to the liver, altering the hepatic pathological consequences.[7] Due to the incredibly high intake of high-caloric and ultra-processed foods such as alcoholic beverages and/or high-fat diet activates microbe-associated molecular pattern (MAMP), and bacterial metabolites are produced. Even the gut microbiome or its-derived metabolites may translocate to the liver, where they can accumulate to cause serious hepatic features such as inflammation, fibrosis, and liver cancer.[7]

2. The healthy gut microbiome

The gastrointestinal tract (GIT) microbial ecology is defined by the diversity, richness, dynamics, equilibrium, and resilience of healthy individuals. *Firmicutes, Bacteroidetes, Actinobacteria*, and *Proteobacteria* predominate the innate GIT ecosystems of healthy individuals. Several bacterial taxa, such as *Verrucomicrobia, Fusobacteria, Tenericutes*, and *Spirochetes*, have also been detected in the least abundant.[8,9] The gut is not as diverse as those in other microbial communities and shows a high degree of functional redundancy from other bodily locations.[10] The gut microbiome has a vital involvement in the formation of host immunity, digestion, the control of intestinal endocrine and neurological signals, the modification of pharmacological effects and the removal of toxins and the production of various chemicals affecting the host.[11] However, the overall distinguishing healthy robust microbial communities are of high taxonomic diversity, high microbial richness and stable multifunctional center. There is no 'gold standard' human gut microbiota reference that is capable of promoting host metabolic health. This is because of considerable variance in the demographic, ethnicity, sex, age and health state of individuals, especially at the taxonomic level.

3. The detangling metabolic disorder: NAFLD and its prevalence

Almost 50 years ago, a connection between liver fat, liver injury and fibrosis was already established.[12] In 1980, Jurgen Ludwig and colleagues anticipated the presence of macrovesicular steatosis, hepatocyte enlargement, lobular inflammation and pericellular fibrosis in induvial who did not consume alcohol or consumed moderately. This was termed nonalcoholic steatohepatitis (NASH).[12] NASH is already recognized as part of a histologic spectrum defined by a pathologic deposition of fat in the liver, with an overall prevalence of roughly 25–30%.[13] The incidence of NASH is markedly higher in the Middle East, and it is expanding across Asia and the Far East. By 2030, the impact of liver disease is projected to elevate in both Western and Asian countries.[14] Globally, NAFLD's overall prevalence is 25.24%, with considerable global variance. In the Middle East and South American countries, the highest incidences, mainly ultrasound-based, have been recorded (32% and 31%, respectively). Still, the limited number of African studies indicated a considerably lower frequency [13.48%].[15] Most research, however, has shown that NAFLD has a prevalence of 21.0–24.7% in the USA and North America. In a meta-analysis published in 2015, a similar NAFLD prevalence (24%) was identified in Europe. As compared to the United States and Europe, the burden of NAFLD in Asia, a rapidly developing continent, is increasing faster.[13] The increased prevalence of NAFLD in Asia-Pacific locations may be attributed to the recent embrace of Western lifestyle and nutrition patterns, as well as prevalent hereditary factors in many populations. The increased frequency of NAFLD in Asia-Pacific regions can be attributed to the recent incorporation of Western lifestyle and nutrition patterns, as well as prevalent hereditary factors. Obesity and diabetes are well-established risk factors for NAFLD.

However, evidence suggests that in the Asia-Pacific area, a considerable percentage of people with a normal BMI also have NAFLD. It is predicted that the number of cases of non-alcoholic fatty liver and non-alcoholic steatohepatitis (NASH) will continue to rise in future, particularly among fast-ageing populations who are more prone to chronic lifestyles.[16] Many previous reports have relied on earlier models and epidemiology to forecast the future burden of disease in Asia and have eventually recommended, concluded, and encouraged effective allocation to NAFLD and NASH-related morbidity mitigation.[16]

4. The culprit of NAFLD: Gut dysbiosis

The healthy gut microbial ecology can be characterized in terms of richness, abundance, and dynamics.[8] Most gastrointestinal microbes are harmless or beneficial. The gut microbiota secures nutrients and energy through the host and also protects their host against enteric pathogens, and contributes to conventional immunological function.[9] The concept of gut bacteria regulating the homeostasis of the liver is a consequence of the physiological linkage between the gastrointestinal system and the liver, frequently alluded to as the "gut-liver axis."[17] The gut microbiota is an intensive metabolic reactor that collectively generates by-products that traverse to liver by the portal vein. Several microbial-derived metabolites have been suspected of causing hepatocellular inflammation or impairment. For instance, microbiota-derived lipopolysaccharide (LPS), apparently discovered in the serum of obese patients and associated with liver damage, suggests a relationship between LPS and the progression of nonalcoholic steatohepatitis.[18] Fat storage is regulated by gut microbiota as evidenced by increased production of triglycerides in the liver and its increased storage in the adipocytes,[19] and antibiotics treatment significantly decreased lipid accumulation and portal endotoxin level in fructose-fed mice.[20] Endotoxins, peptidoglycans, or even whole bacteria, all of whom have the potential to alter many metabolic pathways in the liver, may cause direct damage.[21] Pathogen-recognition receptors on the plasma membranes of many hepatic cells, notably Kupffer cells (stellate macrophages), sinusoidal cells, biliary epithelial cells, and hepatocytes, among others, are triggered by a constant flow of microbes from the gut into the liver.[6] Many studies have found alterations in the intestinal microbiota may influence liver function, systemic or hepatic inflammation, insulin resistance (IR), and increased fat storage, which leads to obesity and, eventually, liver damage and NASH

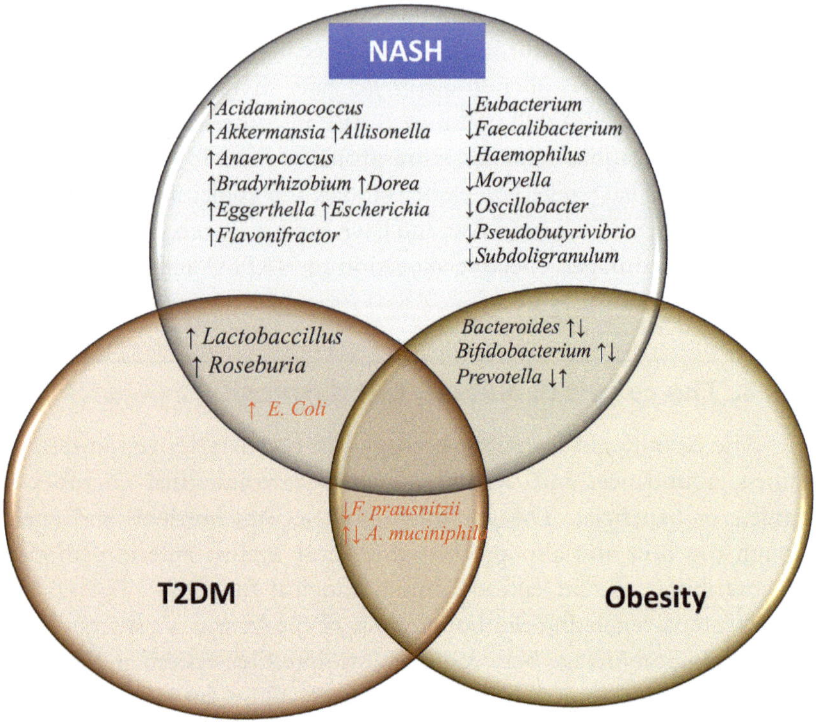

Fig. 1 Signatures of contiguous microbiota species and genera in NAFLD, diabetes, and obesity.

development.[22] In NAFLD patients, there was a higher abundance of phylum *Fusobacteria* and a reduced abundance of *Ruminococcaceae Oscillospira and Ruminococcus,* and *Lachnospiraceae Coprococcus*[23] (Fig. 1). *Lactobacillus, Bifidobacteria,* and *Bacteroides* bacteria may replenish the gut microbial composition and diminish gastrointestinal endotoxemia, thereby mounting a robust anti-inflammatory response and preventing the progression of non-alcoholic fatty liver disease.[24] Several theories have been proposed to explain how gut flora affects NAFLD development.

5. Microbiome signatures and their functional aspect

Gut bacteria and their metabolites transcend a leaky gut barrier to the liver. They induce chronic hepatic inflammation and are correlated with dietary determinants that promote steatosis via many reported pathways. The accumulation of triglycerides in the hepatocytes is a significant feature of NAFLD. Adipose tissue TGs undertake a variety of catabolic processes to produce free fatty acids, which then translocate into hepatocytes for aggregation.[25] Human gut microbiome is a metabolically active component that changes in response to modest changes in its environment. Compared to healthy control patients, the number of Gram-negative pathogens such as *Proteobacteria, Escherichia* spp., and *Enterobacteriaceae* was considerably greater in NASH patients. *Faecalibacterium prausnitzii* and *Akkermansia muciniphila*, which are renowned for their endogenous probiotic nature, are found to be completely lacking.[26,27] In recent years, several different pathways wherein the gut microbiome and their metabolites modulate NAFLD and NASH have been addressed.

5.1 Short-chain fatty acids (SCFAs)

SCFAs, such as acetate, butyrate, propionate, pentanoic acid, and hexanoic acid, are released when complex carbohydrates such as dietary fiber are digested by bacterial species in the gut.[28] Although the majority of SCFAs are used in the gut, few of them can move into the bloodstream of the portal vein via sodium-coupled monocarboxylate transporter 1 (SMCT-1) and monocarboxylate transporter 1 (MCT-1) receptors, and then finally to the liver for energy production through the citric acid cycle. By interacting with the G-protein-coupled receptors GPR41, GPR43, and GPR109A, SCFAs may operate as signaling molecules.[29] For instance, SCFAs can influence fat storage in the hepatic and adipose tissue by regulating insulin levels via GPR43 thereby increasing fat accumulation and energy expenditure in both adipose and liver tissues.[30] SCFAs also affect the biosynthesis of cholesterol in the liver, which is a precursor to bile acids, which play a role in the gut microbiome host relationship.[28] Implementing the use of SCFA might be a potential strategy for treating NASH.

5.2 Bile acids (BA)

After each meal, bile acids are steroidal molecules synthesized from cholesterol in the liver, secreted to the gallbladder, and then delivered to the duodenum. The gut microbiota converts them to secondary bile acids in the gastrointestinal system. The molecules are subsequently reabsorbed into the portal vein, with the majority of them being collected by the liver and recirculated, but some staying as signaling molecules in the entero-hepatic circulation. Bile acids have two purposes in the body: (i) they help in absorption, solubilization and digestion of dietary lipid, and (ii) they are signaling molecules which act through TGR5 and the Farnesoid X receptor (FXR), among other receptors. The bile acid pool's composition and content have an impact on lipids absorption as well as the pool's ability to activate receptors.[31] Other microorganisms may diminish the bile acid pool in humans by suppressing bile acid synthesis in the liver as a result of inflammation. The expression of the intestinal bile acid transporter rises as cirrhosis progresses, resulting in fewer bile acids entering the large intestine owing to reduced bile salt concentrations. Lower levels of primary BAs may lead to fewer 7-dehydroxylating bacteria in the colon. As a consequence, the secondary/primary bile acid ratio in cirrhosis is low.[32]

5.3 Alcohol derived via microbial fermentation

Microbiota can ferment dietary carbohydrates into ethanol, which can subsequently be absorbed into the circulation. The quantity of alcohol generated is proportional to the amount of carbohydrates available in the diet, and it is higher in obese mice.[33] When comparing obese people with NASH to obese people without NASH, there are considerably higher levels of ethanol-producing microbes in the blood. A significant difference between obese and NASH microbiomes was observed exclusively in the phylum, family, and genus of Proteobacteria, Enterobacteriaceae, and *Escherichia* with higher than 1% representation. These bacteria were found in both healthy and obese individuals but have considerably more abundant in individuals with NASH. Blood alcohol concentrations in healthy, obese, and NASH patients all followed a very similar trend.[26] In the liver, alcohol increases oxidative stress and inflammation. Furthermore, alcohol dehydrogenase in the liver converts ethanol to the toxic product acetaldehyde, which forms adducts with proteins and other molecules in cells due to its electrophilic nature, causing cell structure and function to be lost. CYP2E1 is a molecule that metabolites acetaldehyde as well as other compounds, and its activity is connected to the degree of oxidative stress in the liver, which is a significant pro-fibrotic signal. CYP2E1 converts acetaldehyde to acetate. However, this route can become saturated, resulting in a buildup of this hazardous intermediate. In the presence of ethanol, CYP2E1 expression is increased. Patients with NASH had higher CYP2E1 levels, which is intriguing.[34]

5.4 Choline

Choline, a quaternary amine, affects microbial metabolism in the gut.[35] Commensal bacteria like *E. coli* and *Desulfovibrio desulfuricans* can metabolize dietary choline into methylamines such as trimethylamine (TMA), dimethylamine, and monomethyl amine. *In vitro* transformation of choline to TMA has been observed in *Anaerococcus hydrogenalis, Clostridium hathewayi, Clostridium asparagiforme, Clostridium sporogenes, Edwardsiella tarda, Escherichia fergu sonii, Proteus penneri,* and *Providencia rettgeri.*[36] Additionally, dysbiotic microbiota with increased choline to methylamine conversion may contribute to choline deficiency, decreased levels of phosphatidylcholine, and increased levels of TMA in the blood. All these alterations are associated with NAFLD and suggest that the microbiota may play a role in the choline metabolic flux imbalance.[37]

5.5 Lipopolysaccharide (LPS) as endotoxin

Hepatocytes are the most abundant cells in the liver and are constantly exposed to endotoxins (Lipopolysaccharide). Diverse TLRs, such as TLR4, TLR9, TLR2, and TLR5, identify bacterial products and stimulate inflammatory and cytokine responses in different cell types of the liver (Kupffer cells, hepatic stellate cells, and hepatocytes). LPS, being a pathogen-associated molecular pattern (PAMP), can be recognized by the Toll-like receptor 4 (TLR4) and its co-receptors LPS binding protein (LBP) and CD14. Upon activation, it stimulates a downstream inflammatory pathway that includes activation of NFKβ and IRF3 depending on the environment.[38,39] Gut microbiota alterations may be associated with increased gut permeability, bacterial translocation, and the transmission of microbial products from the gut to the liver through the portal vein. A result of altered interaction of different cell types in the liver, such as Kupffer cells and hepatic stellate cells, may also lead to inflammation and fibrosis. Altered expression of TLR4 in the liver may regulate the synthesis of hepcidin, an essential iron-regulatory protein involved in obesity and NAFLD, through the MyD88-IRAK-TRAF6-JNK-AP-1 axis.[40,41]

6. Lean NASH—An intriguing aspect

Although obesity is a global cause of non-alcoholic fatty liver disease (NAFLD), it can also affect people who are not obese. The term "lean NAFLD" is a relatively new concept, and there is much research needed to decipher the mechanism of lean NAFLD individuals in terms of pathogenesis, prognosis, and therapy. This subset of people, referred to as "lean NAFLD" or "non-obese NAFLD," is becoming more common. Asians tend to have a higher prevalence of NAFLD among lean people.[42] A retrospective multi-ethnicity cohort study of 1,090 individuals with biopsy-proven NAFLD found that 125 (~12%) had a BMI of <25 kg/m^2. Interestingly, patients with a BMI of <25 kg/m^2 showed greater lobular inflammation and died at a higher rate (hazard ratio 11.8).[42] Fatty liver has been linked to a higher abundance of *Prevotella* and *Porphyromonas* and ethanol-producing bacteria, as well as a lower abundance of Bacteroidetes and.[26] When comparing patients with NASH to controls, Duarte et al. identified a substantial difference in the abundance *of Bifidobacterium, Faecalibacterium, Lactobacillus* and *Ruminococcus*.[43] higher levels of Lysophosphatidylcholines (lyso-PCs) are associated with a low incidence of obesity because of their anti-inflammatory and insulin-sensitizing properties.[44,45] When compared to healthy controls, the metabolomic analysis revealed reduced levels of lysoPCs in both lean and obese NAFLD patients.[46] Lean individuals with NAFLD, on the other hand, had a greater amount of lysine concentration than their obese counterparts. Increased levels of lysine may be an indication of the disordered metabolic milieu that underpins lean NAFLD people since it is linked to visceral fat formation.[46,47] Future longitudinal prospective studies will definitely be required to define the prognosis in lean NAFLD and to elucidate cellular and molecular mechanisms underlying its progression. *Faecalibacterium prausnitzii* is found mainly in the gastrointestinal tract of healthy people, accounting for 3.5% of total gut microflora. It is capable of producing short-chain fatty acids (butyrate) in the gastrointestinal tract from indigestible food. It also plays a role in modulating the immune response of the host by secreting anti-inflammatory molecules.[48]

6.1 Metabolic characteristics of lean NASH cohorts

With contradictory findings, the fundamental pathophysiologic basis of NAFLD in lean patients is currently unexplained. There does not appear to be a unique mechanism that leads to this complex phenotype. Environmental factors such as food composition and gut microbiota, in combination with genetic susceptibility, may have a significant effect. Polymorphisms in genes that impact lipid metabolism, oxidative stress, IR, and immunological modulation have been identified as NAFLD predisposing factors.[42] Multiple hits acting simultaneously on genetically susceptible lean persons are likely to cause NAFLD (Fig. 2). A retrospective analysis of 646 individuals with biopsy-proven NAFLD revealed that patients with lean NAFLD had lower transaminases, fibrosis stages, and a lower incidence of NASH compared to individuals with a higher BMI but they were more prone to developing the severe liver disease

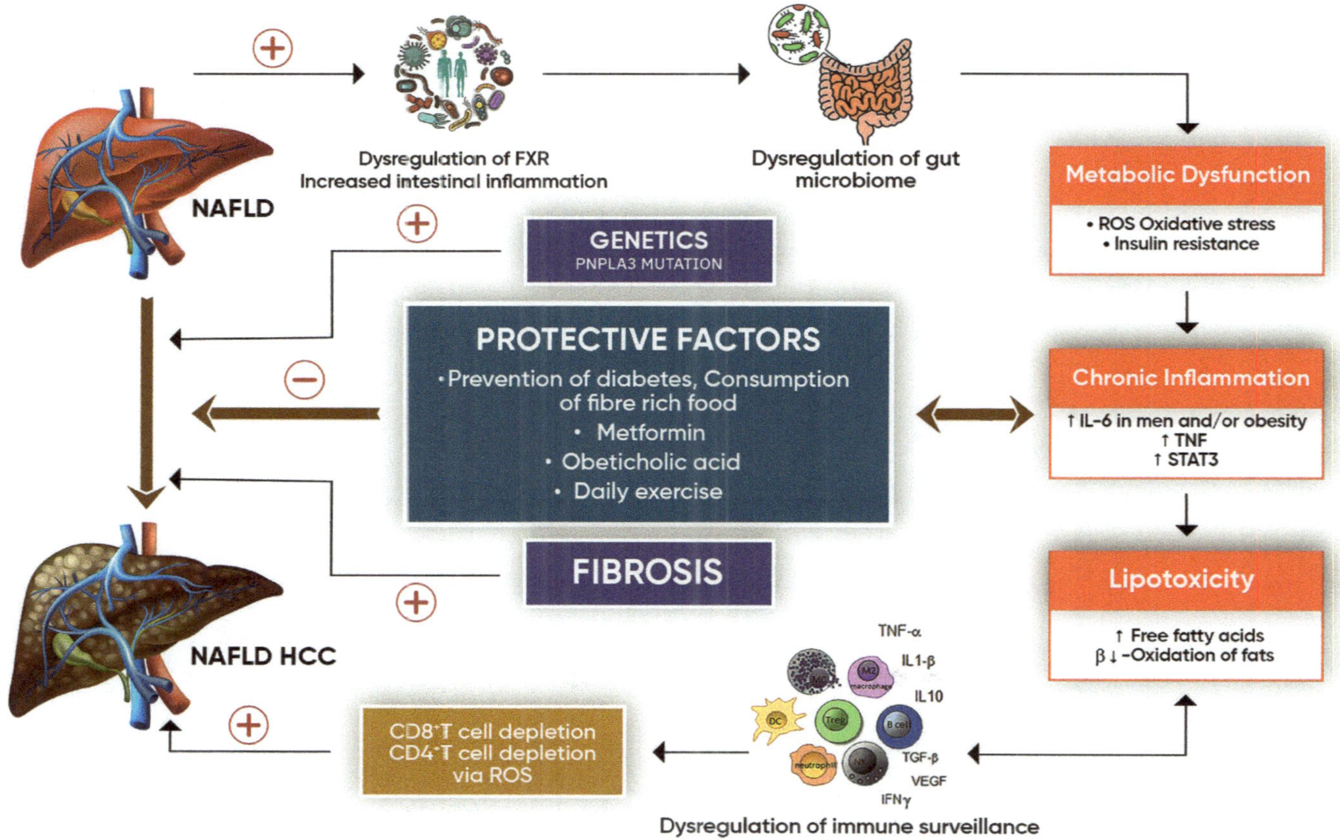

Fig. 2 See figure legend on next page.

(hazard ratio: 2.69; $P=0.007$).[49] Compared to healthy people, individuals with lean NAFLD are more prone to the risk of metabolic syndrome such as hyperlipidemia, arterial hypertension, diabetes, and insulin resistance, which may be related to a more dysfunctional adipose tissue, but not its visceral component.[49]

7. The fungal mycobiome—A synchronized component of the gut microbiome

While bacteria are often associated with the gut microbiome, microbial communities also include archaea, viruses, protists, and fungi. Gastrointestinal fungal communities comprising molds and yeasts, collectively known as the mycobiome, are essential for gut homeostasis and systemic immunity.[50] Fungal cells are 100 times bigger in volume than many bacteria, suggesting substantially higher biomass, and they may play essential roles in sustaining microbial community structure and function while contributing to <0.1% of microbial genes.[51,52] The healthy human gut mycobiome has less diversity and density than bacteria, and it is dominated by Saccharomyces, Malassezia, and Candida.[51] The mycobiome and its significance in the fatty liver are becoming more well-known. Fungal-specific internal transcribed spacer (ITS) amplicon sequencing can be used to characterize mycobiomes.[53,54]

Intestinal fungal overgrowth is related to the translocation of fungal products, as seen by higher plasma levels of 1,3-beta-D-glucan in the latest study of prolonged alcohol administration in a murine model. Furthermore, amphotericin B therapy reduces the increase in intestinal fungi and reduces the plasma level of 1,3-beta-D-glucan. Increased 1,3-beta-D-glucan stimulates Kupffer cells in the liver, prompting them to release IL-1beta, which further damages hepatocytes, causing alcohol-induced inflammation, steatosis, and hepatocyte toxicity[55] (Fig. 3).

Fig. 2 Factors that positively and negatively affect the progression of NAFL to HCC. The release of free fatty acids (FFA) is aided by insulin resistance and metabolic dysfunction. The production of reactive oxygen species (ROS) and the release of pro-inflammatory cytokines are induced by the oxidation of FFAs. Obesity also causes an increase in IL-6 and tumor necrosis factor (TNF), both of which enhance chronic inflammation and activate STAT3, an oncogenic transcription factor. Linoleic acid, a free fatty acid, impairs mitochondrial activity and causes $CD4^+$ T cells to die via selective ROS-mediated apoptosis, lowering immunological surveillance. In nonalcoholic steatohepatitis (NASH), IgA^+ cells increase which inhibits $CD8^+$ T cells and speeds up hepatocarcinogenesis. In nonalcoholic fatty liver disease (NAFLD), the loss of commensal gut flora and dysregulated farnesoid X receptor (FXR) signaling enhances intestinal permeability and hepatic inflammation, hastening hepatocarcinogenesis.

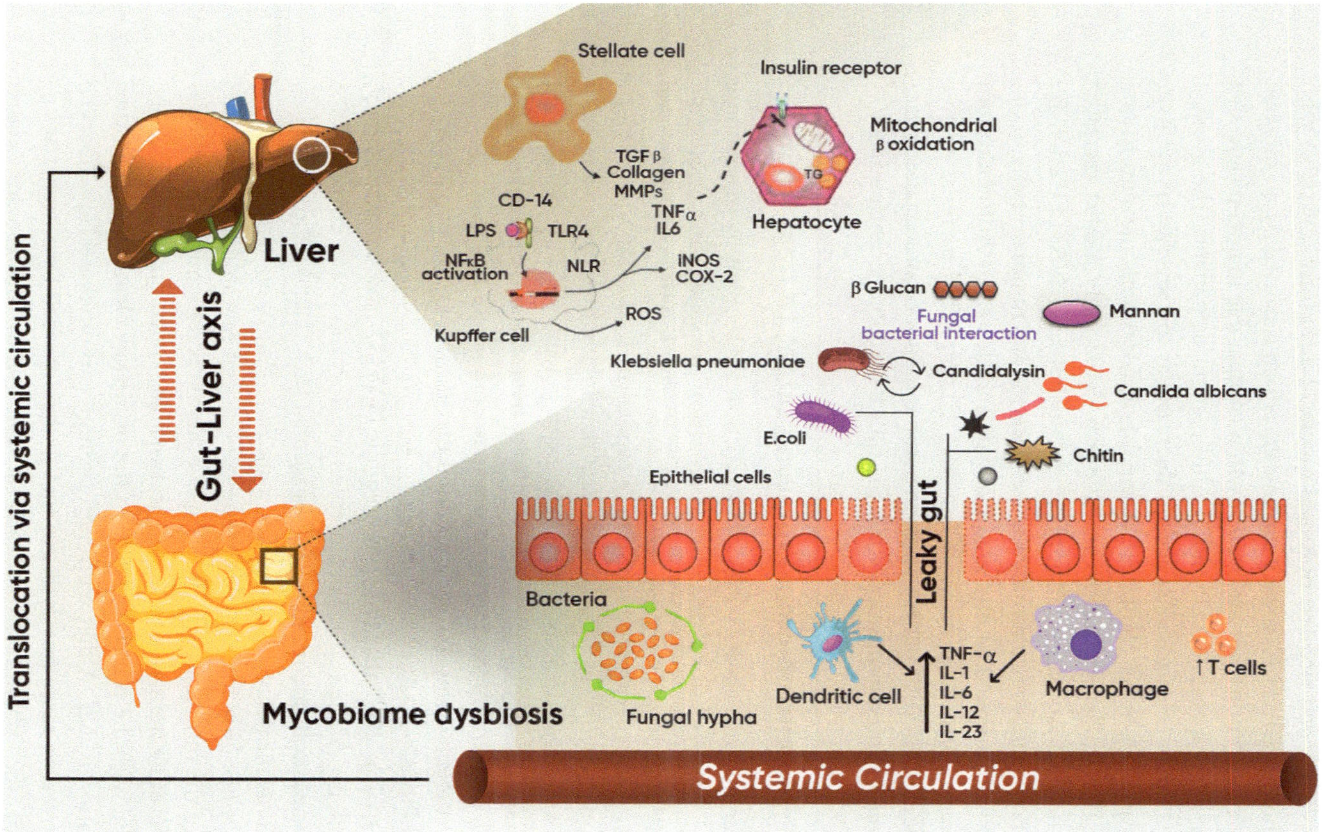

Fig. 3 Schematic of the interplay of liver and gut microbiota in non-alcoholic fatty liver disease. With distinct changed profiles in ALD and NAFLD, intestinal dysbiosis is the foundation of the impaired gut–liver axis. Among some of the significant modifications are: (1) increased microbial invasion and leaky intestinal barrier, and (2) inflammatory signaling activation and toxins accumulation in the liver (particularly in ALD and NAFLD). The portal vein, systemic circulation, and bile acid metabolism connect changes in the gut and liver. Cirrhosis is the result of persistent inflammation and recurring growth inside the liver.

In humans, as compared with healthy controls, alcoholic patients have a reduced variety and richness of fungal species. Furthermore, patients who use alcohol have fecal *Candida* overgrowth, which is caused by an increase in *C. Albicans*. While Candida overgrowth is comparable in non-progressive alcoholic liver disease (ALD) and alcoholic cirrhosis patients, the average relative abundance of *Candida albicans* decreases as the severity of the liver disease rises.[56] A novel combined bacterial-fungal dysbiosis measure having a higher Bacteroidetes/Ascomycota ratio can significantly predict increased 90-day hospitalization in cirrhotic with alcohol and NAFLD etiology. Furthermore, antibiotic exposure reduces bacterial and fungal diversity in cirrhotic, which is linked to increased Candida and decreased Autochthonous (*Ruminococcaceae, Lachnospiraceae, Clostridiales* Cluster XIV) bacterial relative abundance. In cirrhotic patients, proton pump inhibitor treatment affects bacterial diversity and composition while the fungal mycobiota remains stable.[57]

8. The microbiome: Promising target to treat NASH

The gut microbiota serves a key function for humans by maintaining intestinal homeostasis, acting for colonization resistance and protecting against pathogens, improving immune systems growth, and promoting energy and micro-nutrient synthesis. Proposal pathways for the development of non-alcoholic fatty liver disease that highlight the significance of high-fat and sugar diets in combination with gut microbiota dysbiosis can abrogate the development of non-alcoholic liver disease. Oral administration and supplementation of probiotics and prebiotics have assisted treat NAFLD by enhancing the aberrant lipid metabolism and intestinal microbiota dysbiosis.

In a trial in 2019 that enrolled 68 adult NAFLD patients with obesity, published by Ahn et al., The interventional group were cured with a multispecies probiotic mixture (*Bifidobacterium lactis, Lactobacillus acidophilus, Lactobacillus rhamnosus, Lactobacillus paracasei, Pediococcus pentosaceus and Bifidobacterium breve* for 12 weeks. The conclusions revealed that probiotics improved the intrahepatic fat fraction but there was no significant improvement of alanine transaminase (ALT) and aspartate transaminase (AST) when compared with the placebo group.[58] One of the key elements for intestinal modification is diet. A diet with high fat/high carbohydrates/low fiber has been proven to cause gut dysbiosis that is characterized by a higher abundance of the phylum Proteobacteria and genus Bacteroides, classes Erysipelotrichia and Gamma proteobacteria including ethanol-producing

E. coli.[59] Among these dietary habits, the Mediterranean diet and vegetarian/vegan diets have been found to ameliorate gut microbiota dysbiosis by augmenting *Bifidobacterium, Prevotella,* and *Faecalibacterium prausnitzii* while reducing *E. coli* and other gram-negative bacteria.[60–62] A number of evidence have indicated that the use of these intestinal microbiota therapies with dietary supplementation restores the NAFLD-related gut dysbiosis and enhances early biomarkers of NAFLD. This strategy minimizes liver damage, inflammation and NAFLD-related insulin resistance. The etiology of NAFLD is not understood; therefore, therapeutic options are limited and disappointing. NAFLD treatment approaches are generally comprehensive lifestyle changes that also include suggestions to promote physical exercise and make nutritional changes.[63,64] Additional regimens include the antioxidants, anti-inflammatory drugs, lipid-lowering agents, and insulin-sensitizing agents, as well as supplements such as Vitamin C, D and E.[64] A major emphasis needs to be acquired on developing microbiome therapeutics for patients with NAFLD. Manipulation of the microbiome is primarily conducted using medications, prebiotics, probiotics and fecal microbiota transplantation (FMT).

9. Conclusions and perspectives

The growing incidence of NAFLD and NASH has been deemed a worldwide burden. As aforementioned, evidence has emerged on the relationship between gut microbiota and their etiological involvement in the development of NAFLD. Furthermore, various mechanistic studies studying microbial interaction and the development of novel disease pathogenesis therapy regimens might give significant insight into the causes of NAFLD and NASH. High throughput sequencing techniques such as 16S ribosomal RNA gene sequencing for the measurement of species abundance, bacterial and host RNA sequencing for transcript expression, shotgun metagenomic sequencing for the gene abundance measurement and mass spectrometry for metabolomic profiling will be used in the future to recognize abundant bacterial communities and plausible host changes.[65] These robust methodologies can aid in the development of novel medicines that target the inclusion or inhibition of disease-associated metabolites that are specifically adjusted. Furthermore, FMT is a "microbiome restoration" method which has actually been proven to be extremely effective in the treatment of drug-resistant *Clostridium difficile* infections, indicating that it may be useful in treating NAFLD. During NAFLD, the dysbiotic gut microbiota may be analyzed and modulated to assist with targeted treatment. Delineating the mechanistic and molecular basis of gut microbiota-host interactions that drive NAFLD progression might lead to the development of microbiome-targeting therapies for this widespread and chronic disease.

Acknowledgments

The authors are thankful to the Indian Council for Medical Research (ICMR) for providing financial support of this work (Project no. 5/7/1747/CH/Adhoc/RBMCH-2021) and the National Institute of Pharmaceutical Education and Research, Guwahati core fund. Thanks are due to the Indian Council of Medical Research for supporting MJA in the Research Associate position (Grant number 3/1/2(19)/OBS/2019-NCD–II).

Conflict of interest

The authors have no conflicts of interest, financial or otherwise.

References

1. Bäckhed F, Ley RE, Sonnenburg JL, Peterson DA, Gordon JI. Host-bacterial mutualism in the human intestine. *Science*. 2005;307(5717):1915–1920. https://doi.org/10.1126/science.1104816.
2. Grice EA, Segre JA. The human microbiome: our second genome. *Annu Rev Genom Hum Genet*. 2012;13(1):151–170. https://doi.org/10.1146/annurev-genom-090711-163814.
3. Sarkar S, Das B, Banerjee S. Insights into the human gut microbiome and cardiovascular diseases. *J Pract Cardiovasc Sci*. 2018;4(1):10. https://doi.org/10.4103/jpcs.jpcs_18_18.
4. Bäumler AJ, Sperandio V. Interactions between the microbiota and pathogenic bacteria in the gut. *Nature*. 2016;535(7610):85–93. https://doi.org/10.1038/nature18849.
5. Brennan CA, Garrett WS. Gut microbiota, inflammation, and colorectal cancer. *Annu Rev Microbiol*. 2016;70:395–411. https://doi.org/10.1146/annurev-micro-102215-095513.
6. Belkaid Y, Hand T. Role of the microbiota in immunity and inflammation. *Cell*. 2014;157(1):121–141. https://doi.org/10.1016/j.cell.2014.03.011.
7. Ohtani N, Kawada N. Role of the gut-liver axis in liver inflammation, fibrosis, and cancer: a special focus on the gut microbiota relationship. *Hepatol Commun*. 2019;3(4):456–470. https://doi.org/10.1002/hep4.1331.
8. Lozupone CA, Stombaugh JI, Gordon JI, Jansson JK, Knight R. Diversity, stability and resilience of the human gut microbiota. *Nature*. 2012;489(7415):220–230. https://doi.org/10.1038/nature11550.
9. Caporaso JG, Lauber CL, Costello EK, et al. Moving pictures of the human microbiome. *Genome Biol*. 2011;12(5):R50. https://doi.org/10.1186/gb-2011-12-5-r50.
10. Costello EK, Lauber CL, Hamady M, Fierer N, Gordon JI, Knight R. Bacterial community variation in human body habitats across space and time. *Science*. 2009;326(5960):1694–1697. https://doi.org/10.1126/science.1177486.
11. Lynch SV, Pedersen O. The human intestinal microbiome in health and disease. *New Engl J Med*. 2016;375(24):2369–2379. https://doi.org/10.1056/NEJMra1600266.
12. Sanyal AJ. Past, present and future perspectives in nonalcoholic fatty liver disease. *Nat Rev Gastroenterol Hepatol*. 2019;16(6):377–386. https://doi.org/10.1038/s41575-019-0144-8.

13. Mitra S, De A, Chowdhury A. Epidemiology of non-alcoholic and alcoholic fatty liver diseases. *Transl Gastroenterol Hepatol.* 2020;5:16. https://doi.org/10.21037/tgh.2019.09.08.
14. Asrani SK, Devarbhavi H, Eaton J, Kamath PS. Burden of liver diseases in the world. *J Hepatol.* 2019;70(1):151–171. https://doi.org/10.1016/j.jhep.2018.09.014.
15. Younossi ZM, Koenig AB, Abdelatif D, Fazel Y, Henry L, Wymer M. Global epidemiology of nonalcoholic fatty liver disease-meta-analytic assessment of prevalence, incidence, and outcomes. *Hepatology.* 2016;64(1):73–84. https://doi.org/10.1002/hep.28431.
16. Estes C, Chan HLY, Chien RN, et al. Modelling NAFLD disease burden in four Asian regions—2019-2030. *Aliment Pharmacol Ther.* 2020;51(8):801–811. https://doi.org/10.1111/apt.15673.
17. Kolodziejczyk AA, Zheng D, Shibolet O, Elinav E. The role of the microbiome in NAFLD and NASH. *EMBO Mol Med.* 2019;11(2):e9302. https://doi.org/10.15252/emmm.201809302.
18. Imajo K, Fujita K, Yoneda M, et al. Hyperresponsivity to low-dose endotoxin during progression to nonalcoholic steatohepatitis is regulated by leptin-mediated signaling. *Cell Metabolism.* 2012;16(1):44–54. https://doi.org/10.1016/j.cmet.2012.05.012.
19. Bäckhed F, Ding H, Wang T, et al. The gut microbiota as an environmental factor that regulates fat storage. *Proc Natl Acad Sci USA.* 2004;101(44):15718–15723. https://doi.org/10.1073/pnas.0407076101.
20. Bergheim I, Weber S, Vos M, et al. Antibiotics protect against fructose-induced hepatic lipid accumulation in mice: role of endotoxin. *J Hepatol.* 2008;48(6):983–992. https://doi.org/10.1016/j.jhep.2008.01.035.
21. Adams DH, Eksteen B, Curbishley SM. Immunology of the gut and liver: a love/hate relationship. *Gut.* 2008;57(6):838–848. https://doi.org/10.1136/gut.2007.122168.
22. Cani PD, Amar J, Iglesias MA, et al. Metabolic endotoxemia initiates obesity and insulin resistance. *Diabetes.* 2007;56(7):1761–1772. https://doi.org/10.2337/db06-1491.
23. Khan A, Ding Z, Ishaq M, et al. Understanding the effects of gut microbiota dysbiosis on nonalcoholic fatty liver disease and the possible probiotics role: recent updates. *Int J Biol Sci.* 2021;17(3):818–833. https://doi.org/10.7150/ijbs.56214.
24. Xue L, He J, Gao N, et al. Probiotics may delay the progression of nonalcoholic fatty liver disease by restoring the gut microbiota structure and improving intestinal endotoxemia. *Sci Rep.* 2017;7(1):45176. https://doi.org/10.1038/srep45176.
25. Donnelly KL, Smith CI, Schwarzenberg SJ, Jessurun J, Boldt MD, Parks EJ. Sources of fatty acids stored in liver and secreted via lipoproteins in patients with nonalcoholic fatty liver disease. *J Clin Invest.* 2005;115(5):1343–1351. https://doi.org/10.1172/JCI200523621.
26. Zhu L, Baker SS, Gill C, et al. Characterization of gut microbiomes in nonalcoholic steatohepatitis (NASH) patients: a connection between endogenous alcohol and NASH. *Hepatology.* 2013;57(2):601–609. https://doi.org/10.1002/hep.26093.
27. Gerbes A, Zoulim F, Tilg H, et al. Gut roundtable meeting paper: selected recent advances in hepatocellular carcinoma. *Gut.* 2018;67(2):380–388. https://doi.org/10.1136/gutjnl-2017-315068.
28. Høverstad T, Midtvedt T. Short-chain fatty acids in germfree mice and rats. *J Nutr.* 1986;116(9):1772–1776. https://doi.org/10.1093/jn/116.9.1772.
29. Samuel BS, Shaito A, Motoike T, et al. Effects of the gut microbiota on host adiposity are modulated by the short-chain fatty-acid binding G protein-coupled receptor, Gpr41. *Proc Natl Acad Sci USA.* 2008;105(43):16767–16772. https://doi.org/10.1073/pnas.0808567105.
30. Kimura I, Ozawa K, Inoue D, et al. The gut microbiota suppresses insulin-mediated fat accumulation via the short-chain fatty acid receptor GPR43. *Nat Commun.* 2013;4(1):1829. https://doi.org/10.1038/ncomms2852.

31. Pols TWH, Noriega LG, Nomura M, Auwerx J, Schoonjans K. The bile acid membrane receptor TGR5 as an emerging target in metabolism and inflammation. *J Hepatol.* 2011;54(6):1263–1272. https://doi.org/10.1016/j.jhep.2010.12.004.
32. Neimark E, Chen F, Li X, Shneider BL. Bile acid-induced negative feedback regulation of the human ileal bile acid transporter. *Hepatology.* 2004;40(1):149–156. https://doi.org/10.1002/hep.20295.
33. Cope K, Risby T, Diehl AM. Increased gastrointestinal ethanol production in obese mice: implications for fatty liver disease pathogenesis. *Gastroenterology.* 2000;119(5):1340–1347. https://doi.org/10.1053/gast.2000.19267.
34. Zong H, Armoni M, Harel C, Karnieli E, Pessin JE. Cytochrome P-450 CYP2E1 knockout mice are protected against high-fat diet-induced obesity and insulin resistance. *Am J Physiol Endocrinol Metab.* 2012;302(5):E532–E539. https://doi.org/10.1152/ajpendo.00258.2011.
35. Spencer MD, Hamp TJ, Reid RW, Fischer LM, Zeisel SH, Fodor AA. Association between composition of the human gastrointestinal microbiome and development of fatty liver with choline deficiency. *Gastroenterology.* 2011;140(3):976–986. https://doi.org/10.1053/j.gastro.2010.11.049.
36. Romano KA, Vivas EI, Amador-Noguez D, Rey FE. Intestinal microbiota composition modulates choline bioavailability from diet and accumulation of the proatherogenic metabolite trimethylamine-N-oxide. *mBio.* 2015;6(2):e02481. https://doi.org/10.1128/mBio.02481-14.
37. Dumas ME, Barton RH, Toye A, et al. Metabolic profiling reveals a contribution of gut microbiota to fatty liver phenotype in insulin-resistant mice. *Proc Natl Acad Sci USA.* 2006;103(33):12511–12516. https://doi.org/10.1073/pnas.0601056103.
38. Zhai Y, Da SX, O'Connell R, et al. Cutting edge: TLR4 activation mediates liver ischemia/reperfusion inflammatory response via IFN regulatory factor 3-dependent MyD88-independent pathway. *J Immunol.* 2004;173(12):7115–7119. https://doi.org/10.4049/jimmunol.173.12.7115.
39. Abu-Shanab A, Quigley EMM. The role of the gut microbiota in nonalcoholic fatty liver disease. *Nat Rev Gastroenterol Hepatol.* 2010;7(12):691–701. https://doi.org/10.1038/nrgastro.2010.172.
40. Kaden-Volynets V, Basic M, Neumann U, et al. Lack of liver steatosis in germ-free mice following hypercaloric diets. *Eur J Nutr.* 2019;58(5):1933–1945. https://doi.org/10.1007/s00394-018-1748-4.
41. Lee YS, Kim YH, Jung YS, et al. Hepatocyte toll-like receptor 4 mediates lipopolysaccharide-induced hepcidin expression. *Exp Mol Med.* 2017;49(12):e408. https://doi.org/10.1038/emm.2017.207.
42. Younossi ZM, Stepanova M, Negro F, et al. Nonalcoholic fatty liver disease in lean individuals in the United States. *Medicine.* 2012;91(6):319–327. https://doi.org/10.1097/MD.0b013e3182779d49.
43. Soga T, Ohishi T, Matsui T, et al. Lysophosphatidylcholine enhances glucose-dependent insulin secretion via an orphan G-protein-coupled receptor. *Biochem Biophys Res Commun.* 2005;326(4):744–751. https://doi.org/10.1016/j.bbrc.2004.11.120.
44. Hung ND, Kim MR, Sok DE. 2-Polyunsaturated acyl lysophosphatidylethanolamine attenuates inflammatory response in zymosan A-induced peritonitis in mice. *Lipids.* 2011;46(10):893–906. https://doi.org/10.1007/s11745-011-3589-2.
45. Heimerl S, Fischer M, Baessler A, et al. Alterations of plasma lysophosphatidylcholine species in obesity and weight loss. *PLoS One.* 2014;9(10), e111348. https://doi.org/10.1371/journal.pone.0111348.
46. Feldman A, Eder SK, Felder TK, et al. Clinical and metabolic characterization of lean caucasian subjects with non-alcoholic fatty liver. *Am J Gastroenterol.* 2017;112(1):102–110. https://doi.org/10.1038/ajg.2016.318.

47. Yamakado M, Tanaka T, Nagao K, et al. Plasma amino acid profile is associated with visceral fat accumulation in obese Japanese subjects. *Clin Obes.* 2012;2(1-2): 29–40. https://doi.org/10.1111/j.1758-8111.2012.00039.x.
48. van der Hee B, Wells JM. Microbial regulation of host physiology by short-chain fatty acids. *Trends Microbiol.* 2021;29(8):700–712. https://doi.org/10.1016/j.tim.2021.02.001.
49. Hagström H, Nasr P, Ekstedt M, et al. Risk for development of severe liver disease in lean patients with nonalcoholic fatty liver disease: a long-term follow-up study. *Hepatol Commun.* 2018;2(1):48–57. https://doi.org/10.1002/hep4.1124.
50. Kau AL, Ahern PP, Griffin NW, Goodman AL, Gordon JI. Human nutrition, the gut microbiome, and immune system: envisioning the future. *Nature.* 2011;474(7351): 327–336. https://doi.org/10.1038/nature10213.
51. Clemente JC, Ursell LK, Parfrey LW, Knight R. The impact of the gut microbiota on human health: an integrative view. *Cell.* 2012;148(6):1258–1270. https://doi.org/10.1016/j.cell.2012.01.035.
52. Hallen-Adams HE, Suhr MJ. Fungi in the healthy human gastrointestinal tract. *Virulence.* 2016;8(3):352–358. https://doi.org/10.1080/21505594.2016.1247140.
53. Purahong W, Mapook A, Wu YT, Chen CT. Characterization of the Castanopsis carlesii deadwood mycobiome by pacbio sequencing of the full-length fungal nuclear ribosomal internal transcribed spacer (ITS). *Front Microbiol.* 2019;10:983. https://doi.org/10.3389/fmicb.2019.00983.
54. Fouts DE, Szpakowski S, Purushe J, et al. Next generation sequencing to define prokaryotic and fungal diversity in the bovine rumen. *PLoS One.* 2012;7(11):e48289. https://doi.org/10.1371/journal.pone.0048289.
55. Yang AM, Inamine T, Hochrath K, et al. Intestinal fungi contribute to development of alcoholic liver disease. *J Clin Invest.* 2017;127(7):2829–2841. https://doi.org/10.1172/JCI90562.
56. You N, Zhuo L, Zhou J, Song Y, Shi J. The role of intestinal fungi and its metabolites in chronic liver diseases. *Gut Liver.* 2020;14(3):291–296. https://doi.org/10.5009/gnl18579.
57. Bajaj JS, Liu EJ, Kheradman R, et al. Fungal dysbiosis in cirrhosis. *Gut.* 2018;67 (6):1146–1154. https://doi.org/10.1136/gutjnl-2016-313170.
58. Ahn SB, Jun DW, Kang BK, Lim JH, Lim S, Chung MJ. Randomized, double-blind, placebo-controlled study of a multispecies probiotic mixture in nonalcoholic fatty liver disease. *Sci Rep.* 2019;9:5688. https://doi.org/10.1038/s41598-019-42059-3.
59. Agus A, Denizot J, Thévenot J, et al. Western diet induces a shift in microbiota composition enhancing susceptibility to adherent-invasive E. coli infection and intestinal inflammation. *Sci Rep.* 2016;6:19032. https://doi.org/10.1038/srep19032.
60. Mitsou EK, Kakali A, Antonopoulou S, et al. Adherence to the Mediterranean diet is associated with the gut microbiota pattern and gastrointestinal characteristics in an adult population. *Br J Nutr.* 2017;117(12):1645–1655. https://doi.org/10.1017/S0007114517001593.
61. Ruengsomwong S, La-Ongkham O, Jiang J, Wannissorn B, Nakayama J, Nitisinprasert S. Microbial community of healthy thai vegetarians and non-vegetarians, their core gut microbiota, and pathogen risk. *J Microbiol Biotechnol.* 2016;26(10):1723–1735. https://doi.org/10.4014/jmb.1603.03057.
62. De Filippis F, Pellegrini N, Vannini L, et al. High-level adherence to a Mediterranean diet beneficially impacts the gut microbiota and associated metabolome. *Gut.* 2016;65 (11):1812–1821. https://doi.org/10.1136/gutjnl-2015-309957.
63. Chalasani N, Younossi Z, Lavine JE, et al. The diagnosis and management of non-alcoholic fatty liver disease: practice guideline by the American Association for the Study of Liver Diseases, American College of Gastroenterology, and the American Gastroenterological Association. *Hepatology.* 2012;55(6):2005–2023. https://doi.org/10.1002/hep.25762.

64. Eslamparast T, Eghtesad S, Poustchi H, Hekmatdoost A. Recent advances in dietary supplementation, in treating non-alcoholic fatty liver disease. *World J Hepatol.* 2015;7 (2):204–212. https://doi.org/10.4254/wjh.v7.i2.204.
65. Del Chierico F, Nobili V, Vernocchi P, et al. Gut microbiota profiling of pediatric nonalcoholic fatty liver disease and obese patients unveiled by an integrated meta-omics-based approach. *Hepatology.* 2017;65(2):451–464. https://doi.org/10.1002/hep.28572.

CHAPTER NINE

Hepatic drug metabolism and gut microbiome

Tanshi Mehrotra[a] **and Subir Kumar Maulik**[b,*]
[a]Translational Health Science and Technology Institute, Faridabad, Haryana, India
[b]Indian Council Of Medical Research, New Delhi, India
*Corresponding author: e-mail address: skmaulik@gmail.com

Contents

Abstract

This chapter focuses on intestinal microbiota and its effect on drug metabolism. Here, we discussed about different drugs which are metabolized either by some enzymes or gut microbiota and their mechanism. Nowadays, consuming drugs without a doctor's prescription is common. This chapter will make people aware about its negative consequences and how it is related to gut microbiota dysbiosis. Intestinal disorders like inflammatory bowel disease (IBD), colorectal cancer (CRC) and metabolic disorders such as obesity and type 2 diabetes mellitus (T2D) are found to be affected with gut microbiota dysbiosis. To address this issue, we discussed a variety of strategies such as fecal microbiota transplantation (FMT), probiotics and antibiotic stewardship programs which are commonly used to tackle this problem.

Progress in Molecular Biology and Translational Science, Volume 191
ISSN 1877-1173
https://doi.org/10.1016/bs.pmbts.2022.07.005

1. Introduction

The microbiome is popularly known as the second human genome with approximately 10^{13}–10^{14} microbial cells.[1,2] It is well known that all the body sites have host specific microbiota such as respiratory tract, reproductive organs, skin, oral cavity and gastrointestinal tract (GI tract).[3] But, interestingly the human gut in comparison with the whole human body holds approximately 100 trillion microbes with 5000 different species showing the abundance and diversity in a human body.[1] This may be due to higher exposure of the gastrointestinal tract (GI tract) to the environment due to different dietary habits and geographical distribution. Nowadays, it is quite evident that use of drugs without proper prescription by doctors is increasing which is modulating the human gut microbiome drastically. Drugs are typically small chemical compounds that are recognized as foreign to the human body.[4] These drugs when taken orally or parenterally reaches the intestine and affect the gut microbiome either in a positive or negative way. Thus, it is really essential to study how drug metabolizes and how it interacts with intestinal microbiota.[5] Biotransformation is a process of breakdown of organic compounds into simpler forms or metabolites which takes place mostly in the liver, it leads to the excretion of molecules, both endogenous and exogenous. This may lead a drug to be either active, inactive or toxic.[6,7] In hepatic drug metabolism drug enter into the enterohepatic recirculation where the drug is transported from enterocytes to hepatocytes via the portal vein and through bile secretion in intestine where intestinal microbiota can restore the drug activity establishing a bidirectional relationship between intestine and liver.[8] Sometimes if the drug is not metabolized or detoxified appropriately and is circulating in the body for longer durations it may lead to liver toxicity. There are different types of drug metabolizing enzymes cytochrome P450, cytochrome b5 and NADPH-cytochrome P450 reductase. In liver, it is cytochrome P450 which is the major driver for drug metabolism and xenobiotics and constitutes hydrolases, oxidases and reductases.[4,9] It is well known that throughout the GI tract microbial composition is different mostly higher abundance is seen in the colon whereas levels of enzymes like CYP450 are lowest in the colon.[10–12] Prodrugs are transformed into their pharmacologically active form by drug-metabolizing enzymes. In addition, there are also conjugating enzymes and drug transporters. These play crucial roles in the absorption, distribution, metabolism and excretion of drugs, and alterations in their functionality may have a considerable impact on a drug's pharmacokinetics and pharmacodynamics.[13] Thus metabolizing a drug means converting it into a hydrophilic form via enzymes to make it more soluble and fasten the absorption process by the gut.[11,14] The major sites where CYP450 enzymes are found and exert their action on drug metabolism are the small intestinal epithelia and the proximal tubules of the kidneys, smooth endoplasmic reticulum (ER) and mitochondria of the hepatocytes. The nomenclature system for CYP450 enzymes was first made by Nebert about 40 years ago. More than 115 genes and pseudogenes make up the human CYP450 gene family. Adrenal glands, platelets, blood vessels, adipose tissues, reproductive organs and all critical organs such as the brain, heart, lung, kidneys, stomach and liver comprise of CYP450s. The major role in overall metabolism and drug clearance is played by the liver and small intestine. The most prevalent CYP450 enzyme in the human liver is CYP3A4, which is followed by CYP2E1 and CYP2C9.[13] Our main goal in this chapter is to inform public about the benefits and risks associated with gut microbiota and how it affects drug metabolism.

2. Gut microbiota composition and its association with diseases

Human gut microbiota is essential for shaping the gut and promoting a healthy lifestyle. It has a strong impact on different functions of the body like digestion, synthesis of amino acids, immunomodulation and other metabolic functions thus it is very essential to maintain a balance between these microorganisms as a slight misbalance can create havoc in the human body stimulating several diseases.[15] There are a total of six dominant phyla for gut microbiota these are Bacteroidetes, Firmicutes, Fusobacteria, Actinobacteria, Verrucomicrobia and Proteobacteria. Gut microbiota composition varies from proximal to distal axis, i.e., (i) Esophagus—*Pseudomonas, Bacteroides, Gemella, Veillonella, Prevotella, Rothia* sps., *Streptococcus, Megasphaera*; (ii) Stomach—*Streptococcus, Helicobacter pylori, Enterococcus, Lactobacillus, Prevotella*; (iii) Small intestine—*Lactobacillus, Streptococcus, g-Proteobacteria, Bacteroides, Enterococcus, Clostridium*; (iv) Colon—*Clostridium, Prevotella, Bacteroides, Lactobacillus, Eubacterium, Ruminococcus, Porphyromonas, Enterobacterium, Enterococcus, Peptostreptococcus, Streptococcus, Fusobacteria*; (v) Cecum—*Fusobacteria Lachnospira, Ruminococcus, Roseburia, Butyrivibrio, Fecalibacterium.*[16] It is said that in a healthy state Firmicutes and Bacteroidetes are the two dominant phyla whereas in a disease state phylum Proteobacteria is dominant.[11,17] It is well known that dysbiosis in the gut microbiome is a major cause of intestinal disorders such as inflammatory bowel disease (IBD) and colorectal cancer (CRC) and metabolic disorders such as obesity and type 2 diabetes mellitus (T2D). According to the world health organization (WHO) colorectal cancer

(CRC) is the second most common cause of cancer deaths worldwide in 2020 (935,000 deaths) (WHO,[18]). A study reported higher levels of *Fusobacterium nucleatum* (*Fn*) in patients with adenomas and CRC in comparison to healthy controls other bacteria such as Enterotoxigenic *Bacteroides fragilis* (*ETBF*), *E. coli* was also found to induce CRC development.[19] In case of inflammatory bowel disease (IBD), a study reported that phylum Firmicutes (species *F. prausnitzii*) was underrepresented[20] whereas in case of obesity it was proven that a high-fat diet leads to gut microbiota dysbiosis which leads to low levels of *Lactobacillus, Bifidobacterium, Bacteroides* and *Akkermansia* and high levels of Firmicutes (*Clostridium*), *Prevotella* and *Methanobrevibacter.*[21] Several studies have reported gut microbiota dysbiosis as a prominent cause of insulin resistance in Type 2 diabetes mellitus (T2D).[22] It was found that there is a negative association of T2D with genus *Bifidobacterium, Bacteroides, Faecalibacterium, Akkermansia and Roseburia whereas Ruminococcus, Fusobacterium, and Blautia* were positively associated with T2D.[23] Not only metabolic or intestinal disorders even cardiovascular diseases are also considered linked with the gut microbiome. Jie et al. showed the link between the gut microbiome and atherosclerotic cardiovascular disease (ACVD) authors stated that there is an increase in abundance of *Enterobacteriaceae* and *Streptococcus* species when compared with healthy controls.[24] Thus, gut microbiota dysbiosis is a problem which should be considered not only at the clinical level but at the individual level too and people should follow a healthy lifestyle to overcome this. In contrast to the microbiome, the healthy human gut has been found to include a variety of fungi (mycobiome). Only a few kinds of fungi, notably *Candida*, yeasts, and members of its family Dipodascaceae (Galactomyces, Geotrichum, and Saprochaete), can thrive and inhabit the gut.[25] *Malassezia* and the filamentous fungus *Cladosporium* are potential colonizers more work is needed to clarify their role in hepatic drug metabolism. Although *Penicillium* and *Debaryomyces* species, which are frequently found on fermented foods but cannot grow at human body temperature (*Saccharomyces cerevisiae*, a fermentation agent and occasionally probiotic; *Aspergillus* species, ubiquitous moulds) and are not colonizers, they are likely to have an impact on gut ecology. The gut mycobiome seems to be less stable than the bacterial microbiome and is probably influenced by environmental factors.[26]

3. Drug metabolism pathways

3.1 Phase I Pathway

CYP450s are the major group of phase I drug-metabolizing enzymes that chemically alter drugs to produce water-soluble products that can be excreted via the kidney or liver.[27] In the cytoplasm, cofactors, like, NADPH-Cytochrome P450 reductase, cytochrome-reductases are essential to carry out the biotransformation reactions. Iron is necessary for the synthesis of CYP450s and is located in the middle of the binding site for the enzyme and substrate. Thus, the status of each of these regulators influences the overall functional activity of CYP450s, which in turn causes variations in metabolic capacity across and among individuals Table 1 and Fig. 1. This causes differences in the pharmacologic effects (pharmacokinetics and pharmacodynamics) of the same dose of a drug. Several factors can influence CYP450 enzyme expression and function. Drug response is reduced as a result of Induction of CYP450 enzymes which results in increased clearance of drugs that are substrates of CYP450 enzymes. CYP450 inducers could also decrease the risk of hepatotoxicity of certain drugs. Rifampicin (an anti TB drug) and phenobarbital, immunosuppressants like tacrolimus and cyclosporine, sirolimus and everolimus are clinically important inducers of CYP3A. On the other side, CYP450 enzyme inhibition results in decreased activity of the drug-clearing enzyme. CYP450 inhibitors can cause toxicity by dramatically raising the blood levels of different CYP450 enzyme substrates. Ritonavir, an anti-HIV medication, and azole antifungals are examples of CYP450 enzyme inhibitors.

3.2 Phase II pathway

Transferase enzymes combine drugs or their metabolites from phase I pathway with a hydrophilic endogenous substance during phase II drug metabolism Fig. 1 and Table 1. The UDP-glucuronosyltransferases (UGTs), sulfotransferases (SULTs), N-acetyltransferases (NATs), glutathione S-transferases (GSTs), thiopurine S-methyltransferases (TPMTs), and catechol O-methyltransferases (COMTs) are the most prevalent phase II drug-metabolizing enzymes.[28]

Table 1 Drug metabolism pathways.

	Phases of drug metabolism
Phase 1	Drug metabolism greatly depends on hepatic CYP450s
Phase 2	A range of conjugation enzymes expressed in the liver metabolize between 40% and 70% of xenobiotics
Phase 3	Drug transporters aid in both the efflux of xenobiotics and drugs into bile as well as their uptake into hepatocytes

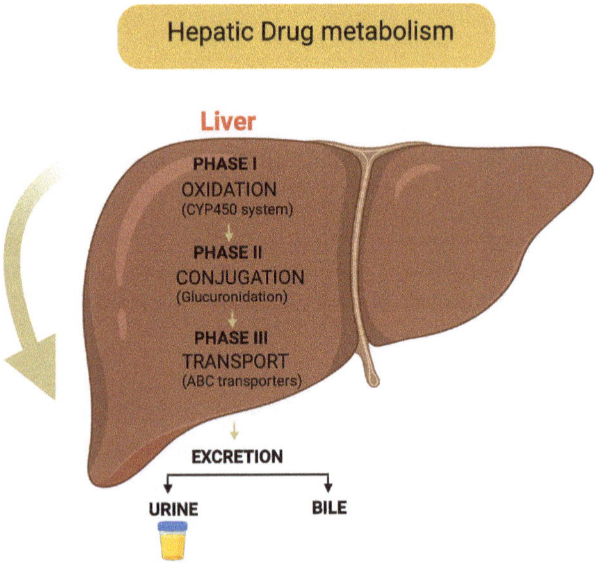

Fig. 1 Overview of drug metabolism pathways.

3.3 Phase III pathway

The ATP-binding cassette (ABC) and solute carrier (SLC) super-families are the two main supergroups that make up the phase III pathway. Transmembrane proteins called drug transporters transport large and/or ionised molecules in and out of the cells Fig. 1 and Table 1.[29] Significant diet/herb–drug interactions may result from how some phytochemicals (bioactive compounds in diverse plants) impact CYP expression and activity. Flavonoids have been shown to interfere with the activities of CYP enzymes in a clinically significant way.[30] According to reports, isoflavones can either induce or inhibit CYP enzymes. Transforming growth factor beta (TGF) and tumor necrosis factor alpha (TNF) and other proinflammatory cytokines either directly or indirectly regulate the expression of CYP in cancer cell lines, animal models and humans.[31–33]

4. Gut microbiota modulating drug metabolism

Intestinal bacteria play a pivotal role in metabolizing endogenous and exogenous substances. Although the major site of biotransformation is the liver only but since drug when administered orally pass through the intestine and thus bacteria residing in the intestine majorly in the colon exert their effect by making a drug transformed into active, inactive or toxic metabolites.[8] The small intestine's gut microbiota release a variety of enzymes that lead to a wide range of bio-transformations, including demethylation, hydrolysis, azo, nitro, sulfoxide, and N-oxide reductions, deacetylation and conjugate hydrolysis which are direct processes. Thus, the gut microbiota act both ways directly and indirectly influencing drug metabolism. By indirectly altering the host's drug metabolism, disposition and competing with metabolites produced by bacteria for pathways involved in xenobiotic metabolism.[34,35] The enzymes of intestinal bacteria are mostly involved in the reduction and deconjugation, as opposed to the primary forms of biotransformation in the liver, which are the oxidation and conjugation. Drugs taken orally undergo the first-pass reaction in the liver and enterohepatic recirculation in the stomach and liver. This generally has an impact on the drugs pharmacokinetics. The bioavailability of oral medications may be impacted by various processes, which may be influenced by the microbial enzyme activity in the gut. Additionally, through altering the metabolism of bile acids, the gut microbiota may influence the bioavailability of medications taken orally as well as the absorption of prodrugs from the upper intestine. Stomach and small intestine has lowest bacterial population whereas colon is the dominant contributor.[36] From the stomach to the large intestine, the abundance and different species of the gut microbiota population declines.[36] Oral drug administration, particularly those that are very inefficiently absorbed from the small intestine, inevitably interacts with the distal gut bacteria.[37] Chloramphenicol is a broad-spectrum antibiotic which is majorly used as an antibacterial drug as it diffuses through the cell wall and can inhibit bacterial protein synthesis.[38] It was discovered that nitro-reduction reactions by bacteria can affect chloramphenicol which is a nitroaromatic drug. Thus, it is one of the major examples of drug metabolism by gut microbiota.[39] Statins are the commonly used drug for cardiovascular diseases. It is used to lower the LDL (Low-density lipoprotein) cholesterol levels in plasma by inhibiting 3-hydroxy-3-methylglutaryl-CoA (HMG-CoA) reductase.[40] Zhao et al. stated that Fluvastatin has a positive association with *Escherichia/Shigella*, *Ruminococcaceae* UCG 014, and *Sutterella* whereas for rosuvastatin (ROS), simvastatin (SIM) and atorvastatin (ATO) gut microbiota composition remained static which showed that all statins can be degraded by gut microbiota.[41] Even the evidences are found that peptide drugs such as insulin and calcitonin can be degraded by some proteolytic enzymes present in the caecal contents of rats.[37] Different types of drugs metabolized by gut microbiota and keystone taxa involved in drug metabolism with common bacterial species are described in Table 2.

Table 2 Describing different types of drugs metabolized by gut microbiota and their mechanism of action.

Drug	Bacterial species/ enzymes	Mechanism of action	Disease/infections/drug type	References
Sulfasalazine	Diazo-reductase enzymes	Reduction of the diazo-bond in the colon produces 5-ASA (5-amino salicylic acid) and sulfapyridine	Ulcerative Colitis [UC]	6,42
Digoxin	*Eggerthella lenta*	Cardiac glycoside reductase (*cgr*) operon inactivates digoxin drug when it binds to Cgr2 (protein related to fumarate reductase) which further leads to reduction	Cardiac drug	6,43
Levodopa (L-dopa)	[a]*Helicobacter pylori*	L-dopa directly interacts with *H. pylori* outer membrane proteins, which are in charge of the organism's adherence to stomach epithelial cells	Parkinson's disease	6,44,45
Metronidazole	*Clostridium perfringens*	By rupturing the imidazole ring, *C. perfringens* causes reductive alteration of metronidazole and produces reduced metabolites such *N*-(2-hydroxyethyl)-oxamic acid and acetamide	Anaerobic infections (bone, joint, central nervous system and respiratory tract infections)	46
Sorivudine	*Bacteroides eggerthii* and *Bacteroides vulgatus*	It is converted into metabolite(E)-5-(2-bromovinyl) uracil (BVU)	Antiviral drug	37,47
Irinotecan	[b]*Escherichia coli*	*E. coli* produces β-glucuronidase, which converts SN-38G (inactive metabolite) to SN-38 (active metabolite) in the intestinal lumen	Gastrointestinal carcinomas and small cell lung cancer	48,49

Brivudine	[a]*Bacteroides* spp.	Metabolized to bromovinyluracil	Antiviral drug	11,44
Prontosil Neoprontosil	Azoreductase enzyme	Conversion of prontosil and neoprontosil to sulfanilamide	Antibacterial drug	50
Olsalazine	Azoreductase enzyme	Olsalazine transformed into two 5-aminosalicylic acid	Ulcerative Colitis [UC]	51
Nitrazepam	Nitroreductase enzyme	Nitroreductase enzyme activates Nitrazepam by nitroreduction to 7-aminonitrazepam by intestinal bacteria which is further converted to 7-acetylaminonitrazepam in liver	Anti-anxiety drug	52
Lactulose	*Lactobacillus* spp., [a]*Bacteroides* spp. and [b]*E. coli*	Intestinal bacteria hydrolyzes Lactulose drug into lactic and acetic acids	Hepatic encephalopathy	44,53
Levamisole	[a]*Bacteroides* and *Clostridium* spp.	Levametabol-I, II, and III are three thiazole ring-opened metabolites of levamisole that are produced by the gut microbiota	Anthelmintic drug and colon cancer	37,44
Balsalazide	*Unknown*	Balsalazide metabolizees to 5-aminosalicylic acid by gut microbiota	Ulcerative Colitis [UC]	54

[a]Represents keystone species.
[b]Represents common bacterial species involved in drug metabolism.

5. Factors affecting the metabolism of drugs

Age, gender, pregnancy, comorbidities, solid organ transplantation, drugs, genetic polymorphism, and the gut flora all have an impact on metabolism. According to reports, the human gut microbiota is a key location for the xenobiotic metabolism, modulating drugs pharmacokinetics. Increased drug metabolism by the gut microbiota is definitely a possibility.[55,56] In mice, it has been demonstrated that gastrointestinal microbes can alter CYP host expression. The pharmacokinetics of xenobiotics are altered by the microbiota, and the latter may also have an impact on the microbiota's viability.[57,58]

6. Causes and imperative measures to tackle gut microbiota dysbiosis

It is well known that gut microbiota composition is not static and vary from individual to individual due to the individual lifestyle and geographical variations. But not only these but some external factors including antibiotic/drugs usage also contribute to dysbiosis Fig. 2.

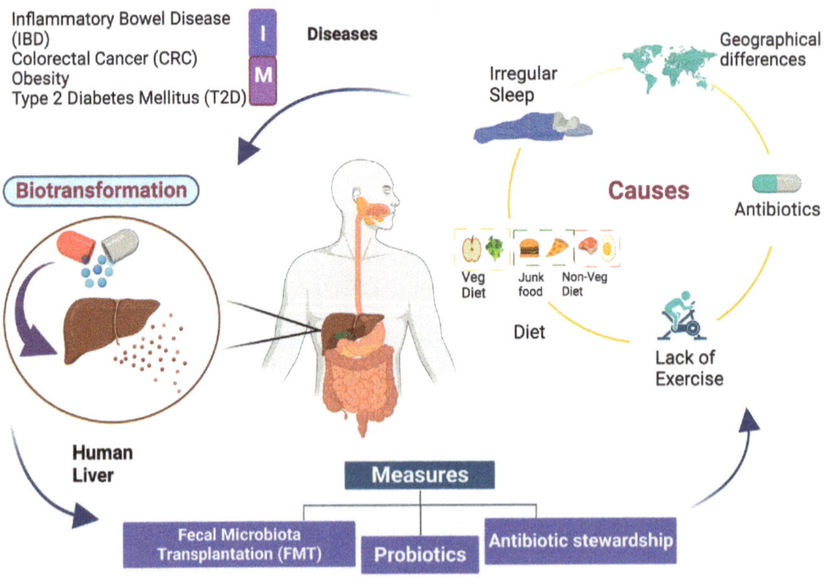

Fig. 2 An overview of causes and imperative measures to tackle gut microbiota dysbiosis and its relation to hepatic drug metabolism.

6.1 Lifestyle

Diet is a very crucial aspect of our life. The food we eat determines whether we become healthy or disease-prone. It is commonly recognized that following a vegetarian or vegan diet has a good impact on the gut microbiome and that these people differ from those who eat omnivorously. It was found that those who follow a plant-based diet or vegans observed have higher counts for Bacteroidetes. Fibrous diet promotes the growth of lactic acid bacteria, such as *Ruminococcus*, *E. rectale*, and *Roseburia* and a decrease in *Clostridium* and *Enterococcus* species levels. Polyphenols which are abundant in plant-based diet leads to an increase in *Bifidobacterium* and *Lactobacillus* species which are majorly used in probiotics.[59] It is well known that sleep has a strong impact on our life and a disturbed sleep pattern leads to stress and anxiety but recently Robert P. Smith et al. reported a linkage between sleep physiology and the gut microbiome. They found Bacteroidetes and Firmicutes phylum were positively associated with sleep efficiency, interleukin-6 concentrations and abstract thinking and taxa *Lachnospiraceae*, *Corynebacterium* and *Blautia* were negatively associated. Thus gut microbiome can be manipulated to improve sleep efficiency.[60] Even exercise plays important role in balancing gut microbiome it was found that by doing exercises regularly, ratio of Bacteroidetes-Firmicutes improves which eventually contributes to weight loss and reduce obesity related problems.[61] The gut microbiome has a major impact on mood disorders, according to most research. In patients, bacteria that make short-chain fatty acids (SCFA) were reduced, whereas bacteria engaged in pro-inflammatory and lipid metabolism were elevated. *Actinobacteria* and *Enterobacteriaceae* were shown to be more abundant in patients with bipolar illness or major depressive disorder, while *Faecalibacterium* was found to be less abundant. As a result, the brain-gut-microbiota axis is critical and should be taken into account.[62] Diet, exercise, sleep and our mood are all important aspects of our lifestyle, and it should be our top priority to eat well and exercise regularly in order to maintain a healthy gut microbiome and a happy lifestyle.

6.2 Geographical variations

Several findings advocate the impact of geographical differences on the gut microbiome. A study reported gut microbiome composition of the hunter-gatherer populations is highly abundant with *Prevotella*, Proteobacteria, Spirochaetes, Clostridiales, *Ruminobacter*. In comparison urban communities has an abundance of *Bacteroides, Bifidobacterium*, and Firmicutes.[63] Even climate has a significant impact, in a study 1020 healthy subjects were taken from 23 populations and six published studies were considered. They found a positive correlation with Firmicutes and a negative correlation with Bacteroidetes with respect to latitude.[64] Yatsunenko et al. investigated 531 individuals fecal samples. The cohort comprises of healthy children and adults from the Amazonas of Venezuela, rural Malawi and US metropolitan areas as well as monozygotic and dizygotic twins. There were notable disparities between US citizens and those from the other two countries in terms of bacterial assemblages and functional gene repertoires.[65] Similarly, a recent study in Malaysia examined gut microbiota composition across different ethnic groups. The subjects were drawn from Malaysia's three dominant ethnic groups: 636 Chinese, 123 Malay and 248 Indians. Interestingly, the authors suggest that the occurrence of the same gut microbiota composition is attributable to shared ethnic origin because all three major ethnic groups lived in Malaysia and had cultural similarities.[66] It has been reported in various studies that rural people suffer from the majority of health problems as a result of poor hygiene and water conditions. In contrast to this Monaghan et al. found *Clostridioides difficile* abundance in urban residents as compare to rural, also urban subjects microbiome was found to be enriched with metabolic pathways which are related to the industrial compounds and third generation cephalosporins and carbapenems resistant genes.[67] To identify gut microbiome composition with respect to the Indian landscape 1000 subjects were involved from overall India and their 16S rRNA gene profiling data from fecal microbiota was analyzed. They found Firmicutes, Bacteroidetes and Proteobacteria phyla with major OTUs. More than 80% prevalence was observed for genera *Prevotella, Faecalibacterium, Bacteroides, Roseburia, Megasphaera, Eubacterium* and *Gemmiger.*[68] As a result, all of these studies suggest how regional differences influence gut microbiome composition.

6.3 Impact of host genetics and environment on gut microbiome

Genome-wide association studies can be used to evaluate the effect of host genetics. This approach focuses on examining genomic markers for disease-associated genetic variants.[69] Investigating the influence of the environment and host genetics on gut microbiota is a larger subject within itself. Turnbaugh et al. reported similar gut microbiota composition between family members but variations were observed at an individual level. When these variations were examined in monozygotic and dizygotic twins a comparable co-variation was observed.[70,71] A study reported microbiome comparison between dizygotic pairs ($n=38$) and monozygotic pairs ($n=18$) who were born with cesarean delivery. Meconium samples were collected and 16S rRNA gene sequencing was done. Similar microbiological classifications are found in the results, although distinct relative abundances of some taxa, like Firmicutes ($P=0.05$, Wilcoxon test), are found at the phylum level. Also, reduced abundances of five genera ($P<0.05$), i.e., *Rheinheimera*, *Proteus*, *SMB53*, *Sphingobium*, and *Megamonas* was observed in dizygotic pairs.[72] Several studies supported host genetics influence microbiome dysbiosis which eventually leads to uncountable diseases. Genome-wide association studies indicated *NOD2* gene mutations are associated with IBD whereas genetic variants *SLC9A2, ELAVL4* and *LINGO2,* has been found to be associated with both obesity and *Blautia* abundance.[73] Several studies have been published in support of host genetics influencing gut microbiome, however, the environment also has a significant impact on gut microbiota. Rothschild et al. looked at genotype and microbiome data from 1046 healthy subjects who shared the same environment but were genetically different. They discovered that the gut microbiome makeup of genetically unrelated persons that share the same household is strikingly similar, and that diet, medicines, and anthropometric measurements account for 20% of inter-person microbiome diversity. They also show that, when compared to models that just employ host genetic and environmental data, microbiome data improves prediction accuracy for numerous human traits, such as glucose and obesity measurements.[74] A study compared the gut microbiome of children adopted as an infant to children who were genetically unrelated but shared the same household and children who were genetically related but raised in different households. They reported common environment is the major driver in shaping the microbiome community during childhood but once it colonizes host genetic factors may influence variations in the abundance of different bacterial species.[75]

6.4 Antibiotics/drugs exposure

Antibiotics are considered as beneficial in treating several infections such as strep throat, whooping cough, urinary tract infection (UTI) but on the other side it exerts selection pressure on bacteria due to this selection pressure they tend to develop antibiotic resistant genes which is a leading problem nowadays. Antimicrobial resistance is predicted to be the cause of 10 million deaths per year by 2050 provided steps were taken.[76–78] It is well known that antibiotics are used to kill pathogenic bacteria but due to broad spectrum antibiotics commensal microbes also get affected or killed unintentionally. Antibiotics can kill bacteria both directly or indirectly one of the major example is vancomycin which is mainly used to treat bacterial infections such as colitis and it is reported that it can lead to a reduction in fecal microbial diversity and gram-positive bacteria. Similarly, *Bifidobacterium adolescentis* uses fructo-oligosaccharides and starch and leads to the production of lactate and acetate which is further utilized by butyrate producing anaerobes as they cannot utilize fructo-oligosaccharides and starch directly. This cross feeding can be affected by the reduction in *B. adolescentis* number due to antibiotics exposure.[79] Thus, we should use antibiotics cautiously as it can lead to gut microbiome dysbiosis. McDonnell et al. analyzed a total of 12 publications out of 4668 and discovered that macrolide exposure reduces richness twice as compared to penicillin. *Bifidobacteria* and *Lactobacillus* were significantly reduced whereas Proteobacteria (*E. coli*) was significantly increased.[80] Antibiotics pipeline is getting narrower day by day, thus reducing its consumption and researches on new alternatives should be our major priority.

7. Measures

Important steps we may take to address the dysbiosis of gut microbiota is fecal microbiota transplantation therapy (FMT) which is one of the most well-known treatments and one effective strategy is to use probiotics and antibiotic stewardship Fig. 2.

7.1 Fecal microbiota transplantation therapy (FMT)

Fecal microbiota transplantation therapy (FMT) is a process of transmitting stool from a healthy donor to the gastrointestinal tract (GI) of the patient to balance the gut microbiota dysbiosis. The concept behind this is to help in reverting the gut microbiota to its earlier state to live a healthy lifestyle.[81] It is used to treat *Clostridium difficile* infection (CDI), Inflammatory bowel disease (IBD), obesity and colorectal cancer. *C. difficile* is an etiological agent of diarrhea and colitis occurs mostly at the age of 65 or above majorly due to overconsumption of antibiotics, hospital-acquired infections or weakened immunity. All these factors accelerate the chances of infection in the gut. It is an anaerobic bacterium and symptoms include nausea, vomiting, stomach pain, fever and diarrhea. An estimated 2,23,900 cases and 12,800 deaths were reported from patients admitted to hospital during 2017 in the United States.[82] Borren et al. reported the estimated prevalence of

CDI with 19.5% for East Asia, 10.5% for South Asia, and 11.1% for the Middle East.[83] Various studies have shown that the cure rates for fecal microbiota transplantation (FMT) for CDI has ranged between 68% to 95% in general population.[84–87] Despite various comorbidities, prolonged antibiotic treatment, or lengthy hospitalization, FMT has a high success rate. Shogbesan et al. reported 88% success rate after a single round of FMT and 93% after multiple FMTs in an immunocompromised population.[88] Comparatively, Kelly et al. reported a 78% therapeutic rate following a single FMT and 89% cure rate with multiple FMT.[89] Both these studies point toward the efficacy of FMT depending on how well the targeted population is and if the patient is stable enough then the cure rates will be high. FMT is a complicated process that requires different skill sets to operate patients suffering from gut microbiota dysbiosis. It is itself very controversial and has lots of ethical challenges associated with it.

7.2 Probiotics

Probiotics are defined as the intake of live microorganisms in an adequate amount that are beneficial to the host health. These probiotics, which are commercially available in the form of yogurt or other additional goods are excellent examples (Probiotics, NCCIH). Most probiotics available in the market are composed of genera *Lactobacillus* or *Bifidobacterium*, and bacteria *Escherichia coli*, *Bacillus subtilis*, *Saccharomyces boulardii* and *Enterococcus faecium*.[90] Till now, we have read success stories of probiotics in restoring gut microbiota and treating several metabolic or intestinal diseases but a study by Bloemendaal et al. supported probiotics positive role in improving cognitive memory during acute stress. They give multistrain probiotics (*Lactobacilli*, *Lactococci*, and *Bifidobacteria*) in healthy female subjects for 28days. Increased abundance in *Ruminococcaceae_UCG-003* was seen after probiotics supplementation, these subjects were found more protected from stress related effects on working memory.[91] Netrotizing Enterocolitis (NEC) is a serious gastrointestinal disease and usually affects newborn infants or premature babies. During this condition, a reduction in Negativicutes and doubling in Gammaproteobacteria was seen. After consuming probiotics relative risk ratio of developing NEC was found to be 0.33 (95% CI, 0.24–0.46). Acute infectious diarrhea is common in children between 6 months to 2 years, for treating acute infectious diarrhea commonly used probiotics are *L. rhamnosus* GG and *L. reuteri*. Antibiotic-associated diarrhea (AAD) uses *Bifidobacillus* as the main component in probiotics for children treatment.[92] Probiotics are considered safe for use but in people with severe illnesses or immune-compromized population risks increases, there are chances of transferring of antibiotic resistant genes from probiotic microorganisms or the contaminants present in the cocktail are some serious harmful effects that should be considered while consuming (Probiotics, NCCIH).

7.3 Antibiotic stewardship

It is a strategy aimed at preventing unintentional use of antibiotics without a doctor's prescription. Antibiotics are the leading cause of antimicrobial resistance (AMR) worldwide, so limiting their use is a top concern. The main goals of antimicrobial stewardship (AMS) programs, according to WHO (World Health Organization), are to optimize antibiotic use by monitoring antibiotic prescribing and dispensing practices, reducing the economic burden on the country, containing the future spread of AMR, promoting research on reviving the existing antibiotics. These are some of the major points to follow. On the other hand, it is the responsibility of local people also to be aware and avoid misuse of antimicrobials.[77] AMR is expected to cost anywhere from $300 billion to >$1 trillion annually by 2050, according to various analyses.[93] According to a report published by the Organization for Economic Cooperation and Development (OECD), implementing AMS programs might save 1.6 million lives and US$ 4.8 billion each year in the 33 OECD countries.[77] Louis B. Rice described several antimicrobial stewardship techniques the foremost is antimicrobial restriction, limiting antimicrobial formularies and restricting the use of particular antibiotics, primarily owing to cost, as new medications are generally expensive. The second is antimicrobial cycling, sometimes known as "crop rotation," which is the practice of avoiding continuous usage of a single antibiotic class and, as a result, the selection of resistance to that class. Third, decision support, which includes prospective audit and feedback, or efforts to provide doctors with the knowledge and skills they need to administer antimicrobials appropriately. Studies reported a reduction in the incidence of *C. difficile* infections (52%) and nosocomial infections after implementing an antimicrobial stewardship program.[94]

8. Conclusion and perspective

Gut microbiota dysbiosis is a major problem and is associated with several intestinal and metabolic disorders and even cardiovascular diseases such as atherosclerotic cardiovascular disease (ACVD), intestinal disorders, i.e., inflammatory bowel disease (IBD), colorectal cancer (CRC) and metabolic disorders such as obesity and type 2 diabetes mellitus (T2D). We attempted to establish a link between drug intake and its influence on the microbiota, or vice versa, which promotes disease. Furthermore, there are multiple leading causes of this, one of which might be concerns associated with an unhealthy lifestyle, which could include disrupted sleep patterns, a poor diet and mental health issues, as well as geographical differences, the impact of host genetics/environment and drug exposure, etc. Thus, in future the goal is to use probiotics and drugs consciously, include antibiotic stewardship programs in national guideline to reduce AMR burden, also consider fecal microbiota transplantation as a great therapy to restore the microbial balance. This chapter will increase your understanding of how the gut microbiota affects drug metabolism and the importance of maintaining microbial balance.

Acknowledgments

T.M. receive research fellowship from University Grants Commission (UGC), Govt. of India. Figures are created with BioRender.com.

Declaration

Authors declare no conflicts of interests.

References

1. Ferranti EP, Dunbar SB, Dunlop AL, Corwin EJ. 20 Things You Didn't Know About the Human Gut Microbiome. *J Cardiovasc Nurs*. 2014;29:479–481. https://doi.org/10.1097/JCN.0000000000000166.
2. Kho ZY, Lal SK. The Human Gut Microbiome – A Potential Controller of Wellness and Disease. *Front Microbiol*. 2018;9:1835. https://doi.org/10.3389/fmicb.2018.01835.
3. Gilbert JA, Blaser MJ, Caporaso JG, Jansson JK, Lynch SV, Knight R. Current understanding of the human microbiome. *Nat Med*. 2018;24:392–400. https://doi.org/10.1038/nm.4517.
4. Benedetti MS, Whomsley R, Poggesi I, et al. Drug metabolism and pharmacokinetics. *Drug Metab Rev*. 2009;41:344–390. https://doi.org/10.1080/10837450902891295.
5. Javdan B, Lopez JG, Chankhamjon P, et al. Personalized Mapping of Drug Metabolism by the Human Gut Microbiome. *Cell*. 2020;181:1661–1679. https://doi.org/10.1016/j.cell.2020.05.001.
6. Crouwel F, Buiter HJC, de Boer NK. Gut Microbiota-driven Drug Metabolism in Inflammatory Bowel Disease. *J Crohns Colitis*. 2021;15:307–315. https://doi.org/10.1093/ecco-jcc/jjaa143.
7. Phang-Lyn S, Llerena VA. *Biochemistry, Biotransformation, StatPearls [Internet]*. StatPearls Publishing; 2021.
8. Sun C, Chen L, Shen Z. Mechanisms of gastrointestinal microflora on drug metabolism in clinical practice. *Saudi Pharm J*. 2019;27:1146–1156. https://doi.org/10.1016/j.jsps.2019.09.011.

9. Sheweita S. Drug-Metabolizing Enzymes: Mechanisms and Functions. *Curr Drug Metab.* 2000;1:107–132. https://doi.org/10.2174/1389200003339117.
10. Albhaisi SAM, Bajaj JS, Sanyal AJ. Role of gut microbiota in liver disease. *Am J Physiol Gastrointest Liver Physiol.* 2020;318:G84–G98. https://doi.org/10.1152/ajpgi.00118.2019.
11. Lucchetti M, Kaminska M, Oluwasegun AK, Mosig AS, Wilmes P. Emulating the gut–liver axis: Dissecting the microbiome's effect on drug metabolism using multiorgan-on-chip models. *Curr Opin Endocr Metab Res.* 2021;18:94–101. https://doi.org/10.1016/j.coemr.2021.03.003.
12. Martignoni M, Groothuis GMM, de Kanter R. Species differences between mouse, rat, dog, monkey and human CYP-mediated drug metabolism, inhibition and induction. *Expert Opin Drug Metab Toxicol.* 2006;2:875–894. https://doi.org/10.1517/17425255.2.6.875.
13. Nebert DW, Gonzalez FJ. P450 genes: structure, evolution, and regulation. *Annu Rev Biochem.* 1987;56:945–993. https://doi.org/10.1146/annurev.bi.56.070187.004501.
14. Almazroo OA, Miah MK, Venkataramanan R. Drug metabolism in the liver. *Clin Liver Dis.* 2017;21(1):1–20.
15. Antushevich H. Fecal microbiota transplantation in disease therapy. *Clin Chim Acta Int J Clin Chem.* 2020;503:90–98. https://doi.org/10.1016/j.cca.2019.12.010.
16. Jandhyala SM, Talukdar R, Subramanyam C, Vuyyuru H, Sasikala M, Reddy DN. Role of the normal gut microbiota. *World J Gastroenterol.* 2015;21:8787–8803. https://doi.org/10.3748/wjg.v21.i29.8787.
17. Rinninella E, Raoul P, Cintoni M, et al. What is the Healthy Gut Microbiota Composition? A Changing Ecosystem across Age, Environment, Diet, and Diseases. *Microorganisms.* 2019;7:14. https://doi.org/10.3390/microorganisms7010014.
18. Cancer [WWW Document], 2021 URL https://www.who.int/news-room/fact-sheets/detail/cancer (accessed 1.22.22).
19. Fan X, Jin Y, Chen G, Ma X, Zhang L. Gut Microbiota Dysbiosis Drives the Development of Colorectal Cancer. *Digestion.* 2021;102:508–515. https://doi.org/10.1159/000508328.
20. Sokol H, Seksik P, Furet JP, et al. Low counts of Faecalibacterium prausnitzii in colitis microbiota. *Inflamm Bowel Dis.* 2009;15:1183–1189. https://doi.org/10.1002/ibd.20903.
21. Amabebe E, Robert FO, Agbalalah T, Orubu ESF. Microbial dysbiosis-induced obesity: role of gut microbiota in homoeostasis of energy metabolism. *Br J Nutr.* 2020;123:1127–1137. https://doi.org/10.1017/S0007114520000380.
22. Sharma S, Tripathi P. Gut microbiome and type 2 diabetes: where we are and where to go? *J Nutr Biochem.* 2019;63:101–108. https://doi.org/10.1016/j.jnutbio.2018.10.003.
23. Gurung M, Li Z, You H, et al. Role of gut microbiota in type 2 diabetes pathophysiology. *EBioMedicine.* 2020;51:102590. https://doi.org/10.1016/j.ebiom.2019.11.051.
24. Jie Z, Xia H, Zhong S-L, et al. The gut microbiome in atherosclerotic cardiovascular disease. *Nat Commun.* 2017;8:845. https://doi.org/10.1038/s41467-017-00900-1.
25. Scanlan PD, Marchesi JR. Micro-eukaryotic diversity of the human distal gut microbiota: qualitative assessment using culture-dependent and -independent analysis of faeces. *ISME J.* 2008;2:1183–1193. https://doi.org/10.1038/ismej.2008.76.
26. Huffnagle GB, Noverr MC. The emerging world of the fungal microbiome. *Trends Microbiol.* 2013;21:334–341. https://doi.org/10.1016/j.tim.2013.04.002.
27. Nelson DR. The Cytochrome P450 Homepage. *Hum Genomics.* 2009;4:59. https://doi.org/10.1186/1479-7364-4-1-59.
28. Achour B, Barber J, Rostami-Hodjegan A. Expression of hepatic drug-metabolizing cytochrome p450 enzymes and their intercorrelations: a meta-analysis. *Drug Metab Dispos.* 2014;42:1349–1356. https://doi.org/10.1124/dmd.114.058834.

29. Giacomini KM, Huang S-M, Tweedie DJ, et al. Membrane transporters in drug development. *Nat Rev Drug Discov*. 2010;9:215–236. https://doi.org/10.1038/nrd3028.
30. Ronis MJJ. Effects of Soy Containing Diet and Isoflavones on Cytochrome P450 Enzyme Expression and Activity. *Drug Metab Rev*. 2016;48:331–341. https://doi.org/10.1080/03602532.2016.1206562.
31. Charles KA, Rivory LP, Brown SL, Liddle C, Clarke SJ, Robertson GR. Transcriptional repression of hepatic cytochrome P450 3A4 gene in the presence of cancer. *Clin Cancer Res*. 2006;12:7492–7497. https://doi.org/10.1158/1078-0432.CCR-06-0023.
32. Elkahwaji J, Robin MA, Berson A, et al. Decrease in hepatic cytochrome P450 after interleukin-2 immunotherapy. *Biochem Pharmacol*. 1999;57:951–954. https://doi.org/10.1016/s0006-2952(98)00372-4.
33. Gorski JC, Hall SD, Becker P, Affrime MB, Cutler DL, Haehner-Daniels B. In vivo effects of interleukin-10 on human cytochrome P450 activity. *Clin Pharmacol Ther*. 2000;67:32–43. https://doi.org/10.1067/mcp.2000.103860.
34. Wilson I, Nicholson J. Gut Microbiome Interactions with Drug Metabolism. *Efficacy and Toxicity Transl Res*. 2016;179. https://doi.org/10.1016/j.trsl.2016.08.002.
35. Wilson ID, Nicholson JK. Gut Microbiome Interactions with Drug Metabolism, Efficacy and Toxicity. *Transl Res*. 2017;179:204–222. https://doi.org/10.1016/j.trsl.2016.08.002.
36. Sender R, Fuchs S, Milo R. Are We Really Vastly Outnumbered? Revisiting the Ratio of Bacterial to Host Cells in Humans. *Cell*. 2016;164:337–340. https://doi.org/10.1016/j.cell.2016.01.013.
37. Sousa T, Paterson R, Moore V, Carlsson A, Abrahamsson B, Basit AW. The gastrointestinal microbiota as a site for the biotransformation of drugs. *Int J Pharm*. 2008;363:1–25. https://doi.org/10.1016/j.ijpharm.2008.07.009.
38. PubChem, n.d. Chloramphenicol [WWW Document]. URL https://pubchem.ncbi.nlm.nih.gov/compound/5959 (accessed 1.10.22).
39. Swanson HI. Drug Metabolism by the Host and Gut Microbiota: A Partnership or Rivalry? *Drug Metab Dispos*. 2015;43:1499–1504. https://doi.org/10.1124/dmd.115.065714.
40. Li H, He J, Jia W. The influence of gut microbiota on drug metabolism and toxicity. *Expert Opin Drug Metab Toxicol*. 2016;12:31–40. https://doi.org/10.1517/17425255.2016.1121234.
41. Zhao C, Hu Y, Chen H, et al. An in vitro evaluation of the effects of different statins on the structure and function of human gut bacterial community. *PLoS One*. 2020;15:e0230200. https://doi.org/10.1371/journal.pone.0230200.
42. Peppercorn MA, Goldman P. The Role of Intestinal Bacteria in the Metabolism of Salicylazosulfapyridine. *J Pharmacol Exp Ther*. 1972;181:555–562.
43. Haiser HJ, Seim KL, Balskus EP, Turnbaugh PJ. Mechanistic insight into digoxin inactivation by Eggerthella lenta augments our understanding of its pharmacokinetics. *Gut Microbes*. 2014;5:233–238. https://doi.org/10.4161/gmic.27915.
44. Banerjee S, Schlaeppi K, van der Heijden MGA. Keystone taxa as drivers of microbiome structure and functioning. *Nat Rev Microbiol*. 2018;16:567–576. https://doi.org/10.1038/s41579-018-0024-1.
45. Niehues M, Hensel A. In-vitro interaction of L-dopa with bacterial adhesins of Helicobacter pylori: an explanation for clinicial differences in bioavailability? *J Pharm Pharmacol*. 2009;61:1303–1307. https://doi.org/10.1211/jpp/61.10.0005.
46. Zemanová N, Lněničková K, Vavrečková M, et al. Gut microbiome affects the metabolism of metronidazole in mice through regulation of hepatic cytochromes P450 expression. *PLoS One*. 2021;16:e0259643. https://doi.org/10.1371/journal.pone.0259643.
47. Nakayama H, Kinouchi T, Kataoka K, Akimoto S, Matsuda Y, Ohnishi Y. Intestinal anaerobic bacteria hydrolyse sorivudine, producing the high blood concentration of 5-(E)-(2-bromovinyl)uracil that increases the level and toxicity of 5-fluorouracil. *Pharmacogenetics*. 1997;7:35–43. https://doi.org/10.1097/00008571-199702000-00005.

48. Tsunoda SM, Gonzales C, Jarmusch AK, Momper JD, Ma JD. Contribution of the Gut Microbiome to Drug Disposition, Pharmacokinetic and Pharmacodynamic Variability. *Clin Pharmacokinet.* 2021;60:971–984. https://doi.org/10.1007/s40262-021-01032-y.
49. Robert J, Rivory L. Pharmacology of irinotecan. *Drugs Today (Barc).* 1998;1998(34): 777–803. https://doi.org/10.1358/dot.1998.34.9.485276.
50. Gingell R, Bridges JW, Williams RT. The Role of the Gut Flora in the Metabolism of Prontosil and Neoprontosil in the Rat. *Xenobiotica.* 1971;1:143–156. https://doi.org/10.3109/00498257109044386.
51. Wadworth AN, Fitton A. Olsalazine. A review of its pharmacodynamic and pharmacokinetic properties, and therapeutic potential in inflammatory bowel disease. *Drugs.* 1991;41:647–664. https://doi.org/10.2165/00003495-199141040-00009.
52. Takeno S, Sakai T. Involvement of the intestinal microflora in nitrazepam-induced teratogenicity in rats and its relationship to nitroreduction. *Teratology.* 1991;44:209–214. https://doi.org/10.1002/tera.1420440209.
53. Elkington SG, Floch MH, Conn HO. Lactulose in the treatment of chronic portal-systemic encephalopathy. A double-blind clinical trial. *N Engl J Med.* 1969;281: 408–412. https://doi.org/10.1056/NEJM196908212810803.
54. Jourova L, Anzenbacher P, Anzenbacherova E. Human gut microbiota plays a role in the metabolism of drugs. *Biomed Pap.* 2016;160. https://doi.org/10.5507/bp.2016.039.
55. Collins SL, Patterson AD. The gut microbiome: an orchestrator of xenobiotic metabolism. *Acta Pharm Sin B.* 2020;10:19–32. https://doi.org/10.1016/j.apsb.2019.12.001.
56. Zimmermann M, Zimmermann-Kogadeeva M, Wegmann R, Goodman AL. Mapping human microbiome drug metabolism by gut bacteria and their genes. *Nature.* 2019;570:462–467. https://doi.org/10.1038/s41586-019-1291-3.
57. Aleksunes LM, Klaassen CD. Coordinated regulation of hepatic phase I and II drug-metabolizing genes and transporters using AhR-, CAR-, PXR-, PPARα-, and Nrf2-null mice. *Drug Metab Dispos.* 2012;40:1366–1379. https://doi.org/10.1124/dmd.112.045112.
58. Selwyn FP, Cheng SL, Klaassen CD, Cui JY. Regulation of Hepatic Drug Metabolizing Enzymes in Germ-Free Mice by Conventionalization and Probiotics. *Drug Metab Dispos.* 2016;44:262–274. https://doi.org/10.1124/dmd.115.067504.
59. Tomova A, Bukovsky I, Rembert E, et al. The Effects of Vegetarian and Vegan Diets on Gut Microbiota. *Front Nutr.* 2019;6:47. https://doi.org/10.3389/fnut.2019.00047.
60. Smith RP, Easson C, Lyle SM, et al. Gut microbiome diversity is associated with sleep physiology in humans. *PLoS One.* 2019;14:e0222394. https://doi.org/10.1371/journal.pone.0222394.
61. Monda V, Villano I, Messina A, et al. Exercise Modifies the Gut Microbiota with Positive Health Effects. *Oxid Med Cell Longev.* 2017;2017:3831972. https://doi.org/10.1155/2017/3831972.
62. Huang T-T, Lai J-B, Du Y-L, Xu Y, Ruan L-M, Hu S-H. Current Understanding of Gut Microbiota in Mood Disorders: An Update of Human Studies. *Front Genet.* 2019;10:98. https://doi.org/10.3389/fgene.2019.00098.
63. Gupta VK, Paul S, Dutta C. Geography, Ethnicity or Subsistence-Specific Variations in Human Microbiome Composition and Diversity. *Front Microbiol.* 2017;8:1162. https://doi.org/10.3389/fmicb.2017.01162.
64. Suzuki TA, Worobey M. Geographical variation of human gut microbial composition. *Biol Lett.* 2014;10:20131037. https://doi.org/10.1098/rsbl.2013.1037.
65. Yatsunenko T, Rey FE, Manary MJ, et al. Human gut microbiome viewed across age and geography. *Nature.* 2012;486:222–227. https://doi.org/10.1038/nature11053.
66. Dwiyanto J, Ayub Q, Lee SM, Foo SC, Chong CW, Rahman S. Geographical separation and ethnic origin influence the human gut microbial composition: a meta-analysis from a Malaysian perspective. *Microb Genomics.* 2021;7. https://doi.org/10.1099/mgen.0.000619.

67. Monaghan TM, Sloan TJ, Stockdale SR, et al. Metagenomics reveals impact of geography and acute diarrheal disease on the Central Indian human gut microbiome. *Gut Microbes*. 2020;12:1752605. https://doi.org/10.1080/19490976.2020.1752605.
68. Shetty SA. Gut microbiota features of the geographically diverse Indian population (preprint). *Microbiology*. 2018. https://doi.org/10.1101/478586.
69. Genome-Wide Association Studies Fact Sheet [WWW Document], n.d. Genome.gov. URL https://www.genome.gov/about-genomics/fact-sheets/Genome-Wide-Association-Studies-Fact-Sheet (accessed 1.23.22).
70. Dąbrowska K, Witkiewicz W. Correlations of Host Genetics and Gut Microbiome Composition. *Front Microbiol*. 2016;7.
71. Turnbaugh PJ, Hamady M, Yatsunenko T, et al. A core gut microbiome in obese and lean twins. *Nature*. 2009;457:480–484. https://doi.org/10.1038/nature07540.
72. Yang J, Yao S, Cheng K, et al. Comparison of Meconium Microbiome in Dizygotic and Monozygotic Twins Born by Caesarean Section (CS). *Front Microbiol*. 2020;11:1139. https://doi.org/10.3389/fmicb.2020.01139.
73. Hall AB, Tolonen AC, Xavier RJ. Human genetic variation and the gut microbiome in disease. *Nat Rev Genet*. 2017;18:690–699. https://doi.org/10.1038/nrg.2017.63.
74. Rothschild D, Weissbrod O, Barkan E, et al. Environment dominates over host genetics in shaping human gut microbiota. *Nature*. 2018;555:210–215. https://doi.org/10.1038/nature25973.
75. Tavalire HF, Christie DM, Leve LD, Ting N, Cresko WA, Bohannan BJM. Shared Environment and Genetics Shape the Gut Microbiome after Infant Adoption. *MBio*. 2021;12. https://doi.org/10.1128/mBio.00548-21.
76. Antimicrobial resistance [WWW Document], n.d. URL https://www.who.int/news-room/fact-sheets/detail/antimicrobial-resistance (accessed 1.15.22).
77. WHO. Antimicrobial stewardship programmes in health-care facilities in low- and middle-income countries: a WHO practical toolkit. *JAC-Antimicrob Resist*. 2019;1. https://doi.org/10.1093/jacamr/dlz072.
78. WHO 2021 New report calls for urgent action to avert antimicrobial resistance crisis [WWW Document], URL https://www.who.int/news/item/29-04-2019-new-report-calls-for-urgent-action-to-avert-antimicrobial-resistance-crisis (accessed 1.15.22).
79. Zhang S, Chen D-C. Facing a new challenge: the adverse effects of antibiotics on gut microbiota and host immunity. *Chin Med J (Engl)*. 2019;132:1135–1138. https://doi.org/10.1097/CM9.0000000000000245.
80. McDonnell L, Gilkes A, Ashworth M, et al. Association between antibiotics and gut microbiome dysbiosis in children: systematic review and meta-analysis. *Gut Microbes*. 2021;13:1870402. https://doi.org/10.1080/19490976.2020.1870402.
81. Zhang F, Luo W, Shi Y, Fan Z, Ji G. Should We Standardize the 1,700-Year-Old Fecal Microbiota Transplantation? *Off J Am Coll Gastroenterol ACG*. 2012;107:1755. https://doi.org/10.1038/ajg.2012.251.
82. CDC. Your life after C. diff can go back to normal. [WWW Document]. *Cent Dis Control Prev*. 2021; [URL https://www.cdc.gov/cdiff/after.html (accessed 1.22.22].
83. Borren NZ, Ghadermarzi S, Hutfless S, Ananthakrishnan AN. The emergence of Clostridium difficile infection in Asia: A systematic review and meta-analysis of incidence and impact. *PLoS One*. 2017;12:e0176797. https://doi.org/10.1371/journal.pone.0176797.
84. Cammarota G, Ianiro G, Gasbarrini A. Fecal microbiota transplantation for the treatment of Clostridium difficile infection: a systematic review. *J Clin Gastroenterol*. 2014;48:693–702. https://doi.org/10.1097/MCG.0000000000000046.
85. Cammarota G, Ianiro G, Tilg H, et al. European consensus conference on faecal microbiota transplantation in clinical practice. *Gut*. 2017;66:569–580. https://doi.org/10.1136/gutjnl-2016-313017.

86. Nowak A, Hedenstierna M, Ursing J, Lidman C, Nowak P. Efficacy of Routine Fecal Microbiota Transplantation for Treatment of Recurrent Clostridium difficile Infection: A Retrospective Cohort Study. *Int J Microbiol.* 2019;2019:7395127. https://doi.org/10.1155/2019/7395127.
87. Youngster I, Mahabamunuge J, Systrom HK, et al. Oral, frozen fecal microbiota transplant (FMT) capsules for recurrent Clostridium difficile infection. *BMC Med.* 2016;14:134. https://doi.org/10.1186/s12916-016-0680-9.
88. Shogbesan O, Poudel DR, Victor S, et al. A Systematic Review of the Efficacy and Safety of Fecal Microbiota Transplant for Clostridium difficile Infection in Immunocompromised Patients. *Can J Gastroenterol Hepatol.* 2018;2018:1394379. https://doi.org/10.1155/2018/1394379.
89. Kelly CR, Ihunnah C, Fischer M, et al. Fecal microbiota transplant for treatment of Clostridium difficile infection in immunocompromised patients. *Am J Gastroenterol.* 2014;109:1065–1071. https://doi.org/10.1038/ajg.2014.133.
90. O'Toole PW, Cooney JC. Probiotic Bacteria Influence the Composition and Function of the Intestinal Microbiota. *Interdiscip Perspect Infect Dis.* 2008;2008:1–9. https://doi.org/10.1155/2008/175285.
91. Bloemendaal M, Szopinska-Tokov J, Belzer C, et al. Probiotics-induced changes in gut microbial composition and its effects on cognitive performance after stress: exploratory analyses. *Transl Psychiatry.* 2021;11:300. https://doi.org/10.1038/s41398-021-01404-9.
92. Liu Y, Tran DQ, Rhoads JM. Probiotics in Disease Prevention and Treatment. *J Clin Pharmacol.* 2018;58:S164–S179. https://doi.org/10.1002/jcph.1121.
93. Dadgostar P. Antimicrobial Resistance: Implications and Costs. *Infect Drug Resist.* 2019;12:3903–3910. https://doi.org/10.2147/IDR.S234610.
94. Rice LB. Antimicrobial Stewardship and Antimicrobial Resistance. *Med Clin North Am.* 2018;102:805–818. https://doi.org/10.1016/j.mcna.2018.04.004.

CHAPTER TEN

Microbiome-based therapeutics: Opportunity and challenges

Archana Pant[a] and Bhabatosh Das[b,*]
[a]Molecular Genetics Lab, National Institute of Immunology, New Delhi, India
[b]Molecular Genetics Laboratory, Infection and Immunology Division, Translational Health Science and Technology Institute, Faridabad, Haryana, India
*Corresponding author: e-mail address: bhabatosh@thsti.res.in

Contents

Abstract

The autochthonous microbial communities comprising symbionts, commensals, and opportunistic pathogens living throughout the human body profoundly contribute to health by reducing disease susceptibility and maturing host immunity. The community compositions and functional repertoires of microbiomes present in the different body habitats are dynamic. The structural and functional balance of the human microbiome could be modulated by environmental factors, lifestyle, and host genetics. Several functions of the microbial community directly or indirectly modulate host

Progress in Molecular Biology and Translational Science, Volume 191
ISSN 1877-1173
https://doi.org/10.1016/bs.pmbts.2022.07.006

cellular signaling pathways that are associated with energy assimilation, sensing and responding to environmental signals through the neuroendocrine pathways, and resistance against colonization of allochthonous microbiota with disease-causing potential. Both culture-dependent and independent characterizations of microbial community compositions and their functional attributes help us to recognize the importance of microbial diversity in individual's health and identify the microbes and their metabolites associated with health and diseases. Such an in-depth understanding of the human microbiome created avenues of microbiome-based translation research, which led to the discovery and development of large numbers of medical therapies. In this chapter, we discuss the current success of microbiome-based therapies for infectious and metabolic diseases, and the major bottleneck and challenges of translational research with the community of symbiotic microorganisms.

Abbreviations

CD Crohn's disease
FMT fecal microbiota transplant
IBS irritable bowel syndrome
NAFLD non-alcoholic fatty liver disease
SCFAs short-chain fatty acids
UC ulcerative colitis

1. Introduction

Functional potency and genetic variability of the human microbiome living in various parts of the human body are widely vary between individuals, over time, and under different health conditions.[1] Such variabilities are strongly associated with disease susceptibility and modulated by multiple factors including diets, consumption of drugs and alcohol, and exposure to the environment. The restoration of microbiota to its native community composition in certain health disorders has huge potency to reduce disease progression and also to recover health. Autochthonous microbiota and their metabolites interact with several cell surface receptors and modulate cell signaling pathways of their host, which protect individuals from the progression of communicable diseases, and the duration and severity of infectious diseases.[2] Our improved understanding of the microbial functional potency has helped us to think of new methods of targeted therapy for treating different diseases either by utilizing microbiota or their metabolites or targeting specific members of microbes in a complex microbial ecosystem. These new methods involved techniques like fecal microbiota transplant (FMT) to tackle deadly diseases where antibiotics alone were failed to recover the health of the patients, utilization of probiotics, prebiotics,[3] postbiotics and paraprobiotics, probioceuticals/probiotaceuticals[4] to maintain individual's health. In regards to personalized medicines, the microbiome could be a great tool for disease diagnosis and knowing the stage of severity. Still, there are many micro-organisms and related pathways which remain undercover. These discoveries will help the researchers to utilize these readily available resources for microbiome-based therapies. The enormous information and plenty of research on microbiota suggest that it can be a clinically promising resource for capitalization. Modulation of intestinal microflora has a reflective effect on immune system development and function.[5]

2. Homeostasis and dysbiosis of human microbiome

2.1 Homeostasis and dysbiosis in gut microbiome

The resilience of the microbiota within the intestinal tract of healthy individuals is brought by the high microbial diversity, richness and stability.[6] The phyla dominated in a healthy gut comprises of Firmicutes followed by Bacteroidetes, Actinobacteria and Proteobacteria[7] along with the less abundant bacterial phyla like Fusobacteria, Tenericutes, Verrucomicrobia and Spirochetes.[8] The most abundant bacterial families include Ruminococcaceae, Lachnospiraceae, Bacteroidaceae, Prevotellaceae and Rikenellaceae.[9] The small intestine is dominated by Enterobacteriaceae and Lactobacillaceae bacterial families. The bacterial species that are noticed within the healthy individual regardless of geography are *Faecalibacterium prausnitzii*, *Ruminococcus intestinalis*, *Roseburia intestinalis*, *Blautia obeum*, *Clostridium* cluster XIVa, *Dorea formicigenerans*, *Eubacterium hallii*, *Bacteroides uniformis*, *Prevotella copri*, *Collinsella aerofaciens*, *Escherichia coli* and *Bifidobacterium bifidum*.[10] However, during any stimulation due to disease, medication or health issue the homeostasis of the gut flora gets disturbed leading to generation of pathogens and reducing the beneficial bacteria. During obesity, the Firmicutes to Bacteriodetes ratio increases,[11] and during undernutrition, abundance of Proteobacteria increases while that of Firmicutes decreases.[12] There is a decreased abundance of *Clostridia* and raised abundance of *Porphyromonas* and *Fusobacterium* during colorectal cancer.[13] During hypertension the gut microbiota (GM) comprised overgrowth of *Klebsiella*, *Megasphaera*, *Parabacteroides*, *Prevotella*, *Escherichia*, *Alistipes*, *Shigella*, *Enterobacter* etc. while *Roseburia*, *Faecalibacterium*, *Ruminococcus*, *Butyrivibrio*, *Coprococcus*, *Bifidobacterium*, *Akkermansia*, *Bacteroides*, *Sporobacter*, *Eubacterium rectale*, *Clostridium*, *Marvinbryantia*, *Intestinimonas*, *Subdoligranulum*, *Faecalibacterium prausnitzii*, *Anaerotruncus*, etc. were exhausted.

This also leads to alteration in function of GM. SCFA production and metabolism of some amino acids were decreased while LPS biosynthesis, ABC transporters, phosphotransferase system were upregulated.[14]

2.2 Homeostasis and dysbiosis in respiratory tract microbiome

Respiratory tract, a complex organ separated into two sections, the upper (URT) and lower (LRT) respiratory tract. URT comprises the following: anterior nares, nasal passages, paranasal sinuses, nasopharynx and oropharynx, and fragment of larynx above vocal cords. LRT comprises region of larynx below the vocal cords, trachea, smaller airways (i.e., bronchi and bronchioli) and alveoli. Highest bacterial load was observed in the URT. The early life exposure and bacterial acquisition establishes the respiratory tract microbiota (RM). The RM formulation is supposed to affect the morphogenesis of respiratory tract. Studies have reported that germ-free mice had smaller lungs[15] with lesser mature alveoli.[16] However, the mature alveoli populated back to normal when pups of germ-free mice were inhabited with *Lactobacillus* spp. in their nasal cavities.[16] Microbiota within anterior nares includes lipophilic skin colonizers like *Propionibacterium*, *Staphylococcus* and *Corynebacterium* along with *Dolosigranulum*, *Moraxella* and *Streptococcus*. Nasopharynx comprises *Moraxella*, *Corynebacterium and Staphylococcus*, *Haemophilus, Dolosigranulum* and *Streptococcus*. The oropharynx has more diverse bacterial community compared to nasopharynx, characterized by *Streptococcus*, *Rothia, Neisseria* and anaerobes, including *Prevotella, Veillonella* and *Leptotrichia*. The LRT compositions include phyla like Bacteroidetes, Firmicutes, Proteobacteria, Fusobacteria and Actinobacteria and the following genus *Prevotella, Veillonella, Streptococcus, Fusobacteria* and others (less presence of pathogens like *Haemophilus*).[17] Pathogenic bacteria grow and cause diffuse disease (pneumonia) or local dysfunction (pharyngitis). The nasopharynx also contains many potential pathogens including *H. influenzae*, *S. pneumoniae* and *S. aureus*, which are considered as part of normal microflora. It also carries germs related to acute respiratory infection. The dispersal of *S. pneumoniae* from rhinopharynx can cause meningitis, pneumonia or sepsis.[18]

2.3 Homeostasis and dysbiosis in reproductive tract microbiome

The urogenital tract harbors only 9% of the total human microbiota.[19] The healthy females have a *Lactobacillus* dominance, which provides protective features including lactate production and maintaining vaginal pH to ~3.0–4.5.[20] This environment helps in harboring and enabling *Lactobacilli* to grow, multiply, and dominate within cervico-vaginal niche making it unsuitable for pathogens to fluorish. Healthy vaginal microbiota (VMB) is populated by 90–95% of *Lactobacillus*, where four majorly abundant species include *L. crispatus, L. iners, L. gasseri,* and L. *jensenii.*[21] Studies also correlate the type of pregnancy, i.e., natural or in vitro fertilization (IVF) with the vaginal microbiome composition. Genital microbial dysbiosis can characterize a risk factor in infertility onset and increased risk of sexually transmitted diseases. In males, *Pseudomonas, Lactobacillus* and *Prevotella* could influence the sperm quality. *Pseudomonas* was directly related with total motile sperm count; however, *Prevotella* was inversely associated with sperm concentration. A dysbiosed VMB has low *Lactobacilli* and more anaerobic bacteria like *Mobiluncus, Prevotella, Ureaplasma, Gardnerella* and *Mycoplasma.*[22,23] The seminal fluids from both healthy and infertile male patients show *Enterococcus faecalis, Streptococcus agalactiae, Escherichia coli, Ureaplasma urealyticum.*[24] Patients with low quality sperm had increased abundance of *Ureaplasma, Mycoplasma, Enterococcus, Bacteroides ureolyticus* and *Prevotella*, whereas *Lactobacilli* abundance was correlated with healthy individuals.[25]

3. Microbiome-based therapies

In the last 10 years, the discovery and development of microbiome-based therapeutics against communicable and metabolic diseases made tremendous progress (Fig. 1). Microbiome, microbiota and microbial-derived proteins and metabolites isolated from different body habitats or produced in the in vitro culture conditions have successfully been used to reduce the disease susceptibility, onset and prevent disease progression (Fig. 1) and (Table 1).

3.1 Fecal microbiota transplantation

The fecal microbiome transplantation (FMT), the complete transplantation of intestinal microflora and their metabolites from a healthy donor to the recipient, has proven promising treatment outcomes against *Clostridium difficile* infections.[26–28] It is administered in three different ways. The most common method involved colonoscopy or enema. Colonoscopy is preferred for the FMT delivery to the upper intestinal tract whereas enema is preferred for lower intestinal tract. The second method involves

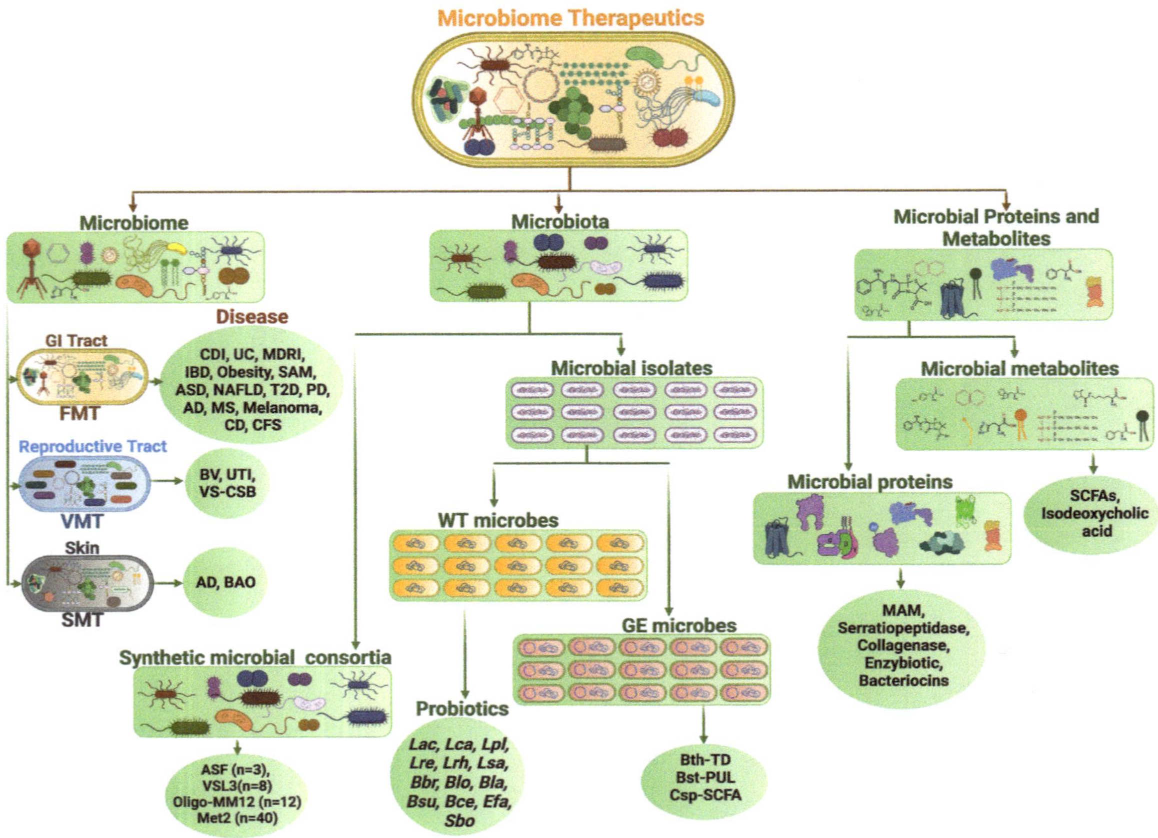

Fig. 1 See figure legend on opposite page.

nasoenteric tube in which a tube is routed from nose to the intestine. A relatively new method involves fecal treatments carried in the pills known as FMT capsules. FMT has established positive treatment effects and considered safe for recurrent *C. difficile* induced colitis even in immunocompromised individuals.[29] Other diseases involving FMT as a treatment are ulcerative colitis (UC), Crohn's disease (CD), obesity, autism and irritable bowel syndrome (IBS). A study documented that FMT from complete response donors who are previously treated with anti-programmed cell death-1 (anti-PD-1) monotherapy for metastatic melanoma to the refractory metastatic melanoma patients with reinduced anti-PD-1 therapy was potentially effective, feasible and safe. Also, in some patients, FMT amplified the intratumoral immune activity, which was deciphered into objective clinical responses. The results of this study support the GM modulation to overcome immunotherapy resistance.[30]

3.2 Probiotics

Probiotics are the "live microorganisms," provided in adequate quantity in order to confer health benefits to the host.[31] Alternative definition is the live microbial feed supplements that beneficially affect the host animal by improving its intestinal microbial balance.[32] Now, there are many well-characterized *lactobacilli* and *bifidobacterial* stains available which are

Fig. 1 Human microbiome, microbiota associated with the human body, and microbial-derived molecules being used in clinical practice or tested for therapeutic potency. Both naturally occurring and genetically engineered bacterial species have been explored for the therapeutic potency against colonization of antibiotic-resistant bacterial pathogens, reducing local and systemic inflammation and improving metabolic activities. CDI, *Clostridium difficile* infection; UC, Ulcerative colitis; MDRI, Multidrug-resistant infection; IBD, Inflammatory bowel disease; SAM, Severe acute malnutrition; ASD, Autism spectrum disorder; NAFLD, Non-alcoholic fatty liver disease; *T2D,* Type 2 diabetes; PD, Parkinson's disease; AD, Atopic dermatitis; MS, Multiple sclerosis; CD, Crohn's disease CFS, Chronic fatigue syndrome; BV, Bacterial vaginosis; UTI, Urinary tract infections; VS-CSB; Vaginal seeding of C-Section baby; BAO, Bad underarm odor; FMT, Fecal microbiota transplantation; VMT, Vaginal microbiota transplantation; SMT, Skin microbiota transplantation; *Oligo-MM12; Oligo-Mouse-Microbiota; Lac; Lactobacillus acidophilus; Lca, Lactobacillus casei; Lpl, Lactobacillus plantarum; Lre, Lactobacillus reuteri; Bbr, Bifidobacterium breve; Blo, Bifidobacterium longum; Bsu, Bacillus subtilis; Bce, Bacillus cereus; Efa, Enterococcus faecium; Sbo, Saccharomyces boulardii;* Bth-TD, *Bacillus thuringiensis*-Tryptophan decarboxylase; ASF, Altered Schaedler's flora; Bth-TD, *Bacteroides thetaiotaomicron*-Tryptophan decarboxylase; *Bst-PUL, Bacteroides stercoris*-Polysaccharide utilization loci; *Csp-SCFA, Clostridium sporogenes*-Short-chain fatty acid; MAM, Microbial Anti-Inflammatory Molecule.

Table 1 Commonly used probiotics and prebiotics across the world.

Microbiome-based therapeutics	Therapeutic type	Company/Product name
L. acidophilus DDS-1	Probiotic	Nebraska Cultures, Inc. (Lincoln, NE)
L. acidophilus NCFM	Probiotic	Rhodia, Inc. (Madison, Wisconsin, USA)
Lactobacillus brevis KB290	Probiotic	Kagome Co., Ltd. (Tochigi, Japan)
L. acidophilus R0011 *L. rhamnosus* R0052	Probiotic	Institute Rosell (Montreal, Canada)
L. acidophilus SBT-2062 *B. longum* SBT-2928	Probiotic	Snow Brand Milk Products Co., Ltd. (Tokyo, Japan)
L. paracasei CRL 431 *B. lactis* Bb-12	Probiotic	Chr. Hansen (Horsholm, Denmark)
L. casei DN014001	Probiotic	Danone Le Plessis-Robinson (Paris, France)
L. casei Shirota *B. breve* strain Yakult	Probiotic	Yakult (Tokyo, Japan)
Streptococcus thermophilus MN-ZLW-002	Probiotic	Inner Mongolia Mengniu Dairy Industry Co. Ltd., (Hohhot, China)
L. fermentum RC-14 *L. rhamnosus* GR-1	Probiotic	Urex Biotech Inc. (London, Ontario, Canada)
L. johnsonii La1 (same as Lj1)	Probiotic	Nestlé (Lausanne, Switzerland)
L. reuteri SD2112 (same as MM2)	Probiotic	BioGaia (Raleigh, North California, USA)
L. plantarum 299 V *L. Rhamnosus* 271	Probiotic	Probi AB (Lund, Sweden)
L. rhamnosus GG	Probiotic	Valio Dairy (Helsinki, Finland)
L. salivarius UCC118	Probiotic	University College (Cork, Ireland)

L. rhamnosus LB21 *Lactococcus lactis* L1A	Probiotic	Essum AB (Umeå, Sweden)
B. lactis HN019 (DR10)	Probiotic	New Zealand Dairy Board
B. longum BB536	Probiotic	Morinaga Milk Industry Co., Ltd. (Zama-City, Japan)
L. paracasei F19	Probiotic	Arla Dairy (Stockholm, Sweden)
L. acidophilus LB	Probiotic	Lacteol Laboratory, (Houdan, France)
L. casei DN 114	Probiotic	Danone, Paris, France
L. crispatus CTV05	Probiotic	Gynelogix, Boulder, Colo.
L. delbrueckii subsp. *bulgaricus* 2038	Probiotic	Meiji Milk Products (Tokyo, Japan)
S. boulardii	Probiotic	Biocodex Inc. (Seattle, Washington, USA)
Inulin and FOS	Prebiotic	Performance lab
Xylooligosaccharides (XOS)	Prebiotic	Peak BioBoost
Inulin	Prebiotic	Align DualBiotic Prebiotic
Inulin, peptidase, lipase, and apple pectin	Prebiotic	Biohm Prebiotic Supplement
Fructooligosaccharides and xylooligosaccharides	Prebiotic	JustThrive Precision Prebiotic
Inulin, FOS, starch, and dietary fiber	Prebiotic	HyperBiotics Prebiotic
Tributyrin, *Bacillus coagulans*, Sunfiber (a Fiber extract), microcrystalline cellulose, hydroxypropyl methylcellulose, silica, and magnesium stearate	Prebiotic, probiotic, and post-biotic	Gundry MD BioComplete 3

From Nagpal R, Yadav H, Kumar M, Jain S, Yamashiro Y, Marotta F. Probiotics, prebiotics and synbiotics: An introduction. In: Semih Ötleş, ed. *Probiotics and Prebiotics in Food, Nutrition and Health.* New York, NY: CRC Press; 2014:1–24; and https://www.peninsulaclarion.com/national-marketplace/15-best-prebiotic-supplements-to-use-top-prebiotics-list-2021/.

capable of decreasing or treating the gut infection risks.[33] Probiotic consumption has demonstrated several health benefits which include gut health enhancement by GM modulation, development of immune system and stimulating its functions, nutrients production and enhancing their bioavailability, strengthening digestion and improvising lactose metabolism and lowering the risk of certain diseases like heart disease, diabetes, blood pressure, urinary tract infection etc.[34–43] The features that are required for an ideal probiotic are survival, colonization and proliferation and activity at the desired site. It should not elicit the immune response[44,45] and capable of withstanding low pH and high concentration of both deconjugated and conjugated bile acids. It should preferably be of human origin[46] and compatible with the food manufacturing processes. Several studies support the benefits contributed by probiotics in individual's health and disease. *Lactobacillus species* like *L. acidophilus* and *L. casei* yield lactic acid as the principal end product of fermentation. *Bifidobacteria* produce acetic and lactic acids.[47] The other acids produced by probiotic bacteria are hippuric and citric acid. Lactic acid bacteria also produce antimicrobial substances like diacetyl, hydrogen peroxide and bacteriocin which provide hostile environments for pathogens. *L. acidophilus, L. gasseri, L. confusus, Streptococcus thermophilus, Bifidobacterium breve* and *B. longum* were antigenotoxic toward N′-nitro-N-nitrosoguanidine (MNNG)[48] and 1,2-dimethylhydrazine (DMH).

3.3 Prebiotics

These are the group of non-digestible food components that are digested only by GM thereby selectively enhancing the growth or activity of one or few bacteria within gut and hence bettering the individual's health. Beneficial GM ferment these substances and obtain their survival energy. This way prebiotics have selective influence on the GM.[49,50] This fermentation results into short-chain fatty acids (SCFAs) including butyric acid, lactic acid and propionic acid are liberated into the circulating blood and affect the gut along with other distal organs. Inulin, fructo-oligosaccharides (FOS), *trans*-galacto-oligosaccharides (TOS), galacto-oligosaccharides (GOS), resistant starch (RS), pectic oligosaccharide (POS) and non-carbohydrate oligosaccharides are the different prebiotics with positive effects on human health. The bacterial genera that are stimulated by different prebiotics include *Lactobacilli, Bifidobacteria, Eubacterium rectale, Ruminococcus bromii, Bacteroides thetaiotaomicron, Bacteroidetes, Enterobacteria* and *Firmicutes.*

The various features of prebiotic are non-absorbable within GI tract and fermented by GM, resistant to acidic pH of stomach and hydrolysis by mammalian enzymes and it should selectively augment the growth as well as activity of the gut bacteria to improve host's health.[51]

3.4 Postbiotics, paraprobiotics and probioceuticals/ probiotaceuticals

(i) Postbiotics can be "non-living products" from bacteria or metabolites from microflora that have biotic activity within the host. (ii) Paraprobiotics/ ghost or inactivated probiotics are "non-living bacterial cells" which are either intact or broken or crude cell extracts provided either orally or topically in acceptable quantities and pose a health benefit to the consumers. (iii) Probioceuticals/probiotaceuticals are defined as factors derived from probiotics like reuterin from *Lactobacillus reuteri*.

Examples include postbiotics that are derivatives of: flavonoids (daidzein, norathyriol, desaminotyrosine, equol daidzein), terpenoids (paeoni lactone glycosides, paeonimetabolin I, II, III, genipin, paeoniflorin), phenol (valerolactones, equol, enterodiol, enterolactone, 8-prenylnaringenin, urolithins), etc.[52,53]

4. Microbiome as therapeutics in metabolic diseases

4.1 Obesity

Obesity, a metabolic disease commonly manifested by the accumulation of excessive body fat, is a major health concern across the globe and leading to almost >70% of premature deaths. The GM manipulation is evolving as a prominent way to tackle this metabolic disorder. Shifting the obesity-associated GM towards the beneficial commensals by consumption of probiotics may be an incredible therapeutic agent as probiotics are known to reduce the visceral fat mass and body mass index (BMI). In a multicentric RCT with 87 individuals, there were substantial reductions in waist circumference (1.8% decline) and abdominal visceral fat (4.6% decline) upon consuming the fermented milk carrying *Lactobacillus gasseri* daily for a duration of 12 weeks.[54] FMT from healthy individuals to the obese people could be another mode to treat obesity. A study by Ridaura and coworkers on germ-free mice supported the concept that a defined lean microbiota along with a healthy diet may be a potential mode for treating obesity and associated disorders.[55] The mice receiving FMTs from human twins discordant for obesity had shown a significant increase in body mass and adiposity even when they were given less fat and more fiber rich-diet. However, their counterparts who received FMT from lean donors have shown opposite phenotypes. Also, co-housing of mice with FMT from obese individuals along with the mice receiving FMT from lean individuals displayed less adiposity when compared with mice not co-residing with lean mice. The phenotypes in this study were diet-dependent as prevention of weight gain was abolished in mice given with low-fiber/high-saturated fat diet.[56] The associated disorders with obesity include a major risk of insulin resistance (IR) and type 2 diabetes (T2D) and impaired lipid metabolism.[57,58] A study utilized FMT from wild boar to understand the consequences on obesity related symptoms which include VLDL, LDL, TChol and serum TG level which were markedly increased in HFD group. The wild boar has leaner muscle and less fat and these phenotypes were assumed to be associated with GM. The fecal suspension from wild boar alleviated the enhanced serum levels of TChol, TG and amended the insulin sensitivity compared to HFD mice. The data from this study suggested that FMT exerted health advantages by blocking obesity in mice. The distinct diversity and structure of jejunum associated microbiota was found after HFD and FMT treatment where Lachnospiraceae and its genus *Lachnoclostridium* were linked with obesity. This study also found enhanced jejunal m6A levels and increased *Lactobacilli* after FMT.[59] *Dendrobium officinale*, a medicinal food homology plant is known for its various health benefits including anti-obesity effects. Although, the probable mechanisms remain unclear. A study has discussed about probable mechanisms by which *D. officinale* dietary fiber (DODF) exerts health beneficial effects to obesity and associated metabolic syndrome. DODF treatment lessened accumulation of lipid droplets, reduced weight gain, promoted WAT browning and hepatic glycogen synthesis in HFD-induced obese mice. This treatment also enhanced relative abundance of beneficial bacteria with reduced Firmicutes/Bacteroides ratio and enhanced microbial diversity. FMT from DODF treated mice to the obese mice partially restored these positive effects. Both DODF and FMT treatments markedly reduced *Bilophila* and amplified *Bifidobacterium*, *Muribaculum* and *Akkermansia* in HFD-induced obese mice along with low level of serum LPS raised due to HFD. Consequently, DODF can be a potential prebiotic to check and control obesity and other related metabolic disorders.[60] One study, by Sugama and coworkers, assessed the part of GM on SCO-792-induced body weight (BW) decrease in HFD-induced obese (DIO) mice. SCO-792, an oral enteropeptidase inhibitor, reported to show anti-obesity and anti-diabetic effects in mice. The SCO-792 administration considerably reduced the food consumption and BW in DIO mice. GM clearance through antibiotics confirmed SCO-792-induced BW decline is mediated through GM. In SCO-792 treated mice, the GM composition was reformed with reduced Firmicutes and enhanced Verrucomicrobia along with *Akkermansia muciniphila*, a bacterium important for host metabolism. Fecal metabolomics revealed enhanced amino acid levels and SCFAs including propionate, and bile acids (BA). The colonic immunoglobulin A (IgA) concentration was also elevated upon SCO-792 treatment.[61] Synbiotics are dietary supplements having the combination of probiotics and prebiotics[62] and could be a therapeutic alternative. Obesity and NAFLD progress have strong correlation with GM modification.[63,64] In a study, a model of juvenile obesity-related NAFLD was developed to estimate the significance of administering the flavonoid quercetin along with *Akkermansia muciniphila* on GM composition, BA flow and hepatic lipid and BA metabolism gene regulation. The synbiotic administration for 3 weeks significantly reduced the insulin resistance index. Administration of *A. muciniphila* lowered white adipose tissue and enhanced plasma leptin concentration. The relative expression of Ppard (involved in adipogenesis and differentiation) and Plin2 (involved in lipid storage) genes within synbiotic treated group was similar to control group while significantly higher expression was observed in group with dietary intervention. Many genes like *Dgat2, Cebpa* and *Srebp2* upregulated during HFD were significantly reduced during synbiotic treatment. The treatment led to enhancement of Verrucomicrobia and Cyanobacteria phylum and *Akkermansia, Oscillospira* genus while lowering the abundance of Actinobacteria phylum, *Blautia, Coprobacillus, Lactobacillus* and *Lactococcus*.[65]

4.2 Diabetes

Different dietary macronutrients like proteins, carbohydrates and lipids are rich sources of hundreds of microbial-derived metabolites including indole, hydrogen sulfide, phenylacetic acid, acetate, butyrate, propionate, succinate, ferulic acid, resveratrol, cathecin, and linoleic acid in the gut. These metabolites directly interact with insulin signaling pathways. SCFAs enhance glucose uptake and lipid oxidation within the skeletal muscle. Within liver, they reduce gluconeogenesis and lipogenesis, enhance the lipid oxidation through activation of AMP-activated protein kinase and phosphatidylinositol 3-kinase-serine/threonine-protein kinase B

(PI3K-AKT-PKB). In adipose tissue, these induces thermogenesis along with adipogenesis, hinder lipid breakdown and diminish inflammation. Hence, energy expenditure increases and fat oxidation ensues within entire body.[66] A meta-analysis study found that probiotic/synbiotic supplementation to the women during pregnancy with gestational diabetes (GD) had several health advantages like the microbiome composition, enhanced colostrum adiponectin levels, better metabolic regulation and maternal and infant health. The beneficial outcomes were more pronounced with synbiotics supplementation compared to probiotics as the viability of probiotics improves due to nutritional and energy supply. The two bacteria *Bifidobacterium lactis Bb12* and *Lactobacillus acidophilus* were widely used amid the studies inspected in this meta-analysis. In clinical trials, daily probiotic doses given were from 10^5 to 10^{14} CFU during pregnancy or post-delivery and the recommended average dose was $\geq 10^7$ CFU/day to regulate of GD. Also, four clinical trials had shown that $\geq 10^7$ CFU/day had no advantage on pregnancy outcomes.[67–70] These supplements significantly decreased insulin resistance (serum insulin levels (INS), FBS levels, HOMA-IR and HOMA-B measures), inflammation markers (IL-6 and TNF) and lipid profile (serum-, HDL-, VLDL- and total cholesterol levels). Although, plasma TAC and MDA levels were significantly increased compared to control. The SCFAs produced by the probiotics results in enhanced GLP-1 secretion, thereby improving glucose levels via various modes: (a) regulation of specific genes expression important for glucose metabolism like genes for leptin and grehlin, glucose-6-phosphatase, glucose transporter type 4 and PPAR-gamma genes,[71] (b) insulin secretion stimulation and delayed gastric draining,[72] (c) lowering toll-like receptor activity, to further increase insulin sensitivity in muscle.[73] The differences in lipid profile were not significant for LDL-C, HDL, TG, Chol, and total Chol/HDL but for VLDL ($P = 0.02$), without or less heterogeneity among various studies. The probable mechanisms for probiotic associated regulation of lipid homeostasis may be (a) cholesterol assimilation into the probiotic cell walls, (b) the bile-salt hydrolase (BSH) activity,[74] (c) cholesterol to coprostanol conversion,[75] (d) production of SCFAs,[76] (e) Suppressing the nuclear factor (NF)-kappa light-chain expression,[77] (f) Alleviating the pro-inflammatory cytokines expression, and (g) altering the fatty-acid oxidation energy pathways.[78,79] Researches on anti-diabetic (T2D) activity of dietary fibers (DF) revealed the DF intake help in enhancing the abundance of *Roseburia, Ruminococcus bromii, Bacteroides, Eubacterium rectale, Prevotella* and *Bifidobacterium* and suppress a number of LPS producing gram-negative bacteria like Enterobacteriaceae and *Desulfovibrio*.[80] Also, DF promoted SCFA production, which can further bind to GPR and increase the PPY and GLP-1 levels into the intestinal epithelial L-cells. This improves IR, energy intake or expense, regulates appetite and lipid oxidation.[81] The antidiabetic properties of polyphenols were suggested to be associated to altered markers of gut barrier function like occludin and ZO-1.[82] An ethanolic extract of *Lessonia nigrescens*, high in flavonoids and phenolics exhibits hypoglycemic property by reducing Firmicutes and enhancing the Bacteroidetes in the gut.[81] Grape pomace extract which is a combination of polyphenols like flavanols, anthocyanins and flavanol glycosides, betters the gut barrier function, impaired glucose tolerance, inflammation of adipose tissue, fat mass gain and IR. Effects are correlated with increased abundance of *Prevotellaceae, Erysipelotrichaceae, Allobaculum,* and *Roseburia* and reduction of *Clostridium* sensu stricto, *Desulfovibrionaceae, Streptococcaceae and Lactococcus*.[83] Capsaicin, an active component of chili peppers, ameliorate insulin tolerance and glucose homeostasis in obese diabetic mice by enhancing Firmicutes/Bacteroidetes ratio and *Roseburia* abundance, which could reduce proinflammatory cytokines like IL-6 and TNF-α.[84] Berberine, the most widely studied alkaloid found in several medicinal plants, has the following effect on DM: (1) Modulation of Firmicutes to Bacteroidetes ratio and thus enhancing SCFA in feces.[85] (2) Regulation of BA cycle and then the ileal FXR signaling pathway.[86] (3) Reducing LPS in plasma and lessening the gut permeability, thereby lowering the highly expressed inflammatory markers.[87] and (4) Regulation of branched chain amino-acid biosynthesis and catabolism in adipose tissue and liver.[88,89] A meta-analysis by Nejad and coworkers illustrated that probiotics use seemed to be helpful in managing postprandial blood glucose levels and lowering the inflammatory markers. Compared to control group, probiotics significantly improved the fasting blood glucose levels by −12.15 mg/dL, IR by 1.63 units and the glycated hemoglobin A1c levels by 0.31% and significantly lowered the high sensitivity C-reactive protein level with a mean difference of 0.26 mg/L.[90] In an observational study by Su and coworkers had assessed the safety and efficacy of three formula diet containing probiotics, prebiotics and whole grains (PPW preparation) and FMT combination treatment process for T2D. Formula 1, a combined dietary fiber powder formulation (15 g/bag) containing resistant dextrin, inulin, galactooligosaccharide (GOS), fructo-oligosaccharide (FOS) and xylooligosaccharide. Formula 2, a pre-processed whole grain powder mixture formulation (25 g/bag) comprising organic wheat, highland barley and oat. Formula 3, a synbiotics powder formulation (4 g/bag) having resistant dextrin, fermented fruit, GOS, stachyose, and vegetable freeze-dried powder, *Bifidobacterium lactis* HN019, *Lactobacillus acidophilus* NCFM, and *Lactobacillus paracasei* Lpc-37. The total probiotics per bag was 1.00E + 10. The study was carried for 90 days PPW dietary intervention with or without FMT found the alterations in the physical conditions of 13 individuals with T2D. Both diet-only and FMT combination interventions were linked through blood glucose control with reduced fasting blood glucose, glycated hemoglobin and significant weight loss. Systolic blood pressure was also reduced significantly in both groups. The DF group has shown effects faster compared to diet-only group; however, the effects were same in both the groups. Weight loss was exhibited quickly in DF group. The overall GM diversity was reduced with the interventions. The GM in T2D individuals shifted from *Bacteroides* to *Prevotella* following the treatment along with significant rise in valuable bacteria like *Bifidobacterium* along with significant decline in harmful bacteria like *Bilophila.*[91] In T1D families, elevated inflammatory state which is consistent with PRR activation was identified.[92,93] In a pilot study by Cabrera and coworkers, a multistrain probiotic supplement was provided to nondiabetic siblings of T1D patients with an aim to reduce this innate state. They found the probiotic supplement well-tolerated and safe with increased participant adherence rate without adverse events and significantly reduced systemic inflammation. The probiotic supplementation was not overtly changing the GM. Through LEfSe analyses, post-supplement stool was enriched with several bacterial families like Lachnospiraceae, Lactobacillaceae and Bifidobacteriaceae. After probiotic supplement, there were noticeably higher levels of butyrate in the participants. The probiotic supplement was established to carry *Bifidobacteria, Lactobacillus* and *Streptococcus* through 16S sequencing. 11 differentiating characters linked to the pre-supplement samples comprised Porphyromonadaceae, Mogibacteriaceae and Peptostreptococcaceae family members with an absolute LDA score >2 and 5 differentiating characters linked to the post-supplement samples included Bifidobacteriaceae and Lactobacillaceae family (LDA score >2) along with Lachnospiraceae (LDA score = 3.1, $P = 0.025$). Importantly, the abundance of Lachnospiraceae and Veillonellaceae were found lower in T1D progressors.[94,95]

4.3 Non-alcoholic fatty liver disease

Non-alcoholic fatty liver disease (NAFLD) is the state of hepatic cells with extra fat that leads to many liver ailments extending from hepatic steatosis to non-alcoholic steatohepatitis (NASH). To elucidate the effect of probiotics as an intervention for NAFLD, *L. plantarum* MA2, a probiotic bacterium and its mode of action were investigated in a study. In this study, methionine and choline-deficient (MCD) induced NAFLD rats were fed the bacterium and markers for liver damage, lipid metabolism, intestinal barrier along with microflora composition were inspected. The strain supplementation improved the phenotype and lipid metabolism along with lipid deposition and pathological liver impairment of the MCD-induced NAFLD rats. However, body weight was not restored. The strain also protected gut mucosal barrier system in NAFLD rats by upregulating the claudin-1 expression. Upon supplementation, the species richness was significantly restored with higher Firmicutes:Bacteroidetes ratio and lower Proteobacteria. In supplemented rats, the dominant genera were reestablished to Ruminococcaceae, *Lactobacillus* and [Eubacterium]_coprostanoligenes cluster alike controls but different from NAFLD group consisting of Lachnospiraceae (unclassified genus), *Escherichia-Shigella, Bacteroides*, and [Eubacterium]_coprostanoligenes_group and *Blautia*. This intervention had markedly ameliorated the gut microecology of NAFLD rats. Liver inflammation was downregulated by expressing inflammation-related pathway proteins. The expression of TLR4, MyD88, AP1, inflammatory markers like TNF-α, IL-6, IL-1, and p65 protein along JNK kinase activity were greatly turned down in MA2 group compared with NAFLD group suggesting that *L. plantarum* MA2 can improve excess liver inflammation in NAFLD rats by remarkably preventing the transcription of TLR4-MYD88 and its downstream pathways.[96] Another study aimed to examine the regulatory role of probiotics on GM and finding probable mechanism of regulating the bile acid receptor farnesoid X receptor/growth factor 15 (FXR/FGF15) signaling pathway in NAFLD rats. The study groups included a normal diet control group (NC), a high fat diet group (HFD), a HFD + probiotic group (HFD-P) and a HFD + atorvastatin group (HFD-A). The results had shown significant reduction in the inflammatory cell infiltration and liver's fatty degeneration and with great improvement in hepatic inflammation in HFD-P and HFD-A groups. Levels of TG, AST, ALT and TBA in the HFD-P and HFD-A groups were reduced and in HFD-P group the CHOL level was decreased compared with the HFD group. There was low similarity of microbial communities among the four groups. Probiotics improved the GM structure in HFD-induced NAFLD. HFD-P group has shown lower abundance of Bacteroidia whereas HFD-A group has higher abundance of Clostridia with increased

Desulfovibrionaceae in both HFD-P and HFD-A groups. FGF15 was markedly higher in the probiotic group than that of the HFD group whereas no substantial difference was found between HFD-A and HFD group. The mRNA and/or protein expressions of FXR and FGF15 were substantially higher in HFD-P group than in HFD group, suggesting that probiotics may affect bile acid metabolism via upregulating the FXR/FGF15 pathway expression.[97] Cassiae Semen (CS) is a commonly used Chinese herbal medicine with confirmed health benefits in reducing fevers, loosen bowels to relieve constipation and improving the vision.[98–101] However, there were no proper investigations on its inter-connection between CS effects on the liver and its effect on the GM. One study suggested that CS prohibited NAFLD by lowering fat accumulation and liver inflammation, regulating GM disorder, improving liver and gut barrier injury and metabolic endotoxemia. They used CS extract (CSE) (10 g/kg), total aglycone extract of CS (TA) (10 g/kg), rubrofusarin-6-β-gentiobioside (RG) (20 mg/kg) and aurantio-obtusin (AO) (20 mg/kg). Compared to other groups (TA, RG, and AO), the effect of CSE on lipid accumulation was somewhat enhanced. However, no significant difference between CSE and TA suggested the enzymatic hydrolysis of glycosides into aglycones had comparable effect with glycosides of CS on its hepatoprotective effect. FMT from CSE treated mice into HFD-fed mice restored recipient GM composition back to normal and reduced blood lipids, metabolic endotoxemia, liver inflammatory reactions, serum transaminases, along with enhanced gut mucosal protein expression and improved the gut barrier damage. The genera enriched in HFD group were reversed upon compound/extract intervention, with CSE impacting significant reduction in *Klebsiella, Erwinia, Trabulsiella* and *Morganella* abundance. The genera which were reduced in the HFD group (*Adlercreutzia* from Coriobacteriaceae, *Dehalobacterium* from Dehalobacteriaceae, *Rikenella* from Rikenellaceae, *Odoribacter* from Odoribacteraceae, *Mucispirillum* from Ferribacteraceae, *Dorea* and *Lachnobacterium* in Lachnospiraceae, *Clostridium* from Clostridiaceae and *Oscillospira* and *Ruminococcus* in Ruminococcaceae) enhanced after administration of extract/compound, especially *Oscillospira, Dehalobacterium,* and *Ruminococcus* upon CSE treatment. In addition, the genera from Lachnospiraceae, i.e., *Ruminococcus* and *Coprococcus* were markedly enhanced after CSE treatment and *Akkermansia* was increased after RG treatment. Bacteria reduced during NAFLD, i.e., *Akkermansia, Bifidobacterium, Blautia, Allobaculum, Coprobacillus, Oscillospira, Desulfovibrio, Coprococcus, Roseburia* and *Prevotella*[102–107] were reversed after CS treatment or FMT, signifying important role in NAFLD relief. Though, *Bilophila, Mucispirillum, Lactobacillus* and *Paraprevotella,*[102–105,108] were more during NAFLD, but lower during intervention. The functional prediction of GM also emphasized bacteria regulated biosynthetic capability for primary and secondary bile acids and glycerolipid metabolism by CSE administration, indicating that CSE induced GM composition, may reduce lipid aggregation in HFD mice, encourage lipid digestion and absorption, and lessened lipotoxicity. Compared with the control, out of 80/81 pathways downregulated in HFD group were recovered and 75/76 pathways were reduced after CSE administration. In TA, RG, and AO treated groups, 74, 67 and 80 pathways were increased and 57, 45, and 70 were decreased, respectively. The amino acid, carbohydrate, lipid, and energy metabolism pathways, cell motility and sporulation in cellular processes and signaling were mainly enhanced in CSE treated group. The three-level metabolic function pathways involved in glycerolipid metabolism and biosynthesis of primary and secondary bile acids were enhanced, whereas unsaturated fatty acid biosynthesis was reduced following CSE treatment, explaining the potential hypolipidemic properties of CSE. Also, the metabolic pathways related to LPS biosynthesis and LPS biosynthetic proteins were reduced upon CSE treatment. Therefore, this study suggests the use of CS as a prebiotic to treat NAFLD.[109] Another prebiotic and a synthetic disaccharide of galactose and fructose, Lactulose, is utilized routinely as first-line treatment to manage hepatic encephalopathy. It boosts the growth of *Lactobacillus* and *Bifidobacterium* at clinically relevant dosages.[110] However, its effects on GM composition may be variable.[111,112]

5. Microbiome as therapeutics in infectious diseases

5.1 Diarrhea

Modulation of GM as a therapy to treat severe diarrhea caused due to various reasons has been supported by various studies. Methotrexate (MTX), a chemotherapeutic drug has been shown to have detrimental effects on host microbiota as well as GI tract leading to mucosal barrier injury (MBI) as major clinical obstacle with no effective treatment. MBI reduces gut absorptive capability and manifesting diarrhea.[113,114] Antibiotic (vancomycin, metronidazole and imipenem)-induced microbial dysbiosis (ABX) worsens MTX-induced MBI and promotes pathogen expansion. ABX greatly enhanced MTX60-induced mortality in rats. However, this study suggested that administration of FMT after MTX resulted in increased microbial diversity as well as richness. Although, MBI effects were not mitigated by FMT, however, food intake was increased during peak mucosal injury. This study also suggested that FMT lessens the lethal effects of ABX on MTX induced diarrhea with its efficacy dictated by the colonization of Muribaculaceae (S24–7) a key regulator of mucosal injury and repair. This rodent bacterial family belonging to the phylum Bacteroidetes has established role in regulating FMT efficacy in rodents.[115] Its abundance is enhanced after prebiotic intervention with selenium and vitamin B2,[116] both of which are beneficial in preventing MBI resulted from chemotherapy.[117–119] Another study has suggested shifting FMT from last resort therapy to forefront treatment for *C. difficile* infection (CDI) which involved severity ranging from self-controlled diarrhea to deadly pseudomembranous colitis, sepsis or death. Antibiotics administration may cause the dysbiosis dominated by Proteobacteria[120] and may cause resistance resulting into relapsed or refractory CDI. The meta-analysis of 37 studies suggested FMT to be more efficient in rectifying the recurring and refractory CDI than vancomycin with 92% clinical resolution across all studies. The incidences of improvement had with consecutive FMTs performed after first ones failed.[121] A systematic review by Iqbal and coworkers, covering 6 studies having dataset from 341 patients reviewed efficacy of FMT as ingestible capsules. The first treatment was effective in 285 patients without recurrence. A second and a third FMT treatment was required to resolve the condition in 28 and 3 patients, respectively, with a resolution rate of 93%.[122] In a retrospective study of 59 patients, Roshan and the coworkers have performed FMT by first treating with colonoscopic infusion, followed by enema the subsequent day and found that 58 patients got rid of *C. difficile* infection. It was verified using toxin and culture tests over next 4–8 weeks. Symptoms like diarrhea, abdominal pain, and blood in stool were also significantly reduced.[123] The longest and largest study by Perler and coworkers have reported that FMT is a long-term cure for rCDI with no adverse effects associated with FMT or dysbiosis.[124,125]

5.2 Inflammatory bowel disease

FMT has found to be promising in improving the condition in irritable bowel syndrome (IBS). A study has accessed the effectivity of FMT as capsules in lessening diarrhea-predominant IBS (IBS-D) by performing a randomized, placebo-controlled, double blind, crossover trial in 48 participants with moderate to severe IBS-D. The IBS-SSS, IBS-QOL and BSFS scores were improved in both placebo and FMT recipients at week 12.

Although, at week 12, no difference was observed between these groups. They reported a higher clinical response rate of 61% in placebo recipients compared to 50% response rate in FMT recipients at week 12. Though, the difference was non-significant. The intestinal flora of FMT recipients was more similar to donors following FMT. However, donor similarity and α-diversity were not different between FMT non-responders and responders. The Bacteroidetes is to Firmicutes ratio was higher at baseline within the responders than non-responders. In responders, *Prevotella* was predominated in the Bacteroidetes phylum.[126] IBD is correlated with decreased gut microbial density which is further linked with intestinal Treg cells proportion in specific pathogen-free mice.[127,128] Another study by Britton and the coworkers colonized the C57BL/6 germ-free mice with defined microbiotas transplanted (DMT) from IBD patients and assessed their immunophenotypic response to DMT with healthy donor (HD) microbiota transplanted 3 weeks later. Following DMT, the intestinal immune landscape within the mice was constantly reshaped. The study demonstrated an increased Treg cells population with potential immunoregulatory RORγt+ Treg subset, following HD-FMT to IBD microbiota colonized mice. Enhanced RORγt+ Treg cells accompanied by restoring microbial density following DMT. The humans receiving FMT for ulcerative colitis (UC) possess higher microbial density after treatment which results in improved endoscopic outcomes, signifying a shift toward a more tolerance promoting mucosal niche. Fascinatingly, the enhanced RORγt+ Treg cells were not associated with increases in alpha diversity, which is often linked with healthy state and the most frequently used metrics for GM dysbiosis. This advocates that microbiota density could be a valuable biomarker for non-invasive monitoring of FMT as it is linked to the tolerogenic Treg compartment. The study also found that DMT using HD microbiota repressed Th17 cells population in the gut of mice colonized with a microflora obtained from a Crohn's disease (CD) patient donor. Interestingly, a specific strain, *E. coli* A6, was found to be the primary driver of RORγt+ and IL-17A+ Th17 cell augmentation in mice inhabiting this donor consortia and reduced Th17 cells was linked with the same consortium lacking this strain.[129]

5.3 Bacterial vaginosis

Vaginal microbiota (VMB) dominated by *Lactobacillus* plays a substantial role in managing female reproductive tract health and protects against urogenital infections.[130–132] The major players include *Lactobacillus acidophilus, L. jensenii, L. gasseri, and L. crispatus*[133,134] along with *L. iners*, the recently emerged predominant species,[135,136] however, its role is still debatable.[137] *Lactobacilli* depletion and higher microbial diversity within vagina intensifies the threat of poor gynecologic and obstetric outcomes. Bacterial vaginosis (BV), a medical ailment, associated with VMB dysbiosis and higher microbial diversity with more anaerobes and low H_2O_2-producing *Lactobacilli*.[138,139] It occurs frequently even after antibiotic treatment, may be due to persistent dysbiosis and *Lactobacilli* incapable to repopulate the vagina. Globally, 15–50% of reproductive-age women are affected by BV, globally with a recurrence rate of 20–75% in 3 months after standard antibiotic regimen.[140,141] Metronidazole selectively kills the BV-associated bacteria leaving vaginal *Lactobacilli* which have inherent resistant to nitroimidazoles. This therapy alone is not capable to restore *Lactobacillus*-dominated community and dysbiosis can continue. Replacing the dysbiotic VMB with protective *L. crispatus* CTV-05 is a promising approach. This strain, isolated from a healthy female vagina[134,136] is the active component of LACTIN-V. It is the first VMB-based live biotherapeutic product (LBP), established under an Investigational New Drug application with the FDA as an adjuvant therapy to avoid relapse of BV and rUTI after antimicrobial therapy. CTV-05 is a H_2O_2 producer and a glucose homofermenter to produce lactic acid (both l and d isomers). It is capable of adhering the epithelial cells and colonizes the vagina.[136,142,143] CTV-05 antagonizes many urogenital pathogens in vitro and demonstrates outstanding preclinical and clinical safety record. Original formulation of LACTIN-V was vaginally administered gelatin capsule carrying 5×10^8 CFU/capsule. In a phase 2 clinical trial in 100 females receiving a standard drug therapy for uncomplicated cystitis, after that LACTIN-V or placebo pills regularly for next 5 days and then once in a week for next 10 weeks. Fascinatingly, women who received placebo attained comparably high endogenous *L. crispatus* but not comparable protection against rUTI, signifying CTV-05 superiority over endogenous strains.[144] The other formulation is LACTIN-V applicator at three doses extending from 5×10^8 to 2×10^9 CFU/applicator (NCT00537576).[145] Upto 2×10^9 CFU/applicator doses were well tolerated and safe.[146] Another study by Sagie and the group reported the feasibility of using vaginal microbiota transplant (VMT) as a long-term therapy for antibiotics-non-responsive, repeated and non-tractable BV. In this study, they observed no adverse effects related with VMT, however, they say the potential risks with VMT cannot be exclude completely. The risks may include gynecologic and obstetric problems, sperm transfer leading to unintended pregnancy or transmitting the undetected pathogens with the vaginal fluid. Five patients suffering from intractable BV, reverting after repeated, prolonged and diverse antibiotic attempts and needing constant care and antibiotic therapy to remain symptom free were recruited in a study. Three healthy and premenopausal volunteers, with no BV past 5 years or any recurrent BV history were recruited as donors. VMT was executed 1 week after antibiotic treatment completion. The vaginal fluid was collected from the upper half of the vagina and cervical fornices, whereas avoiding the cervix from donors from the 7th day of menstrual cycle. The fluid was assessed by microscopy and pH. It was then diluted with 1 mL of sterile saline and relocated to the recipient's posterior fornix, without using speculum. VMT was repeated where symptoms recurred or if one or more Amsel criteria reappeared positive during follow-ups. Four out of five patients had long-lasting positive effects in their Amsel scores, whiff test, microscopic vaginal fluid appearance, discharge and vaginal fluid pH after 1–3 VMT sittings. The fifth patient had partial remission, manifesting as a subjective reduction of symptoms, negative whiff test, cumulative Amsel scoring of 0–1 and an intermediate microscopic vaginal fluid appearance. The characterization of the genus-level changes linked to VMT was performed. The follow up of recipient's microbiota before and after VMT suggested a speedy compositional shift after VMT, along with recovery in all Amsel criteria. This compositional shift was maintained and remained different from the baseline BV composition for approximately 5–21 months after successful VMT. The post-VMT microbiota was dominated by members of genus *Lactobacillus* with a reduction in members of genus *Bifidobacterium*, *Gardnerella*, *Fannyhessea* and *Prevotella*.[147]

5.4 Respiratory infections

The intestinal microbiota maintains the health by regulating immunity of GI tract as well as the distal organs probably by regulating extra-intestinal T cells population, SCFAs production, regulating systemic inflammation and developing oral immune tolerance through Tregs.[148] The blood and lymphatic system can carry cytokines and immune cells induced by gut commensals and their metabolites and modulate lung immune and inflammatory responses thereby influencing respiratory health and disease.[149] Respiratory infections, especially with antibiotic resistance are serious threats to the world as the infection might raise through aerosols and spread very rapidly. One such serious threat is antibiotic resistant tuberculosis (TB). High dosing of anti-TB drugs over a very long duration causes serious side effects. Emergence of MDR, XDR, extreme XDR (XXDR) and total drug resistant (TDR) strains is a global public health threat. In this regards, alternative treatment strategies are needed to overcome health threats and bring positive treatment outcomes. Studies on probiotics suggest that anti-*Mtb* immunity within lungs can be restored by supplementing with *Lactobacillus*.[150] In a study by Lee and coworkers, a probiotic, PMC201, identified to be *L. crispatus* strain, with an anti-TB property, was isolated from a healthy Korean woman's vagina. It was applied as immunomodulators to develop new anti-TB therapy. The researchers have measured the *Mtb* inhibitory effect of this strain within macrophages. Both the drug resistant and drug sensitive were inhibited by this probiotic strain at a concentration that was non-cytotoxic. The strain displayed synergy with isoniazid against XDR *Mtb* infected macrophages. This study suggested that PMC201 administration has reduced the level of pro-inflammatory molecule, nitric oxide (NO), suggesting its role in anti-inflammation due to reduction in intracellular *Mtb*.[151,152] This study advocated that PMC201 may be a probable candidate to treat drug-resistant tuberculosis.[153] A study by Wunderink et al. suggested bacteriophage based therapy could be promising against pneumonia caused by MDR bacteria.[154] One study showed that the effectiveness of an intravenous concoction of four bacteriophages was similar to teicoplanin and lowered the bacterial load and enhanced the survival of mice having pneumonia by MRSA.[62,155] Probiotic supplementation also resulted in a reduced frequency of microbiologically confirmed ventilator-associated pneumonia (VAP) along with shortened durations at ICUs[156] and hospitals.[157] The clinical advantages of probiotic supplementation on ventilated patients with reduction in VAP incidences were also suggested in a meta-analysis study.[149,158]

6. Conclusion and future perspective

The major outcomes of human microbiome research are capitalized on therapeutics and diagnostics discovery by modulating microbiome composition and detecting microbial origin biomolecules. Understanding of baseline microbiome in a healthy population and its variations across geographies, ethnicity, diet, age, and lifestyle tremendously helps in exploring the contribution of microbiome dysbiosis to disease susceptibility and disease progression, respectively. The autochthonous microbial communities or their metabolites are used to test their potential to prevent or recover from an infectious and metabolic disease. In addition, Biotech startups and genetic engineers have manipulated the genome of commensal microbiota for optimal production of therapeutically important proteins that can regulate host cellular signaling pathways and reduce systemic or local inflammation. Correcting microbiome imbalance in different body habitats and reducing disease susceptibility and increasing drug efficacy against communicable and metabolic diseases will shape the future clinical practice across the globe.

The bottleneck in utilizing microbiome as therapeutics is to precisely identify the cause-effect correlation and develop formulation with optimal viable numbers of sustainable microbiota which can colonize the desired body sites and provide expected benefits by modulating the composition or functions of the existing local microbial community. Correctly identifying disease-causing microbial taxa or their functional potency, reproducing the effect in a well-accepted animal model, and conducting clinical trials remain challenging for several preliminary findings with massive potential for therapeutics development. However, recent development in multiomics platforms, a computational algorithm for the integration of different omics data, expansion of organoid research, and commercial interest in live therapeutics open new opportunities for microbiome-based therapeutics in infectious and metabolic disease management.

References

1. Huttenhower C, Gevers D, Knight R, et al. Structure, function and diversity of the healthy human microbiome. *Nature*. 2012;486(7402):207–214. https://doi.org/10.1038/nature11234.
2. Hsiao A, Ahmed AMS, Subramanian S, et al. Members of the human gut microbiota involved in recovery from vibrio cholerae infection. *Nature*. 2014;515(7527):423–426. https://doi.org/10.1038/nature13738.
3. Davani-Davari D, Negahdaripour M, Karimzadeh I, et al. Prebiotics: Definition, types, sources, mechanisms, and clinical applications. *Foods*. 2019;8(3):1–27. https://doi.org/10.3390/foods8030092.
4. Nataraj BH, Ali SA, PV B, Yadav H. Postbiotics-parabiotics: The new horizons in microbial biotherapy and functional foods. *Microb Cell Fact*. 2020;19(1):1–22. https://doi.org/10.1186/s12934-020-01426-w.
5. Brown J, de Vos WM, Distefano PS, et al. Translating the human microbiome. *Nat Biotechnol*. 2013;31(4):304–308. https://doi.org/10.1038/nbt.2543.
6. Lozupone CA, Stombaugh JI, Gordon JI, Jansson JK, Knight R. Diversity, stability and resilience of the human gut microbiota. *Nature*. 2012;489:220–230.
7. Das B, Ghosh TS, Kedia S, et al. Analysis of the gut microbiome of rural and urban healthy Indians living in sea level and high altitude areas. *Sci Rep*. 2018;8:10104.
8. Caporaso JG, Lauber CL, Costello EK, et al. 2011 moving pictures of the human microbiome. *Genome Biol*. 2011;12:50.

9. Arumugam M, Raes J, Pelletier E, et al. Enterotypes of the human gut microbiome. *Nature*. 2011;473:174–180.
10. Yatsunenko T, Rey FE, Manary MJ, et al. Human gut microbiome viewed across age and geography. *Nature*. 2012;486:222–227.
11. Turnbaugh PJ, Ley RE, Mahowald MA, Magrini V, Mardis ER, Gordon JI. An obesity-associated gut microbiome with increased capacity for energy harvest. *Nature*. 2006;444:1027–1031.
12. Ghosh TS, Sen Gupta S, Bhattacharya T, et al. Gut microbiomes of Indian children of varying nutritional status. *PLoS One*. 2014;9:e95547.
13. Ahn J, Chen CY, Hayes RB. Oral microbiome and oral and gastrointestinal cancer risk. *Cancer Causes Control*. 2012;23:399–404.
14. Guo Y, Li X, Wang Z, Yu B. Gut microbiota dysbiosis in human hypertension: A systematic review of observational studies. *Front Cardiovasc Med*. 2021;8. https://doi.org/https://doi.org/10.3389/fcvm.2021.650227.
15. Wostmann BS. The germfree animal in nutritional studies. *Annu Rev Nutr*. 1981;1:257–279.
16. Yun Y, Srinivas G, Kuenzel S, et al. Environmentally determined differences in the murine lung microbiota and their relation to alveolar architecture. *PLoS One*. 2014;9:e113466.
17. Man WH, de Steenhuijsen Piters WAA, Bogaert D. The microbiota of the respiratory tract: Gatekeeper to respiratory health. *Nat Rev Microbiol*. 2017;15(5):259–270.
18. Santacroce L, Charitos IA, Ballini A, et al. The human respiratory system and its microbiome at a glimpse. *Biology*. 2020;9(10):1–16. https://doi.org/10.3390/biology9100318.
19. Sirota I, Zarek SM, Segars JH. Potential influence of the microbiome on infertility and assisted reproductive technology. *Semin Reprod Med*. 2014;32(1):35–42.
20. García-Velasco JA, Budding D, Campe H, et al. The reproductive microbiome - clinical practice recommendations for fertility specialists. *Reprod Biomed Online*. 2020;41(3):443–453.
21. Riganelli L, Iebba V, Piccioni M, et al. Structural variations of vaginal and endometrial microbiota: Hints on female infertility. *Front Cell Infect Microbiol*. 2020;10:350.
22. Srinivasan S, Morgan MT, Fiedler TL, et al. Metabolic signatures of bacterial vaginosis. *MBio*. 2015. https://doi.org/10.1128/mBio.00204-15.
23. Verstraelen H, Swidsinski A. The biofilm in bacterial vaginosis: Implications for epidemiology, diagnosis and treatment. *Curr Opin Infect Dis*. 2013;26(1):86–89.
24. Moretti E, Capitani S, Figura N, et al. The presence of bacteria species in semen and sperm quality. *J Assist Reprod Genet*. 2009;26(1):47–56.
25. Farahani L, Tharakan T, Yap T, Ramsay JW, Jayasena CN, Minhas S. The semen microbiome and its impact on sperm function and male fertility: A systematic review and meta-analysis. *Andrology*. 2021;9(1):115–144.
26. Gopalakrishnan V, Spencer CN, Nezi L, et al. Gut microbiome modulates response to anti-PD-1 immunotherapy in melanoma patients. *Science*. 2017;359:97–103.
27. Matson V, Fessler J, Bao R, et al. The commensal microbiome is associated with anti-PD-1 efficacy in metastatic melanoma patients. *Science*. 2018;359:104–108.
28. Routy B, Chatelier EL, Derosa L, et al. Gut microbiome influences efficacy of PD-1-based immunotherapy against epithelial tumors. *Science*. 2017;359:91–97.
29. Lin SC, Alonso CD, Moss AC. Fecal microbiota transplantation for recurrent Clostridium difficile infection in patients with solid organ transplants: An institutional experience and review of the literature. *Transpl Infect Dis*. 2018;20(6):e12967.
30. Baruch EN, Youngster I, Ben-Betzalel G, et al. Fecal microbiota transplant promotes response in immunotherapy-refractory melanoma patients. *Science*. 2021;2021(371):602–609 [https://www.science.org].

31. FAO/WHO (2002) Guidelines for the evaluation of probiotics in foods. Food and Agriculture Organization of the United Nations and World Health Organization Expert Consultation Report. Food and Agricultural Organization of the United Nations and World Health Organization Working Group Report (online).
32. Fuller R. Probiotics in man and animals. *J Appl Bacteriol.* 1989;66:65–378.
33. Salminen SJ, Gueimonde M, Isolauri E. Probiotics that modify disease risk. *J Nutr.* 2005;135:1294–1298.
34. Kumar M, Behare PV, Mohania D, Arora S, Kaur A, Nagpal R. Health-promoting probiotic functional foods: Potential and prospects. *Agro Food Ind Hi Tech.* 2009;20:29–33.
35. Kumar M, Mohania D, Poddar D, et al. A probiotic fermented milk prepared by mixed culture combination reduces pathogen shedding and alleviates disease signs in rats challenged with pathogens. *Int J Probiotics Prebiotics.* 2009;4:211–218.
36. Kumar M, Kumar A, Nagpal R, et al. Cancer-preventing attributes of probiotics: An update. *Int J Food Sci Nutr.* 2010;61:473–496.
37. Kumar M, Verma V, Nagpal R, et al. Anticarcinogenic effect of probiotic fermented milk and chlorophyllin on aflatoxin-B1 induced liver carcinogenesis in rats. *Br J Nutr.* 2011;107:1006–1016.
38. Kumar M, Verma V, Nagpal R, et al. Effect of probiotic fermented milk and chlorophyllin on gene expressions and genotoxicity during AFB1-induced hepatocellular carcinoma. *Gene.* 2011;490:54–59.
39. Nagpal R, Kaur A. Synbiotic effect of various prebiotics on in-vitro activities of probiotic lactobacilli. *Ecol Food Nutr.* 2011;50:63–68.
40. Nagpal R, Yadav H, Puniya AK, Singh K, Jain S, Marotta F. Potential of probiotics and prebiotics for symbiotic functional dairy foods. *Int J Probiotics Prebiotics.* 2007;2:75–84.
41. Nagpal R, Kumar A, Arora S. In-vitro probiotic potential of lactobacilli from indigenous milk products. *Int J Probiotics Prebiotics.* 2010;5:103–110.
42. Yadav H, Jain S, Sinha PR. Formation of oligosaccharides in skim milk fermented with mixed dahi cultures, Lactococcus lactis ssp. diacetylactis and probiotic strains of lactobacilli. *J Dairy Res.* 2007;74:154–159.
43. Yadav H, Jain S, Sinha PR. Anti-diabetic effect of probiotic dahi containing lactobacillus acidophilus, lactobacillus casei and Lactococcus lactis bacteria in high fructose diet fed rats. *Nutrition.* 2007;72:62–68.
44. Toma MM, Pokrotnieks J. Probiotics as functional food: Microbiological and medical aspects. *Acta Univ Latviensis Biol.* 2006;710:117–129.
45. Ohashi Y, Ushida K. Health-beneficial effects of probiotics: Its mode of action. *Anim Sci J.* 2009;80:361–371.
46. Collins JK, Thornton G, Sulliva GO. Selection of probiotic strains for human application. *Int Dairy J.* 1998;8:487–490.
47. Desjardins ML, Roy D. Growth of Bifidobacteria and their enzyme profiles. *J Dairy Sci.* 1990;73:299–307.
48. Pool-Zobel BL, Neudecker C, Domizlaff I, et al. Lactobacillus and Bifidobacterium mediated antigenotoxicity in the colon of rats. *Nutr Cancer.* 1996;26:365–380.
49. Bouhnik Y, Raskine L, Simoneau G, et al. The capacity of nondigestible carbohydrates to stimulate fecal bifidobacteria in healthy humans: A double-blind, randomized, placebo-controlled, parallel-group, dose-response relation study. *Am J Clin Nutr.* 2004;80:1658–1664 [CrossRef] [PubMed].
50. Flint HJ, Scott KP, Louis P, Duncan SH. The role of the gut microbiota in nutrition and health. *Nat Rev Gastroenterol Hepatol.* 2012;9:577–589 [CrossRef] [PubMed].
51. Gibson GR, Scott KP, Rastall RA, et al. Dietary prebiotics: Current status and new definition. *Food Sci Technol Bull Funct Foods.* 2010;7:1–19.

52. Cavallari JF, Fullerton MD, Duggan BM, et al. Muramyl dipeptide-based postbiotics mitigate obesity-induced insulin resistance via IRF4. *Cell Metab.* 2017;25:1063–1074.
53. Cortés-Martín A, Selma MV, Tomás-Barberán FA, González-Sarrías A, Espín JC. Where to look into the puzzle of polyphenols and health? The postbiotics and gut microbiota associated with human metabotypes. *Mol Nutr Food Res.* 2020;64: [1900952].
54. Kadooka Y, Sato M, Imaizumi K, et al. Regulation of abdominal adiposity by probiotics (lactobacillus gasseri SBT2055) in adults with obese tendencies in a randomized controlled trial. *Eur J Clin Nutr.* 2010;64(6):636–643. https://doi.org/10.1038/ejcn.2010.19.
55. Ridaura VK, Faith JJ, Rey FE, et al. Gut microbiota from twins discordant for obesity modulate metabolism in mice. *Science.* 2013;341(6150):[1241214].
56. Santos-Paulo S, Costello S, Forster S, Travis S, Bryant R. The gut microbiota as a therapeutic target for obesity: A scoping review. *Nutr Res Rev.* 2021;1–14. https://doi.org/10.1017/S0954422421000160.
57. Kahn SE, Hull RL, Utzschneider KM. Mechanisms linking obesity to insulin resistance and type 2 diabetes. *Nature.* 2006;444:840–846.
58. Martinez-Guryn K, Hubert N, Frazier K, et al. Small intestine microbiota regulate host digestive and absorptive adaptive responses to dietary lipids. *Cell Host Microbe.* 2018;23:458–469.
59. Zhu L, Fu J, Xiao X, et al. Faecal microbiota transplantation-mediated jejunal microbiota changes halt high-fat diet-induced obesity in mice via retarding intestinal fat absorption. *J Microbial Biotechnol.* 2022;15(1):337–352. https://doi.org/10.1111/1751-7915.13951.
60. Zheng H, Ji H, Fan K, et al. Targeting gut microbiota and host metabolism with *Dendrobium officinale* dietary fiber to prevent obesity and improve glucose homeostasis in diet-induced obese mice. *Mol Nutr Food Res.* 2022. https://doi.org/10.1002/mnfr.202100772.
61. Sugama J, Moritoh Y, Yashiro H, Tsuchimori K, Watanabe M. Enteropeptidase inhibition improves obesity by modulating gut microbiota composition and enterobacterial metabolites in diet-induced obese mice. *Pharmacol Res.* 2021;163. https://doi.org/10.1016/j.phrs.2020.105337.
62. Pérez-Cobas AE, Baquero F, de Pablo R, Soriano MC, Coque TM. Altered ecology of the respiratory tract microbiome and nosocomial pneumonia. *Front Microbiol.* 2022;12. https://doi.org/10.3389/fmicb.2021.709421.
63. Juárez-Fernández M, Porras D, García-Mediavilla MV, et al. Aging, gut microbiota and metabolic diseases: Management through physical exercise and nutritional interventions. *Nutrients.* 2021;13:16.
64. Quiroga R, Nistal E, Estébanez B, et al. Exercise training modulates the gut microbiota profile and impairs inflammatory signaling pathways in obese children. *Exp Mol Med.* 2020;52:1048–1061.
65. Juárez-Fernández M, Porras D, Petrov P, et al. The synbiotic combination of akkermansia muciniphila and quercetin ameliorates early obesity and NAFLD through gut microbiota reshaping and bile acid metabolism modulation. *Antioxidants.* 2021;10(12). https://doi.org/10.3390/antiox10122001.
66. Bastos RM, Rangel ÉB. Gut microbiota-derived metabolites are novel targets for improving insulin resistance. *World J Diabetes.* 2022;13(1):65–69. https://doi.org/10.4239/wjd.v13.i1.65.
67. Callaway LK, McIntyre HD, Barrett HL, et al. Probiotics for the prevention of gestational diabetes mellitus in overweight and obese women: Findings from the SPRING double-blind randosmized controlled trial. *Diabet Care.* 2019;42(3):364–371.

68. Lindsay KL, Brennan L, Kennelly MA, et al. Impact of probiotics in women with gestational diabetes mellitus on metabolic health: A randomized controlled trial. *Am J Obstet Gynecol.* 2015;212(4):[496.e1–496.e11].
69. Lindsay KL, Kennelly M, Culliton M, et al. Probiotics in obese pregnancy do not reduce maternal fasting glucose: A double-blind, placebo-controlled, randomized trial (probiotics in pregnancy study). *Am J Clin Nutri.* 2014;99(6):1432–1439.
70. Pellonperä O, Mokkala K, Houttu N, et al. Efficacy of fish oil and/or probiotic intervention on the incidence of gestational diabetes mellitus in an at-risk group of overweight and obese women: A randomized, placebo-controlled, double-blind clinical trial. *Diabetes Care.* 2019;42(6):1009–1017.
71. Miraghajani M, Dehsoukhteh SS, Rafie N, Hamedani SG, Sabihi S, Ghiasvand R. Potential mechanisms linking probiotics to diabetes: A narrative review of the literature. *Sao Paulo Med J.* 2017;135(2):169–178.
72. McNabney SM, Henagan TM. Short chain fatty acids in the colon and peripheral tissues: A focus on butyrate, colon cancer, obesity and insulin resistance. *Nutrients.* 2017;9(12):1348.
73. Hommelberg PP, Langen RC, Schols AM, Mensink RP, Plat J. Inflammatory signaling in skeletal muscle insulin resistance: Green signal for nutritional intervention? *Curr Opin Clin Nutr Metab Care.* 2010;13(6):647–655.
74. Lambert JM, Bongers RS, de Vos WM, Kleerebezem M. Functional analysis of four bile salt hydrolase and penicillin acylase family members in lactobacillus plantarum WCFS1. *Appl Environ Microbiol.* 2008;74(15):4719–4726.
75. Lye HS, Rusul G, Liong MT. Removal of cholesterol by lactobacilli via incorporation and conversion to coprostanol. *J Dairy Sci.* 2010;93(4):1383–1392.
76. Pereira DI, Gibson GR. Cholesterol assimilation by lactic acid bacteria and bifidobacteria isolated from the human gut. *Appl Environ Microbiol.* 2002;68(9): 4689–4693.
77. Shi H, Kokoeva MV, Inouye K, Tzameli I, Yin H, Flier JS. TLR4 links innate immunity and fatty acid-induced insulin resistance. *J Clin Invest.* 2006;116 (11):3015–3025.
78. Wang X, An C, Yang M, et al. Immunization with individual proteins of the Lrp/AsnC family induces protection against brucella melitensis 16M challenges in mice. *Front Microbiol.* 2015;6:1193.
79. Mahdizade Ari M, Teymouri S, Fazlalian T, et al. The effect of probiotics on gestational diabetes and its complications in pregnant mother and newborn: A systematic review and meta-analysis during 2010–2020. *J Clin Lab Anal.* 2022. https://doi.org/10.1002/jcla.24326.
80. Zhao C, Yang C, Chen M, et al. Regulatory efficacy of Brown seaweed Lessonia nigrescens extract on the gene expression profile and intestinal microflora in type 2 diabetic mice. *Mol Nutr Food Res.* 2018;62. https://doi.org/10.1002/mnfr.201700730. PMID: 29210521.
81. Liu G, Liang L, Yu G, Li Q, Pumpkin polysaccharide modifies the gut microbiota during alleviation of type 2 diabetes in rats. *Int J Biol Macromol.* 2018;115:711–717 [PMID: 29702167 DOI:https://doi.org/10.1016/j.ijbiomac.2018.04.127].
82. Van Hul M, Geurts L, Plovier H, et al. Reduced obesity, diabetes, and steatosis upon cinnamon and grape pomace are associated with changes in gut microbiota and markers of gut barrier. *Am J Physiol Endocrinol Metab.* 2018;314:E334–E352.
83. Wang P, Gao J, Ke W, et al. Resveratrol reduces obesity in high-fat diet-fed mice via modulating the composition and metabolic function of the gut microbiota. *Free Radic Biol Med.* 2020;156:83–98.
84. Song JX, Ren H, Gao YF, et al. Dietary capsaicin improves glucose homeostasis and alters the gut microbiota in obese diabetic Ob/Ob mice. *Front Physiol.* 2017;8:602.

85. Zhang W, Xu JH, Yu T, Chen QK. Effects of berberine and metformin on intestinal inflammation and gut microbiome composition in db/db mice. *Biomed Pharmacother.* 2019;118, 109131.
86. Sun R, Yang N, Kong B, et al. Orally administered berberine modulates hepatic lipid metabolism by altering microbial bile acid metabolism and the intestinal FXR signaling pathway. *Mol Pharmacol.* 2017;91:110–122.
87. Liu D, Zhang Y, Liu Y, et al. Berberine modulates gut microbiota and reduces insulin resistance via the TLR4 signaling pathway. *Exp Clin Endocrinol Diabetes.* 2018;126:513–520.
88. Yue SJ, Liu J, Wang AT, et al. Berberine alleviates insulin resistance by reducing peripheral branched-chain amino acids. *Am J Physiol Endocrinol Metab.* 2019;316: E73–E85.
89. Xia F, Wen LP, Ge BC, Li YX, Li FP, Zhou BJ. Gut microbiota as a target for prevention and treatment of type 2 diabetes: Mechanisms and dietary natural products. *World J Diabetes.* 2021;12(8):1146–1163. https://doi.org/10.4239/wjd.v12.i8.1146.
90. Nejad L, Al-Naimi S, Rajic A, Ibrahim F. Probiotics as an additional therapeutic approach in the management of type 2 diabetes mellitus: A systematic review and meta-analysis. *Proc Nutr Soc.* 2022;81. (OCE1). https://doi.org/https://doi.org/10.1017/s0029665122000131.
91. Su L, Hong Z, Zhou T, et al. Health improvements of type 2 diabetic patients through diet and diet plus fecal microbiota transplantation. *Sci Rep.* 2022;12(1). https://doi.org/10.1038/s41598-022-05127-9.
92. Cabrera SM, Henschel AM, Hessner MJ. Innate inflammation in type 1 diabetes. *Transl Res.* 2016;167:214–227. https://doi.org/10.1016/j.trsl.2015.04.011.
93. Cabrera SM, Chen YG, Hagopian WA, Hessner MJ. Blood-based signatures in type 1 diabetes. *Diabetologia.* 2016;59:414–425. https://doi.org/10.1007/s00125-015-3843-x.
94. Kostic AD, Gevers D, Siljander H, et al. The dynamics of the human infant gut microbiome in development and in progression toward type 1 diabetes. *Cell Host Microbe.* 2015;17:260–273. https://doi.org/10.1016/j.chom.2015.01.001.
95. Cabrera SM, Coren AT, Pant T, et al. Probiotic normalization of systemic inflammation in siblings of type 1 diabetes patients: An open-label pilot study. *Sci Rep.* 2022;12 (1):3306. https://doi.org/10.1038/s41598-022-07203-6.
96. Wang Y, Zhang Y, Yang J, Li H, Wang J, Geng W. Lactobacillus plantarum MA2 ameliorates methionine and choline-deficient diet induced non-alcoholic fatty liver disease in rats by improving the intestinal microecology and mucosal barrier. *Foods.* 2021;10 (12). https://doi.org/10.3390/foods10123126.
97. Luo M, Yan J, Wu L, et al. Probiotics alleviated nonalcoholic fatty liver disease in high-fat diet-fed rats via gut microbiota/fxr/fgf15 signaling pathway. *J Immunol Res.* 2021;2264737. https://doi.org/10.1155/2021/2264737.
98. Patil UK, Saraf S, Dixit VK. Hypolipidemic activity of seeds of Cassia tora Linn. *J Ethnopharmacol.* 2004;90:249–252.
99. Li BP, Chen YY, Pan JQ, et al. The hepatoprotective effect of semen cassiae extract by antagonizing insulin resistance and inhibiting oxidative-glycation in rats with nonalcoholic fatty liver disease. *Glob Tradit Chin Med.* 2015;8:1171–1175.
100. Zheng RB, Huang XD, He RR, et al. Effects of Cassia obtusifolia extract on oxidative stress status in STZ-induced diabetic mice. *Chin J Exp Tradit Med Form.* 2012;18:202–206.
101. Zhang L, Ci H. Chemical constituents and clinical application of semen Cassiae for moistening intestines and defecating. *Cap Med.* 2011;18:56.

102. Li Y, Liu T, Yan C, et al. Diammonium glycyrrhizinate protects against nonalcoholic fatty liver disease in mice through modulation of gut microbiota and restoration of intestinal barrier. *Mol Pharm.* 2018;15:3860–3870.
103. Qian M, Lyu Q, Liu Y, et al. Chitosan oligosaccharide ameliorates nonalcoholic fatty liver disease (NAFLD) in diet-induced obese mice. *Mar Drugs.* 2019;17:1–14.
104. Li X, Wang Y, Xing Y, Xing R, Liu Y, Xu Y. Changes of gut microbiota during silybin mediated treatment of high-fat diet-induced nonalcoholic fatty liver disease in mouse. *Hepatol Res.* 2020;50:5–14.
105. Li X, Zeng F, Huang Y, Liu B. The positive effects of Grifola frondose heteropolysaccharide on NAFLD and regulation of the gut microbiota. *Int J Mol Sci.* 2019;20:5302.
106. Lu S, Zhang T, Gu W, et al. Volatile oil of Amomum villosum inhibits nonalcoholic fatty liver disease via the gut-liver axis. *Biomed Res Int.* 2018;1–16.
107. Zhao L, Zhang Q, Ma W, Tian F, Shen H, Zhou M. A combination of quercetin and resveratrol reduces obesity in high-fat diet-fed rats by modulation of gut microbiota. *Food Funct.* 2017;8(12):4644–4656. https://doi.org/10.1039/c7fo01383c [PMID: 29152632].
108. Porras D, Nistal E, Martínez-Flórez S, et al. Protective effect of quercetin on high-fat diet-induced nonalcoholic fatty liver disease in mice is mediated by modulating intestinal microbiota imbalance and related gut-liver axis activation. *Free Radic Biol Med.* 2017;102:188–202.
109. Luo H, Wu H, Wang L, et al. Hepatoprotective effects of Cassiae semen on mice with non-alcoholic fatty liver disease based on gut microbiota. *Commun Biol.* 2021;4(1). https://doi.org/10.1038/s42003-021-02883-8.
110. Bouhnik Y, Attar A, Joly AF, Riottot M, Dyard F, Flourie B. Lactulose ingestion increases faecal bifidobacterial counts: A randomised double-blind study in healthy humans. *Eur J Clin Nutr.* 2014;58:462–466.
111. Sarangi AN, Goel A, Singh A, Sasi A, Aggarwal R. Faecal bacterial microbiota in patients with cirrhosis and the effect of lactulose administration. *BMC Gastroenterol.* 2017;17:125.
112. Sharpton SR, Schnabl B, Knight R, Loomba R. Current concepts, opportunities, and challenges of gut microbiome-based personalized medicine in nonalcoholic fatty liver disease. *Cell Metab.* 2021;33(1):21–32. *Cell Press.* https://doi.org/https://doi.org/10.1016/j.cmet.2020.11.010.
113. Tarricone R, Abu Koush D, Nyanzi-Wakholi B, Medina-Lara A. A systematic literature review of the economic implications of chemotherapy-induced diarrhea and its impact on quality of life. *Crit Rev Oncol Hematol.* 2016;99 [37e48].
114. Diaz C, Neubauer S, Rechenmacher F, Kessler H, Missirlis D. Recruitment of alphanubeta3 integrin to alpha5beta1 integrininduced clusters enables focal adhesion maturation and cell spreading. *J Cell Sci.* 2020;133(1). https://doi.org/10.1242/jcs.232702 [PMID: 31780581].
115. Lleal M, Sarrabayrouse G, Willamil J, Santiago A, Pozuelo M, Manichanh C. A single faecal microbiota transplantation modulates the microbiome and improves clinical manifestations in a rat model of colitis. *EBioMedicine.* 2019;48:[630e41].
116. Pereira AM, Pinna C, Biagi G, et al. Supplemental selenium source on gut health: Insights on fecal microbiome and fermentation products of growing puppies. *FEMS Microbiol Ecol.* 2020;96(11).
117. da Silva Ferreira AR, Wardill HR, Havinga R, Tissing WJE, Harmsen HJM. Prophylactic treatment with vitamins C and B2 for methotrexate-induced gastrointestinal mucositis. *Biomolecules.* 2020;11(1).

118. Jahangard-Rafsanjani Z, Gholami K, Hadjibabaie M, et al. The efficacy of selenium in prevention of oral mucositis in patients undergoing hematopoietic SCT: A randomized clinical trial. *Bone Marrow Transplant.* 2013;48(6):[832e6].
119. Wardill HR, van der Aa SAR, da Silva Ferreira AR, Havinga R, Tissing WJE, Harmsen HJM. Antibiotic-induced disruption of the microbiome exacerbates chemotherapy-induced diarrhoea and can be mitigated with autologous faecal microbiota transplantation. *Eur J Cancer.* 2021;153:27–39. https://doi.org/10.1016/j.ejca.2021.05.015.
120. Binyamin D, Nitzan O, Azrad M, Hamo Z, Koren O, Peretz A. The microbial diversity following antibiotic treatment of Clostridioides difficile infection. *BMC Gastroenterol.* 2021;21(1):166.
121. Quraishi MN, Widlak M, Bhala N, et al. Systematic review with meta-analysis: The efficacy of faecal microbiota transplantation for the treatment of recurrent and refractory Clostridium difficile infection. *Aliment Pharmacol Ther.* 2017;46(5):479–493.
122. Iqbal U, Anwar H, Karim MA. Safety and efficacy of encapsulated fecal microbiota transplantation for recurrent Clostridium difficile infection: A systematic review. *Eur J Gastroenterol Hepatol.* 2018;30(7):730–734.
123. Roshan N, Clancy AK, Borody TJ. Faecal microbiota transplantation is effective for the initial treatment of Clostridium difficile infection: A retrospective clinical review. *Infect Dis Ther.* 2020;9(4):935–942.
124. Perler BK, Chen B, Phelps E, et al. Long-term efficacy and safety of fecal microbiota transplantation for treatment of recurrent Clostridioides difficile infection. *J Clin Gastroenterol.* 2020;54(8):701–706.
125. Dolai S, Ng J, Hazan S, Dawson V, Borody TJ. Faecal microbiota transplantation as primary treatment for Clostridioidoes difficile infection-evidence for change. *J Exp Pathol.* 2021;2:2.
126. Aroniadis OC, Brandt LJ, Oneto C, et al. Faecal microbiota transplantation for diarrhoea-predominant irritable bowel syndrome: A double-blind, randomised, placebo-controlled trial. *Lancet Gastroenterol Hepatol.* 2019;4(9):675–685. https://doi.org/10.1016/S2468-1253(19)30198-0.
127. Contijoch EJ, Britton GJ, Yang C, et al. Gut microbiota density influences host physiology and is shaped by host and microbial factors. *Elife.* 2019;8:e40553.
128. Vandeputte D, Kathagen G, D'hoe K, et al. Quantitative microbiome profiling links gut community variation to microbial load. *Nature.* 2017;551:507–511.
129. Britton GJ, Contijoch EJ, Spindler MP, et al. Defined microbiota transplant restores Th17/RORγt + regulatory T cell balance in mice colonized with inflammatory bowel disease microbiotas. *Proc Natl Acad Sci U S A.* 2020;117(35):21536–21545.
130. Smith SB, Ravel J. The vaginal microbiota, host defence and reproductive physiology. *J Physiol.* 2017;595:451–463.
131. Kroon SJ, Ravel J, Huston WM. Cervicovaginal microbiota, women's health, and reproductive outcomes. *Fertil Steril.* 2018;110:327–336.
132. Anahtar MN, Gootenberg DB, Mitchell CM, Kwon DS. Cervicovaginal microbiota and reproductive health: The virtue of simplicity. *Cell Host Microbe.* 2018;23:159–168.
133. Giorgi A, Torriani S, Dellaglio F, Bo G, Stola E, Bernuzzi L. Identification of vaginal lactobacilli from asymptomatic women. *Microbiologica.* 1987;10:377–384.
134. Antonio MA, Hawes SE, Hillier SL. The identification of vaginal lactobacillus species and the demographic and microbiologic characteristics of women colonized by these species. *J Infect Dis.* 1999;180:1950–1956.
135. Zhou X, Bent SJ, Schneider MG, Davis CC, Islam MR, Forney LJ. Characterization of vaginal microbial communities in adult healthy women using cultivation-independent methods. *Microbiology.* 2004;150:2565–2573.
136. Ravel J, Gajer P, Abdo Z, et al. Vaginal microbiome of reproductive-age women. *Proc Natl Acad Sci U S A.* 2011;108:4680–4687.

137. Petrova MI, Reid G, Vaneechoutte M, Lebeer S. Lactobacillus iners: Friend or foe? *Trends Microbiol.* 2017;25:182–191.
138. Eschenbach DA, Hillier S, Critchlow C, Stevens C, DeRouen T, Holmes KK. Diagnosis and clinical manifestations of bacterial vaginosis. *Am J Obstet Gynecol.* 1988;158:819–828.
139. Srinivasan S, Hoffman NG, Morgan MT, et al. Bacterial communities in women with bacterial vaginosis: High resolution phylogenetic analyses reveal relationships of microbiota to clinical criteria. *PLoS One.* 2012;7:e37818.
140. Faught BM, Reyes S. Characterization and treatment of recurrent bacterial vaginosis. *J Womens Health (Larchmt).* 2019;28:1218–1226.
141. Sobel JD, Kaur N, Woznicki NA, et al. Prognostic indicators of recurrence of bacterial vaginosis. *J Clin Microbiol.* 2019;57:[e00227-19].
142. Kwok L, Stapleton AE, Stamm WE, Hillier SL, Wobbe CL, Gupta K. Adherence of lactobacillus crispatus to vaginal epithelial cells from women with or without a history of recurrent urinary tract infection. *J Urol.* 2006;176:2050–2054.
143. Hemmerling A, Harrison W, Schroeder A, et al. Phase 2a study assessing colonization efficiency, safety, and acceptability of lactobacillus crispatus CTV-05 in women with bacterial vaginosis. *Sex Transm Dis.* 2010;37:745–750.
144. Stapleton AE, Au-Yeung M, Hooton TM, et al. Randomized, placebo-controlled phase 2 trial of a lactobacillus crispatus probiotic given intravaginally for prevention of recurrent urinary tract infection. *Clin Infect Dis.* 2011;52:1212–1217.
145. Hemmerling A, Harrison W, Schroeder A, et al. Phase 1 dose-ranging safety trial of lactobacillus crispatus CTV-05 for the prevention of bacterial vaginosis. *Sex Transm Dis.* 2009;36:564–569.
146. Lagenaur LA, Hemmerling A, Chiu C, et al. Connecting the dots: Translating the vaginal microbiome into a drug. *J Infect Dis.* 2021;223:S296–S306. https://doi.org/10.1093/infdis/jiaa676.
147. Lev-Sagie A, Goldman-Wohl D, Cohen Y, et al. Vaginal microbiome transplantation in women with intractable bacterial vaginosis. *Nat Med.* 2019;25(10):1500–1504. https://doi.org/10.1038/s41591-019-0600-6.
148. Samuelson DR, Welsh DA, Shellito JE. Regulation of lung immunity and host defense by the intestinal microbiota. *Front Microbiol.* 2015;6:1085.
149. Chunxi L, Haiyue L, Yanxia L, Jianbing P, Jin S. The gut microbiota and respiratory diseases: New evidence. *J Immunol Res.* 2020;2020. https://doi.org/10.1155/2020/2340670.
150. Negi S, Pahari S, Bashir H, Agrewala JN. Gut microbiota regulates mincle mediated activation of lung dendritic cells to protect against mycobacterium tuberculosis. *Front Immunol.* 2019;10:1142.
151. Zerin T, Lee M, Jang WS, Nam KW, Song HY. Ursolic acid reduces mycobacterium tuberculosis–induced nitric oxide release in human alveolar A549 cells. *Mol Cells.* 2015;38:610–615.
152. Mahmud HA, Seo H, Kim S, et al. Thymoquinone (TQ) inhibits the replication of intracellular mycobacterium tuberculosis in macrophages and modulates nitric oxide production. *BMC Complement Altern Med.* 2017;17:279.
153. Lee Y, Seo H, Kim S, et al. Activity of lactobacillus crispatus isolated from vaginal microbiota against mycobacterium tuberculosis. *J Microbiol.* 2021;59(11):1019–1030. https://doi.org/10.1007/s12275-021-1332-0.
154. Wunderink RG. Turning the phage on treatment of antimicrobial-resistant pneumonia. *Am J Respir Crit Care Med.* 2019;200:1081–1082. https://doi.org/10.1164/rccm.20190 8-1561ED.
155. Prazak J, Iten M, Cameron DR, et al. Bacteriophages improve outcomes in experimental Staphylococcus aureus ventilator-associated pneumonia. *Am J Respir Crit Care Med.* 2019;200:1126–1133. https://doi.org/10.1164/rccm.201812-2372OC.

156. Peled JU, Devlin SM, Staffas A, et al. Intestinal microbiota and relapse after hematopoietic-cell transplantation. *J Clin Oncol.* 2017;35(15):1650–1659.
157. Mahmoodpoor A, Hamishehkar H, Asghari R, Abri R, Shadvar K, Sanaie S. Effect of a probiotic preparation on ventilator-associated pneumonia in critically ill patients admitted to the intensive care unit: A prospective doubleblind randomized controlled trial. *Nutr Clin Pract.* 2019;34(1):156–162.
158. Weng H, Li JG, Mao Z, et al. Probiotics for preventing ventilator-associated pneumonia in mechanically ventilated patients: A meta-analysis with trial sequential analysis. *Front Pharmacol.* 2017;8:717.

Index

Note: Page numbers followed by "*f*" indicate figures and "*t*" indicate tables.

C

I

K

L

M

FSC
www.fsc.org
MIX
Papier | Fördert
gute Waldnutzung
FSC® C083411

Kolibri 360 GmbH
Ferdinand-Jühlke-Straße 7
99095 Erfurt, Deutschland
produktsicherheit@kolibri360.de